PMI-PBA® Exam Practice Test and Study Guide

Best Practices and Advances in Program Management Series

Series Editor
Ginger Levin

PMI-PBA® Exam Practice Test and Study Guide

Brian Williamson

PfMP, PgMP, PMP, PMI-RMP, PMI-SP, PMI-ACP, PMI-PBA, 6σ, ITIL Expert, CFE

CRC Press
Taylor & Francis Group
Boca Raton London New York

CRC Press is an imprint of the
Taylor & Francis Group, an **informa** business

AN AUERBACH BOOK

Parts of the following publications are reprinted with permission of the Project Management Institute, Inc., 14 Campus Boulevard, Newtown Square, PA 19073-3299 USA, a worldwide organization advancing knowledge in portfolio, program, and project management and business analysis.

Business Analysis for Practitioners: A Practice Guide, 2015.

Requirements Management: A Practice Guide, 2016.

A Guide to the Project Management Body of Knowledge (PMBOK ®), 2013

PMI Professional in Business Analysis (PMI-PBA®) Examination Content Outline, 2013

PMI Professional in Business Analysis (PMI-PBA®) Handbook, 2016

PMI's Pulse of the Profession®, Requirements Management: A Core Competency For Project and Program Success, 2014

"PfMP," "PgMP," "PMP," "PMI-SP," "PMI-RMP," "PMI-ACP," "PMI-PBA," and "PMI" are registered trademarks of the Project Management Institute, Inc., which is registered in the United States and other nations.

"ITIL" is a trademark of AXELOS Limited, which is registered in the United Kingdom and other nations.

"6σ" is a trademark of Motorola Trademark Holdings, LLC., which is registered in the United States.

CRC Press
Taylor & Francis Group
6000 Broken Sound Parkway NW, Suite 300
Boca Raton, FL 33487-2742

First issued in paperback 2022

© 2018 by Taylor & Francis Group, LLC
CRC Press is an imprint of Taylor & Francis Group, an Informa business

No claim to original U.S. Government works

ISBN 13: 978-1-03-247655-1 (pbk)
ISBN 13: 978-1-138-05447-9 (hbk)

DOI: 10.1201/9781315166803

Visit the Taylor & Francis Web site at
http://www.taylorandfrancis.com

and the CRC Press Web site at
http://www.crcpress.com

Dedication

To my parents, thank you for the endless support and encouragement; to my father, thank you for the mentorship, setting the bar, and all your contributions to our country; to my children, remember to always dream big, and be the change that others seek in the world; and to my wife, thank you for allowing me the time to pursue my passions.

Contents

List of Tables

List of Illustrations

Preface

Analysis is a critical skill in terms of both business and project execution. The concepts, tools, and techniques are widely applicable and sufficiently robust, in that they can be applied to nearly every industry, microvertical, and even personal life. Having held various positions in business for over 20 years, I cannot begin to stress strongly enough the importance of business analysis and the Project Management Institute's Professional in Business Analysis (PMI-PBA®) credential. For a moment, consider the cartoon below, which was adopted for this book based on a March 1973 University of London Computer Center Newsletter and then put online on September 9, 2003, in a blog titled "Typical Project Life"; it metaphorically explains the gaps between business requirements and execution.

Figure P.1 Tree swing cartoon. (Licensed under Creative Commons.)

I have often referred to *Business Analysis for Practitioners: A Practice Guide* and *Requirements Management: A Practice Guide,* publications from the Project Management Institute (PMI), as the missing puzzle pieces to the *PMBOK® Guide,* otherwise known as *A Guide to the Project Management Body of Knowledge.* This book is unlike other exam preparation publications on the market; my goal is that it's akin to a reference guide, something that can be used long after you pass the PMI-PBA® Exam. One of my mantras is that ***knowing what to do*** must be translated into ***doing what you know.*** This book will help you prepare for the exam by instilling knowledge and encouraging critical thinking. As a result, the skills attained will lead to improved project success and outcomes, and you'll have a much stronger understanding of the material, along with the tools and techniques of business analysis.

On to the preparation for the PMI-PBA® Exam. Based on my experience in both developing and delivering training content for other PMI exams, this book will attempt to address all your questions and concerns. Two of the most sought-after study aids are memory maps and practice questions. The systematic use of memory maps helps aid in the efficient recall of information and can boost confidence during the exam. Well-crafted practice questions are fantastic study aids that can be used track your progress as you learn new concepts, introduce you to the complex sentence structure that is likely to appear on the exam, and concentrate your studies by domain, essentially preparing you to pass the very challenging PMI-PBA® Exam in the allotted four hours.

In addition to study hints and exam topics, this book provides reference to tools and techniques that should be incorporated into your work immediately to help avoid situations as shown in Figure P.1: Tree swing cartoon. For each of the five domains outlined in the *PMI Professional in Business Analysis (PMI-PBA)® Examination Content Outline* 2013 (the ECO), I have prepared 20 questions to test your knowledge. Also included is challenging 200-question practice exam, which is representative of the actual exam.

To enhance your studies, we also include a timed, online simulated exam. At the end of the simulated exam, you will see your score according to the number of questions you answered correctly. Because PMI does not sell or publish past exams, the best option is to develop these practice tests based on actual experience. These questioned are crafted to foster learning and reinforce content; they are not obscure or overly complicated, but rather are representative of the actual exam.

As with Taylor & Francis's other exam practice tests and study guide books, we have included a plainly written rationale for each correct answer along with a supporting reference. Each question is also anchored back to the ECO, which is based on the PMI's Role Delineation Study (RDS), which is the basis for the creation of the exam. Also included is a method for predicting your overall exam score. This technique is explained in Section 8.2.9.1, on page 190; please review this material before attempting your first domain quiz.

The PMI-PBA® Exam is very challenging, but by focusing your studies as outlined in this book, you will significantly increase your goal of passing the exam on your first attempt. Most important, you will be able to make a measurable impact for your stakeholders. On that note, ***good luck on the exam and in all your future endeavors.***

Brian

Brian Williamson
April, 2017

Acknowledgments

I would like to thank Dr. Ginger Levin for all her contributions to the field of project, program, and portfolio management—she personally supported this project and previewed the book; John Wyzalek, my publisher, who was my guide and shepherd through the entire publishing process and who also programmed the online exam; Theron Shreve of DerryField Publishing Services for guiding the production; Susan Culligan, who masterfully edited and typeset this entire book; and Julie Martin at the Project Management Institute, who guided me through the nuances of intellectual property licensing and permissions.

Thank you all, for your patience, dedication, and assistance; you've helped me to fulfill a life-long dream. My hope is that all business analysts and project practitioners benefit from this book.

Acknowledgements

About the Author

Brian Williamson is a strategic portfolio, program, and project management executive, with over 20 years of experience managing and implementing large-scale enterprise software applications. He has extensive experience in the microverticals of financial management, supply chain management, and human capital management spanning several industry verticals. Over the course of his career, he has been very fortunate to have worked with a number of industry leaders including Praxair, Western Connecticut Health Network, Avaap, PMO Advisory, PwC (PricewaterhouseCoopers), Deloitte Consulting, Ernst & Young, Infor, and Oracle, to name just a few. He is an expert at business analysis, strategic planning, leading transformational change, and helping organizations align their portfolios to their mission, vision, goals, and objectives.

At the time of this publishing, Brian is counted among the less than 20 people globally to hold all the renewable certifications from the Project Management Institute: Professional in Business Analysis (PMI-PBA)®, Portfolio Management Professional (PfMP)®, Program Management Professional (PgMP)®, Project Management Professional (PMP)®, Risk Management Professional (PMI-RMP)®, Scheduling Professional (PMI-SP)®, and Agile Certified Professional (PMI-ACP)®.

Brian also holds the designation of ITIL® Expert in IT Service Management, having earned the following credentials from AXELOS: Foundation in IT Service Management (v2011), Release Control and Validation (RCV), Service Offerings and Agreements (SOA), Operational Support and Analysis (OSA), Service Operation (SO), and Managing Across the Lifecycle (MALC). He is also a Certified Fraud Examiner (CFE) by the Association of Certified Fraud Examiners (ACFE) and has received extensive training in Six Sigma. Brian holds two Masters degrees from Polytechnic University, New York: Management of Technology (MoT), and Information Systems Engineering (ISE); a Bachelor of Science degree from Teikyo Post University in Waterbury, Connecticut; and an Associate degree in Specialized Business from the Art Institute of Philadelphia.

As a consultant, trainer, and coach, Brian has mentored many industry professionals using his own material and methodology aligned to the PMI Standards and Practice Guides. Brian has also coauthored and contributed to a number of books on portfolio, program, and project management, and now business analysis. He resides with his family in a small country town in Connecticut, USA.

How to Best Use This Exam Practice Test and Study Guide

[This guide is repeated in Chapter 8, Section 8.2.1, on page 182.]

This *PMI-PBA® Exam Practice Test and Study Guide* is aligned with the following PMI publications: *Business Analysis for Practitioners: A Practice Guide, Requirements Management: A Practice Guide,* and the *PMI-PBA® Examination Content Outline* (ECO); in addition, it references key content from *A Guide to the Project Management Body of Knowledge (PMBOK®)*. Each of the chapters, as outlined below, corresponds to a domain as defined in the 2012 Role Delineation Study (RDS) of Business Analysts. For further information on the RDS, please refer to Appendix A: Role Delineation Study (RDS) Process, in the ECO.

Introductory Chapter

The introductory chapter offers further context to foundational elements, which are carried throughout all the domains, and provides a necessary baseline for the subject material.

Needs Assessment

The Needs Assessment domain includes the tasks necessary to understand the opportunities or problems of the sponsoring organization; the tools and techniques to determine the value proposition; the creation of project goals and objectives linked to the organizational pillars; and a comprehensive stakeholder assessment. The activities in this domain produce deliverables that are used throughout the project, including the situation statement, the solution scope statement, the business case, and various stakeholder assessment artifacts.

Planning

The Planning domain contains the elements that focus on the preparation required to manage the business analysis activities throughout the lifecycle of the project.

Analysis

The Analysis domain is the single most important domain on the exam. It contains those elements that focus solely on the tasks and activities related to the identification, detailed specification, approval, and validation of product requirements.

Traceability and Monitoring

The Traceability and Monitoring domain contains those tasks and activities that focus on managing the lifecycle of the requirements. This domain focuses on establishing the requirements baseline and the continual monitoring, tracking, and communicating the status of requirements. Furthermore, the domain encompasses change control and the management of issues, risks, and decisions.

Evaluation

The Evaluation domain contains those tasks and activities that validate test results, analyze and communicate gaps, and obtain stakeholder signoff, concluding with an evaluation of the deployed solution.

Based on our experience with adult students, we are advocating the following approach to prepare for the exam:

Step 1 Read the *PMI Professional in Business Analysis (PMI-PBA)® Handbook.*

Step 2 Read *Requirements Management: A Practice Guide* in its entirety.

Step 3 Read *Business Analysis for Practitioners: A Practice Guide* in its entirety.

Step 4 Read a domain chapter in this *PMI-PBA® Exam Practice Test and Study Guide.*

Step 5 Reread the corresponding domain chapter in *Requirements Management: A Practice Guide* and *Business Analysis for Practitioners: A Practice Guide.*

Step 6 Read the supporting chapter in the *PMBOK® Guide.*

Step 7 Practice the exercises at the end of the relevant chapter of this book, then take the 20-question domain quiz.

Step 8 Review and understand the rationale for correct and incorrect answers.

Step 9 Repeat this process for all the domains.

Step 10 Take the final practice exam on paper.

Step 11 Review and understand the rationale for correct and incorrect answers.

Step 12 Use our online exam bank to simulate the actual exam. To access the practice test, please send a message to **pbaexam@certpreptests.com** and include in the subject line the phrase: Test Access. The time allotted for the test is four hours, and a timer displays the remaining time for taking the test. Once you have finished the test or time has run out, the percentage of correct answers per domain is displayed, as well as all of the test questions and answers, which you can review.

65-Day Study Plan

For your convenience, we've created a simple 65-day study plan (see Table 8.1 on page 184) designed to improve your retention as you learn new concepts and theories, along with the practical application of the tools and techniques associated with business analysis. At the same time, this will condition your mind and improve your focus as you prepare for the four-hour, 200-question PMI-PBA® Exam.

There is no substitute for hard work.

— Thomas Edison

Chapter 1

Introduction

Congratulations on deciding to pursue the Project Management Institute's Professional in Business Analysis (PMI-PBA)® Certification. This is without doubt one of the most important certifications a person involved with projects and supporting business operations could pursue. From the senior portfolio manager to a college intern, if you are involved with assessing the needs of an organization, eliciting requirements, or measuring delivered solution value, you will derive value from this book and certainly from the PMI-PBA® certification. As organizations institute lean principles and become more efficient, it is not uncommon for project managers to also fulfill the role of business analysts. Likewise, in small organizations that do not have a formal project management office (PMO) or a standardized project delivery methodology, it is not uncommon to expect business analysts to manage projects. Both can be ingredients for a difficult and sometimes challenging road ahead and, unfortunately, in many cases can lead to either project failure or unfulfilled stakeholder expectations. Even for those individuals who are not directly involved with projects, the tools and techniques that we will explore can improve upon a broad range of activities.

Consider, for a moment, a project that is run incredibly well and is delivered on time and under budget. Is it safe to assume the project was a success? As a reminder, projects can only be considered a success upon sign-off from the customer. This credential bridges the gap between project management (which focuses on managing the project-related activities) and requirements delivery (which focuses on ensuring that the appropriate solution was selected and managed to benefits realization).

In addition to using this book to help you pass the PMI-PBA® Exam, the material can help you ensure that the customer's problem or opportunity is clearly understood; all proposed solutions are vetted and anchored to the sponsoring organizations' goals and objectives; the requirements, once approved, are traced and tracked until delivery; and the delivered solution is measured against the initial project justification and value proposition.

Because the various practice guides cited in this book, as well as the *PMBOK® Guide*,[1] discuss most of the activities, the tools and techniques, and the outputs of each task in each domain, this book will highlight the pertinent information to help you pass the very challenging PMI-PBA® Exam. At the conclusion of this chapter, please take a few moments to complete the exercises. They are designed to help you relate concepts and theories to your organization—a key step in bridging knowledge to practice.

---

[1] PMI (2013). *A Guide to the Project Management Body of Knowledge (PMBOK®)*. Newtown Square, IL: Project Management Institute.

1.1 History of Business Analysis

Business analysis and requirements management can be tracked to the dawn of ages; consider for a few moments just a few of the interesting facts about the Great Pyramid of Giza (GPoG). The pyramid is widely believed to have been commissioned in the Fourth Dynasty by the Egyptian Pharaoh Khufu, and it is the oldest of the Seven Wonders of the Ancient World.

1.1.1 Facts About the Great Pyramid of Giza

Built. Approximately 4,500 years ago, between 2589 and 2504 BC, and originally stood at 481 feet, with a base of 756 feet. Construction duration has been estimated at between 10 and 20 years. To construct the pyramid in 20 years would require installing approximately 800 tons of stone each day, which equates to a block every five minutes.

Resources. The outer stones were white limestone from Tura, and the quarries of Aswan produced the granite stones for the King's Chamber, whereas the balance of the material is believed to have come from the plateau and Giza quarries. Those involved the construction, estimated at between 4,000 and 6,700 primary laborers and 10,000 to 20,000 secondary resources, included architects, artists, engineers, quarrymen, transporters, masons, polishers, skippers on the Nile, scribes, stone layers, cooks, and doctors, to name just a few.

Position. The pyramid is precisely aligned with the constellation Orion and true north to only a 3/60 degree of error. The Great Pyramid of Giza is also located at the center of planet Earth's land mass.

Exterior Design. It was originally covered with highly polished limestone, designed to reflect the Sun's light, so that it could be visible from the Moon. The outer mantle consists of 144,000 casing stones, all polished flat to an accuracy of 1/100 of an inch, which at the time could be visible from Israel. The four faces of the pyramid are concave, resulting in eight sides.

Interior Design. The interior temperature is a constant 68°F, the average temperature of the Earth. The granite coffer, inside the King's Chamber, was installed during construction. There was a swivel door at the entrance to the pyramid. At the time of construction, the descending passage of the pyramid was aligned to the north star, and the southern passage in the King's Chamber was aligned to a star in the constellation Orion.

Reflecting on this magnificent and impressive structure and relating it to the PMI-PBA® credential, two concepts should begin to surface: *project* versus *product* management.

The focus of *project management* is on *"doing the work right"*; consider some of the knowledge areas and process groups above, along with the collaboration between project managers and business analysts. The following case study is purely hypothetical, and while based on facts pertaining to the Great Pyramid, it should only be considered interpretively and not as historical reference.

1.1.2 Project Integration

Project Manager. Contributes to the development of the charter for the project; once he or she is appointed as the project manager (PM), then work is done on the project plan. Next, the key tasks are in project execution, monitoring, and controlling. Once the project is complete, the project manager handles administrative and contract closure.

Business Analyst. Collaborates with the PM to prepare the scope statement and charter; provides input to the PM as the project plan and other artifacts are created.

If there are changes, the business analyst (BA) ensures that any items considered are linked to the organizational pillars (which are further explained in Section 1.2) and performs an impact assessment of the change.

Case Study. During the planning and construction phases, the GPoG PM and BA would collaborate closely on all product-related activities. This would include creation of the key plans, monitoring and controlling the work, and addressing change requests if they occur.

1.1.3 Management of Scope

Project Manager. A scope management plan, a work breakdown structure (WBS), and a WBS dictionary are key documents prepared by the project manager and constitute the project's scope baseline. If changes are made, the PM considers their impact to the scope baseline and changes these documents if necessary.

Business Analyst. To ensure that the final deliverable is in line with expectations, the BA will collaborate with the key stakeholders during the preparation of the scope statement and charter, using the business case as the foundation.

During integrated change control, the BA ensures that any items considered are linked to the approved product scope and the needs of the organization and that any approved items are added to the trackers.

Case Study. The scope statement clearly articulates the boundaries for the project and product. For example, there will be one pyramid, with a descending passage aligned to the north star and a southern passage in the King's Chamber aligned to a star in the constellation Orion. The pyramid shall have a height of 481 feet and a base of 756 feet.

1.1.4 Management of Time

Project Manager. Creates the schedule management plan for project-related activities. Using a work plan, defines and sequences project activities; estimates activity resources and durations; develops and controls the project schedule.

Business Analyst. Collaborates with the PM on an integrated approach and plan related to product activities. Key aspects include assessing the availability of resources, establishing predecessor and successor task dependencies, understanding the interrelationship of product work to project work, addressing staffing requirements, assessing the complexity of tasks, and understanding the nature of risks.

Case Study. The Aswan quarries are located approximately 934 km south of Giza up the Nile River, whereas the Tura quarries are to the south of Cairo on the eastern shore of the Nile River, about 13–17 km from Giza. The Giza quarries lie only a few hundred meters south from the pyramid of Khufu.

The GPoG BA will be focused on collaborating with the key resources responsible for the construction of the pyramid to ensure that their work effort is aligned to that of the overall project.

1.1.5 Management of Cost

Project Manager. Creates the cost management plan for project-related activities; estimates costs, determines budget, and controls costs.

Business Analyst. Collaborates with the PM and financial analysts on the preparation of estimates.

Case Study. Cost is only one of the mysteries of the Great Pyramid; more than likely there was a mixture of paid laborers and volunteers. In traditional terms, the BA would want to ensure that the initial estimates were in line with actual expenses incurred for items related to the solution. Any variation from plan to actual would be brought to the attention of the Steering Committee. In addition, any projected cost variances from the feasibility study to actual bidding would require approval before proceeding.

1.1.6 Management of Quality

Project Manager. Creates the project quality management plan; performs quality assurance and controls quality.

Business Analyst. Collaborates with the PM on the quality requirements for the product and creates inspection validation checks in the work plan.

Case Study. The GPoG project quality team might preform a QA audit, looking at the pyramid's work plan, stakeholder communication, and perhaps stakeholder engagement, whereas the BA could audit the quality of work: for example, the pyramid's alignment to the constellation Orion or the accuracy of the outer mantle casing stones.

1.1.7 Management of Human Resources

Project Manager. Plans HR management; acquires, develops, and manages the project team.

Business Analyst. Collaborates with the PM to determine resource needs for each phase of the project and that all areas are represented. Builds the responsibility assignment matrix and assesses capability and capacity of the organization.

Case Study. The GPoG PM needed to first develop a plan for how to manage all the HR aspects of the project, including hiring, training, and managing over 6,700 resources, spanning the roles of architects, artists, engineers, quarrymen, transporters, masons, skippers on the Nile, scribes, cooks, doctors, and perhaps many more roles.

The BA would be especially concerned with roles directly involved in creating the end product; for example, the architects, artists, engineers, quarryman, transporters, stone layers, and stone polishers.

1.1.8 Management of Communication

Project Manager. Plans communications management; manages and controls communications during the project.

Business Analyst. Collaborates with the PM, focusing on four distinct aspects of communication: (a) method, (b) medium, (c) message, and (d) audience, with each ensuring that the communication was clear, concise, and concrete.

Case Study. In modern times, PMs could use status reports, newsletters, SharePoint, etc. Consider for a moment if there were 10,000 stakeholders identified for the GPoG project. That would equate to 49,995,000 communication channels (formula: $n\,(n-1)\,/\,2$).

The PM would be concerned with ensuring that a consistent message was delivered to all stakeholders involved or impacted by the project: for example, they wanted to communicate an improved method for transporting material from the quarries to all haulers.

However, the BA would concentrate the messaging around the product: for example, a new method was discovered to measure the stones with a greater degree of accuracy.

In both cases, remember to address:

✓ WHO	✓ Stakeholders
✓ WHAT	✓ Methods
✓ WHEN	✓ Frequency
✓ WHERE	
✓ WHY	
✓ HOW	✓ Techniques

When crafting the communication, remember to tailor the message for the recipient. I often suggest that authors put themselves in the position of the recipient and address, "What's in it for me?" (sometimes referred to as WIIFM).

1.1.9 Management of Risk

Project Manager. Plans project risk management; identifies risks; performs qualitative and quantitative analysis; and controls project risks.

Business Analyst. Collaborates with PM on risk identification, analysis, and response, with a concentration on product risk versus project risk.

Case Study. Risks include both opportunities and threats. Potential positive responses include enhancement, exploitation, sharing, or acceptance, as well as cautionary responses of avoidance, transference, mitigation, and acceptance. The BA would want to understand the risk profile (risk tolerant, risk seeking, or risk adverse) when considering potential responses to product risks on the GPoG project. In our case study, the quarry might be able to double output if a new tool were used. Or perhaps they'd like to understand the implications for a 100-year storm, which might impact production.

Remember, risks are events that *could* occur, whereas issues *have* occurred and are presently affecting the project. Issues may (or may not) have been first identified as a risk.

1.1.10 Management of Procurements

Project Manager. Plans, conducts, controls, and closes all project-related procurements.

Business Analyst. Collaborates with PM on procurements directly related to the product. Although the PM holds the final authority to select vendors, the BA has influence to ensure that the proposed vendors are capable of delivering a solution within the specified tolerance of the product.

Case Study. The GPoG PM would need to coordinate purchases, including items such as tools, boats, clothing, food and water, medical supplies, and the stones.

The BM would be concerned with the granite quarried from the Aswan quarries and the limestone from the Tura and Giza quarries.

1.1.11 Engagement of Stakeholders

Project Manager. Identifies all project stakeholders; plans, manages, and controls stakeholder engagement over the duration of the project.

Business Analyst. Collaborates with the PM on stakeholder identification and analysis and elaborates details for those stakeholders involved with tasks associated with each of the BA knowledge areas.

Case Study. Whereas the PM's concern is very broad, the BA's focus would be much narrower. Although they would both engage with the sponsor to discuss the project, the BA's focus would be on the product deliverable, whereas the PM would address all facets of project delivery. Both could rely on stakeholder assessment tools such as power, interest, influence diagrams or salience mappings.

The focus of business analysis as relates to project management is on meeting stakeholder expectations and delivering sustained value through requirements management. As outlined in Section 1.5 of *Business Analysis for Practitioners: A Practice Guide,*[2] "Business analysis is the application of knowledge, skills, tools, and techniques to:

- Determine problems and identify business needs
- Identify and recommend viable solutions for meeting those needs
- Elicit, document, and manage stakeholder requirements in order to meet business and project objectives
- Facilitate the successful implementation of the product, service, or end result of the program or project"

Now from the product perspective, reflect on the hypothetical business analyst's role and tasks related to the Great Pyramid of Giza.

1.1.12 Needs Assessment

- ❖ How did the construction of the Great Pyramid align with goals and objectives of the Old Kingdom during the Fourth Dynasty?
- ❖ What were the expected benefits associated with building the pyramid?
- ❖ Who were the stakeholders, and how should they be engaged over the duration of the project?
- ❖ Business cases provide clarity and justification for the product scope and are constantly updated over the life of the project. Was one even created or were the cost/benefits ever considered?

The outcome of the Needs Assessment domain is:

- ✓ A clearly defined situation statement
- ✓ An accompanying analysis that establishes the needed capabilities (financial, human resources, and equipment), goals, and objectives of the product
- ✓ The results of a feasibility study
- ✓ Assumptions, constraints, risks, dependencies
- ✓ A stakeholder analysis
- ✓ The overall construction approach

---

[2] PMI. *Business Analysis for Practitioners: A Practice Guide.* January, 2015. ISBN: 9781628250695. PMI Product ID: 00101570601.

1.1.13 Planning

❖ What were the underlying rationale and driving factors influencing the construction of the Great Pyramid? Was this clearly communicated to all of Egypt?

❖ To ensure that all of Pharaoh Khufu's needs and requirements were addressed, they would need to establish a means to trace requirements starting at ideation through final walk-through of the finished pyramid. Could this have been accomplished by memory, or perhaps a checklist was used. When you consider all the details of the finished product, a detail tracker would have been critical throughout the duration of construction.

❖ Considering the timeline and significance of the Great Pyramid, plans would have been formulated to address the approach to both product and project requirements management. How was this to be documented and communicated?

❖ Over the duration of the project, some 10 to 20 years, elements of the project evolved, possibly impacting multiple aspects of the pyramid. To ensure continual alignment, a change management approach would be vital to the success of the project; how would this be addressed? Was there a hierarchical change management approach, a governing body, or did Pharaoh Khufu have the one and only vote regarding changes?

❖ The writing medium in the Fourth Dynasty was stone, clay, and papyrus, scribed using black iron gall ink and carbon black ink, in what has been suggested as a pictorial hieroglyphic writing system. How did the scribes implement a document and version control system to ensure that all documentation was preserved over the duration of the project?

❖ While planning the Great Pyramid, there were probably many conversations with Pharaoh Khufu to fully understand the undertaking. At any time, did they discuss the rationale to clearly understand and document metrics and success criteria?

The culmination of the planning tasks is a comprehensive plan and approach for managing the product-related activities over the duration of the project. Please keep in mind, it is entirely feasible and realistic to presume that some of the above elements evolved over time. For example, what was the first change request, and did the process evolve as a result?

1.1.14 Analysis

❖ The business analyst, architects, and engineers would listen to the ideas from Pharaoh Khufu and categorize the solution requirements accordingly.

❖ Once Pharaoh Khufu had identified the need for a pyramid, the team would work to clarify the concept and decompose the elements into manageable construction objects.

❖ With all the objects decomposed, there would be meetings to discuss the aspects, elements, and objects that would be accepted, deferred, or rejected.

❖ With all the facets agreed to, the scribes could establish the scope baseline for the pyramid. This would represent the body of the work.

❖ With the baseline created, Pharaoh Khufu would be in position to authorize the creation of the requirements documents.

❖ Architects, engineers, and masons could now begin the process of working with the scribes to document the requirements for the pyramid.

❖ With the completion of each requirements document, the scribes and business analyst could ensure that each aligned with the previously identified needs of the kingdom.

❖ How would Pharaoh Khufu know if the constructed pyramid was aligned to the initial vision? Evaluation metrics and acceptance criteria would be established to provide for both qualitative and quantitative analysis. The business analyst would need to ensure that any proposed changes were addressed in accordance with the previously established plan.

For the construction of the pyramid, these are some of the most important activities, resulting in the formalization of measurable requirements, which are decomposed into manageable work packages. These activities focus on the identification, detailed specification, approval, and validation of pyramid requirements.

1.1.15 Traceability and Monitoring

❖ Starting with the initial conversations with Pharaoh Khufu, identified requirements would have been added to the checklist or matrix and traced back to the needs, goals, and objectives of the Old Kingdom.

❖ With the cutting, transport, and laying of stones, the team would be monitoring these construction objects, continually reviewing documentation to ensure that the work in progress was in alignment with the specified needs and requirements.

❖ As construction evolved, the checklist or matrix would be updated, and completed items that addressed needs would have been noted.

❖ The pyramid's construction was resource intensive over a very long duration. Establishing a means and forum for communication would be vital to ensure that all stakeholders were apprised of status, issues, risks, and decisions, especially Pharaoh Khufu, as the executive sponsor. How would this be conducted effectively, as the project involved nearly the entire country of Egypt?

❖ Architects, engineers, masons, stone layers, etc. would be managing changes in each of their respective trades, which is both customary and anticipated in projects. How would this have been done effectively and in an integrated manner?

Throughout the construction period, the GPoG business analyst would be tasked with monitoring, tracking, and communicating the overall status of the pyramid, in addition to product-related work packages. In the event that changes were proposed or issues/risks emerged, the business analyst would be prepared to address them in an integrated and timely manner. You'll recall from Section 1.1.1: Facts About the Great Pyramid of Giza, they were laying a block every five minutes; with this cadence, there was virtually no time for lengthy debate.

1.1.16 Evaluation

❖ Validation and verification exercises produce both qualitative and quantitative measures. While stones were being set, one every five minutes, was quality control performed to verify that tolerances were in alignment with the specification?

❖ During walkthroughs, were there any identified gaps in the cost–benefit analysis, the feasibility study, the solution scope statement, or the project charter, and if so, how were those items addressed?

❖ As the sponsor, did Pharaoh Khufu, his designee, or perhaps his heir, sign off on completed construction?

❖ Based on the cost–benefit analysis, the solution scope statement, and project charter, did the constructed pyramid fulfill the value proposition?

The Great Pyramid of Giza is the only one of the Seven Wonders of the Ancient World that still stands today. Furthermore, it is one of the most massive structures on earth and is a tremendous feat of architecture, engineering, masonry, and sheer will power. The true intent remains unknown, as 21st-century amateur archaeologists can only assume it has more than fulfilled its original value proposition.

Without the knowledge of business analysis, could the ancient Egyptians in the Fourth Dynasty (2589 BC) have:

✓ Accurately understood Pharaoh Khufu's problems and needs?
✓ Identified and recommended possible solutions to Khufu?
✓ Elicited, documented, communicated, and managed Pharaoh Khufu's requirements, goals, and objectives over the estimated 10 to 20 years of construction?
✓ Monitored the progress, reported status to Khufu, and ensured that the final product fulfilled the value proposition?

These ancient business analysts were ultimately responsible for facilitating the construction of the Great Pyramid of Giza, which has not only stood the test of time, but has earned itself a place in Earth's history as an engineering marvel and one of the Seven Wonders of the Ancient World.

1.2 Organizational Pillars

Both on the exam and in practice, one of the key principles of business analysis is ensuring that the proposed, and ultimately selected, solution is aligned with the organization's key pillars. The most common and widely known pillars are an organization's mission, vision, and its strategic plan, which comprise the frameworks for effective decision making. However, as outlined in Table 1.1: The Organizational Pillars, you'll find there are additional elements that serve to represent an organization. Please take a few moments to familiarize yourself with the seven organizational pillars.

Understanding these pillars can often be difficult and challenging, if not related to something tangible. Consider for a moment an analogy to a recreational youth club, XYZ Soccer, in Table 1.2.

1.3 Origin of Requirements

In the simplest form, business processes drive requirements. *Requirements Management: A Practice Guide,*[3] on page 13, shows the direction and synchronization of requirements originating at the portfolio and continuing to the program and project levels, where a product, service, or result is defined that meets requirements. However, both in practice and on the exam, this synchronization could be bidirectional, as the discovery of problems and opportunities is far more likely to occur from the area's most familiar with the process.

---

[2]PMI (2017). *Requirements Management: A Practice Guide.* Newtown Square, PA: Project Management Institute.

Table 1.1 The Seven Organizational Pillars

Pillar	Description
Mission	The intent of the mission statement is to articulate why the organization exists.
Vision	A vision statement expresses where the organization wants to be in a defined time period.
Strategy	Typically expressed in the form of a strategic plan, the strategy outlines how the organization will realize its vision and fulfill its mission.
Objectives	Objectives can be both short-term and long-term milestones designed to evaluate if the organization is effectively executing its strategic plan.
Goals	Goals are short-term and long-term outcomes that position an organization to achieve its objectives.
Critical Success Factors	CSFs are elements (metrics) that help the organization to achieve its goals.
Key Performance Indicators	A KPI is a quantifiable (measurable) performance indicator tied to a specific critical success factors.

Note: Section 2.4.1.1, page 16 of *Business Analysis for Practitioners: A Practice Guide*, states, "Figure 2-1 shows an example of the hierarchical relationship between goals . . ." As outlined above, you'll note a slight variation to this context. On the exam, I would advocate that you follow the approach outlined in this book, in which objectives are achieved by successfully completing one or more goals. Consider the example in Table 1.2—if the objective is to win the match, it would be achieved by scoring goals.

Table 1.2 The Organizational Pillars in Practice

Pillar	Description
Mission	XYZ Soccer is a non-profit educational organization whose mission is to provide a safe and positive environment for the youth of our community to learn about soccer. It's our mission to help all players, regardless of skill, age, or background, to develop a lifelong interest in the game.
Vision	To be the leader in teaching young players about sportsmanship, fair play, respecting others (both on and off the field), and teaching our participants about the fundaments of soccer.
Strategy	• Teamwork is more important than individual skills. • Teams will demonstrate fairness of play and sportsmanship at all times (both on and off the field). • Coaches, players, and parents will demonstrate respect at all times to referees and to all players, coaches, and spectators. • Parents and siblings will support the players in an effort to develop their skills.
Objectives	In a season, the team will strive to win an equal number of games than it loses.
Goals	In any game, the team will score one more goal then their opponent to win the match.
Critical Success Factor 1	Each player will be provided with time during a game to practice and demonstrate their skills.
Key Performance Indicators 1	Each player will log a minimum of 10 minutes per half of game play.
Critical Success Factor 2	The team will embody respect, courteousness, and encouragement regardless of the outcome of the game.
Key Performance Indicator 2	At the conclusion of each game, one player from the opposing team will be recognized for their contribution to the game.

1.4 Organizational Structures

To further explain the origin of requirements, we will explore two important concepts. The first concept is organizational structures or hierarchies. In large organizations, there can be five or more levels, whereas in smaller organizations there can be fewer. Please take a moment to familiarize yourself with Table 1.3: Organizational Structures.

Table 1.3 Organizational Structures

Level 0	Board of Directors
Level 1	Enterprise
Level 2	Corporate
Level 3	Business Unit
Level 4	Functional
Level 5	Operations

Because portfolios encompass programs, projects, and operational activity, strategic opportunities and problems can originate from the Board of Directors (Level 0) to functional areas (Level 4), whereas tactical maneuvers to address problems or capitalize on opportunities can originate within operations (Level 5).

In both cases, the proposed solutions must be anchored to the organizational pillars. This is further illustrated in Table 1.4: Origin of Requirements.

Table 1.4 Origin of Requirements

	Level 0	Board of Directors	
Strategic	Level 1	Enterprise	Tactical
	Level 2	Corporate	
	Level 3	Business Unit	
	Level 4	Functional	
	Level 5	Operations	

1.5 Key Artifacts

The tasks, processes, tools, and techniques available to business analysts to aid in the journey from understanding the needs of the business to evaluating the value proposition are guided by set of key artifacts.

At the end of each domain chapter, the book will highlight the key artifacts. For the exam, please familiarize yourself with each artifact, taking note as to when it's created and its key contents.

1.6 Requirement Types

Business Analysis for Practitioners: A Practice Guide references six specific categories of business requirements (see Table 1.5). All too often, project teams focus solely on the functional requirements, which are a subset of solution requirements. By ignoring the other categories, up to 80% of a customer's requirements might be overlooked. Please take a moment to familiarize yourself with structure of requirements.

Table 1.5 Requirement Types

1.	Business Requirements	Business requirements describe the higher-level needs of the enterprise. Typically, in the form of a business requirements document (BRD), they describe the purpose/intent of the component and metrics to evaluate its impact on the organization. BRDs contain the component goals, which are then linked to the strategic plan. The BRD focuses on the entire enterprise, not specific organizational levels.
2.	Stakeholder Requirements	Stakeholder requirements describe the needs of a stakeholder or group of stakeholders and how they will benefit from and interact with the component.
3.	Solution Requirements	Solution requirements describe the characteristics of the component (e.g., features, functions), which are further decomposed into functional and nonfunctional requirements.
3a.	Functional Requirements	Functional requirements describe the behaviors or functions of the component—a statement of conformity. Described in Yes/No terms, they are often categorized as solution utility.
3b.	Nonfunctional Requirements	Nonfunctional requirements describe the environmental conditions of the component. They are typically characterized as quality or additive requirements and are described as "must haves," often categorized as solution warrantee requirements.
4.	Transition Requirements	Transition requirements describe the temporary capabilities that must exist for the component to transition from a current to a desired state. They cannot be fully defined until the "as-is" and "to-be" solutions are completely defined.

Table 1.6: Examples of Requirements Types, highlights a few samples for each classification.

Table 1.6 Examples of Requirement Types

1.	Business Requirements	Background Key stakeholders Current state design Future state design	Preliminary scope Success criteria Constraints Data models and diagrams
2.	Stakeholder Requirements	Objectives Critical success factors	Goals Key performance indicators
3. 3a.	Solution Requirements Functional Requirements	Authentication Reporting requirements Interfaces	Data conversions Regulatory requirements Extensions
	System Functions		
3b.	Nonfunctional Requirements	Availability Capacity Continuity	Performance Security and compliance Service level management
	Quality & Behavior Attributes		
4.	Transition Requirements	Data conversion	Training

☛ *Tip:* Remember, on the exam, *functional requirements* describe how the component works, whereas *nonfunctional requirements* describes the quality of the component. Requirements are then decomposed into *specifications* (e.g., functional specifications, technical specifications, etc.). Functional requirements are often know as RICE objects (reports, interfaces, conversions, and extensions). I often advocate expanding this to include application configuration/build and workflows. Figure 5.3 (page 117) provides a hierarchical and working view of Table 1.6.

! *Hint:* On the exam, you can expect at least one question on the subject of Planguage, which comprises key words to address concerns with ambiguous and incomplete nonfunctional requirements. This tool is further discussed in the Evaluation domain.

Introductory
Chapter Exercises

Chapter Exercise: The Seven Organizational Pillars

INSTRUCTIONS: **Step 1:** Take 10 minutes and complete the table below based on your organization.

Step 2: After completing the table below, take a few moments to validate your answers. Consider, for a moment, how well your responses align with the published information about your company. In your role as a business analyst, it is paramount that you have a solid understanding of these pillars, especially as you approach the experience-based questions on the exam.

Table 1.7 The Organizational Pillars

Mission	
Vision	
Strategy	
Objectives	
Goals	
Critical Success Factors 1	
Key Performance Indicators 1	
Critical Success Factor 2	
Key Performance Indicator 2	

Exercise: Organizational Structure

INSTRUCTIONS: Take a few minutes to document the structure of either your organization or that of a client.

Level 0: Board of Directors	
Level 1: Enterprise	
Level 2: Corporate	
Level 3: Business Unit	
Level 4: Functional	
Level 5: Operations	

Exercise: Great Pyramid of Giza Requirements

INSTRUCTIONS: Referring to Section 1.1.1: Facts About the Great Pyramid of Giza (page 2), write down one or two requirements Egyptian Pharaoh Khufu may have expressed for each of the categories below.

1.	Business Requirements	
2.	Stakeholder Requirements	
3.	Solution Requirements	
3a.	Functional Requirements	
3b.	Nonfunctional Requirements	
4.	Transition Requirements	

Exercise: Types of Requirements

INSTRUCTIONS: For an initiative that you are either presently working on or have recently completed, write down three requirements for each of the below categories.

1.	Business Requirements	
2.	Stakeholder Requirements	
3.	Solution Requirements	
3a.	Functional Requirements	
3b.	Nonfunctional Requirements	
4.	Transition Requirements	

Chapter 2

Needs Assessment

Study Hints

The Needs Assessment domain includes the business analyst tasks necessary to (a) understand the opportunities or problems of the sponsoring organization; (b) incorporate the tools and techniques to determine the value proposition; (c) establish the project goals and objectives and link them to the organizational pillars; and (d) complete a comprehensive stakeholder assessment. The activities in this domain produce deliverables, which will be used throughout the project, including the situation statement, solution scope statement, business case, and various stakeholder assessment artifacts.

> *You want to be extra rigorous about making the best possible thing you can.*
> *Find everything that's wrong with it and fix it. Seek negative feedback, particularly from friends.*

— Elon Musk

It's important to remember that a needs assessment can be mandated by a sponsoring organization, formally requested by a stakeholder, or recommended by a business analyst. Although typically conducted prior to the start of a project or program, external factors influenced by the market or regulations can necessitate a needs assessment. You'll recall from the introductory chapter, a request for a needs assessment can originate at the portfolio or program level (top down), or occur at the request of an end user (bottom up—for example, operations).

On the exam, it is very likely that you will see questions related to the contents of the business case and charter. Once the business case is created, this artifact will be updated over the course of the project and used to support effective decision making. Furthermore, it serves as a reminder of why the initiative was approved. Once the business case is created, it will be used to draft the scope statement and the artifact that formally authorizes the project—the charter. In the Evaluation domain, the business analysts will refer to the business case to validate the expected versus the actual benefits.

The most important concepts to remember in terms of the business case for the exam are:

a. The business case is an approved document that contains the costs and benefits for the product and the project.
b. The business case is a living document, which is constantly referenced and updated over the life of a project.
c. At the conclusion of the project, the business case is validated based on actual measurements.

The most important concepts to remember in terms of the charter for the exam are that it:

    d. Documents the rationale for the project
    e. Establishes the authority of the project manager
    f. Formally authorizes the project

At the end of this chapter are exercises to help in your recall of the contents of the business case, the scope statement, and project charter. The standards and templates for these artifacts will be provided by the organization's designated project management office (PMO) and are commonly referred to as either *portfolio* or *program process assets* (PPA), as compared to *organizational process assets* (OPA), which extend beyond the PMO.

The *Examination Content Outline,* shown in Figure 2.1, contains five essential business analyst tasks associated with the Needs Assessment domain. Please refer to the practice guides and the *PMBOK® Guide* for additional context related to all the items outlined below.

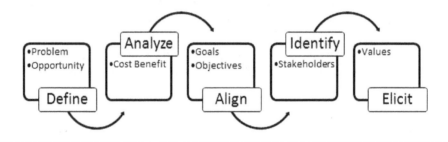

Figure 2.1   Lifecycle of Needs Assessment.

The key artifacts associated with the Needs Assessment domain are shown in Figure 2.2.

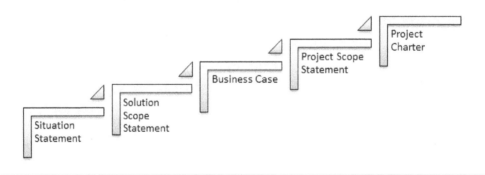

Figure 2.2   Needs Assessment artifacts.

At the end of this chapter are exercises to reinforce the concepts from this domain, the answers to which can be found throughout this exam prep book. This domain represents 18% (or 36) of the questions that will appear on the PMI-PBA® Exam.

Major Topics

2.1 Situation Statement, Solution Scope Statement, Business Case Content

Problem and Opportunity Analysis Techniques
- Brainstorming
- Scenario analysis
- User journey maps
- Value engineering

Root Cause Analysis Techniques
- 5 Whys
- Cause-and-effect diagrams:
  - Ishikawa (fishbone)
  - Interrelationship diagrams
  - Process flows

➢ Additional Concepts

Assessment and Analysis Techniques
- Capability analysis
  - Capability table
  - Affinity diagrams
  - Benchmarking
  - "As-is" process analysis
  - Enterprise and business architectures
  - Capability frameworks
  - Gap analysis

Alternative Option Analysis & Feasibility Assessment
- Cost-effectiveness
- Operational
- Technology/system
- Time
- PESTEL: political, economic, social, technological, environmental, and legal

Data Analysis Tools and Techniques
- Benchmarking
- Pareto analysis
- Trend analysis

RAIDC Analysis Tools and Techniques

- Risks
- Assumptions
- Issues
- Dependencies
- Constraints

Prioritization Tools and Techniques

- Weighted ranking and scoring
- Multi-criteria scoring
- Pair-wise analysis
- MoSCoW
- MultiVoting
- Timeboxing

Quality Tools and Techniques

- Voice of the customer
- SIPOC diagrams

2.2 Solution Value Proposition

Valuation Tools and Techniques

- Cost–benefit analysis
  - Payback period
  - Return on investment
  - Internal rate of return
  - Net present value
- Force-field analysis
- Feasibility studies
  - Political, economic, social, technological, environmental, and legal (PESTEL)
- Kano model
- Net promotor score
- Purpose alignment model
- SWOT analysis
- Value stream map

2.3 Project Goals and Objectives, Solution Alignment to Organizational Goals and Objectives

Goals and Objective Techniques

- SMART goals and objectives
- SWOT analysis

Scope Modeling Techniques
- Goal model and Business Objective Model (BOM)
- Ecosystem map
- Context diagram
- Feature model
- Use-case diagram

2.4 Stakeholder Analysis, Stakeholder Engagement, Communications Management

Stakeholder Analysis Techniques
- Personas
- Role definition (RACI)
- Job analysis
- Skills assessment

➢ **Additional Concepts**
  - Brainstorming
  - Decomposition modeling
  - Interviews
  - Organizational modeling
  - Surveys and focus groups
  - Political and cultural awareness

  Stakeholder Artifacts
  - Stakeholder register
  - Stakeholder communication matrix
  - Stakeholder interest table
  - Power and interest grids
  - Salience diagramming

2.5 Stakeholder Values, Requirements Prioritization Baseline

Elicitation Tools and Techniques
- Brainstorming
- Document analysis
- Focus groups
- Interviewing
- Questionnaires
- Observation
- Research
- Surveys
- Workshop facilitation

➢ **Additional Concepts**
  ○ Facilitation tools and techniques
    ▪ Brainstorming, focus groups, ice breakers, MultiVoting (NGT), scatter diagrams, flow charting, affinity diagramming, consensus decision making, didactic interaction

To reinforce the material from this chapter, on the following pages are a few simple exercises related to the Needs Assessment domain. Using information from a current project or initiative, it's our recommendation that you begin to incorporate these concepts and terminology into your current practice. By doing so, the exam will be easier, and you will ultimately become more proficient in the Needs Assessment domain of business analysis.

Needs Assessment
Chapter Exercises

Test Your Knowledge: Needs Assessment

INSTRUCTIONS: On the exam, it's not enough to simply recall what tool or technique is used in the process; you'll need to understand the fundamentals, the output, and how it relates to business analysis. Below are some of the key concepts related to this chapter; take the time to research each and fill in the blanks.

Q	Practice Area	Example	Explanation
	These diagrams are used to show the causes of specific events.	Ishikawa or fishbone diagram	High-level causes of a problem. Categories can include machines, materials, measurements, stakeholders, and locations.
1	What are three methods to identify stakeholders?		
2	What is a Pareto analysis?		
3	To recommend a viable and appropriate solution, a business analyst may use what nine tools or techniques?		

Q	Practice Area	Example	Explanation
4	To determine required capabilities, what are the three tools or techniques that can be used?		
5	To document an organization's current capabilities, what tools or techniques can be used?		
6	What is the final deliverable of the Needs Assessment domain?		

Q	Practice Area	Example	Explanation
7	What are the five types of feasibility assessments?		
8	What are five commonly used tools for ranking options?		

Q	Practice Area	Example	Explanation
9	When conducting a cost–benefit analysis, what are four commonly used tools?		
10	What are the six types of requirements?		

Memory Game: Stakeholders

INSTRUCTIONS: Make a photocopy of the next two pages, then cut out the pieces along the dotted lines. Without referring to the book, try to match the stakeholders with their roles.

Sponsor	Governance Team	Portfolio Manager
Executive (or group) who provides resources (financial, human capital, equipment) and who are ultimately responsible for delivering on the intended benefits	The group tasked with ensuring that the component's goals are achieved, and that the effort is aligned with the organization's overall strategy	Executive responsible for overseeing the programs, projects, and operational activities within their area of responsibility
Program Manager	Project Manager	Team Member
The individual responsible for managing and overseeing a program	The individual responsible for managing and overseeing a project within the portfolio or program	The resources performing project, program, or operational activities

Funding Organization	Performing Organization	Project Management Office
The group, either internal or external to the organization, that is providing funding	The group, either internal or external to the organization, that is performing the work of the project, program, or operational activities	The group tasked with defining and managing the standards and providing administrative support for the programs and projects
Customers	**Potential Customers**	**Suppliers/ Contractors**
The organization or individual(s) who will use the new product or capabilities and who will determine the overall success/ failure of the initiative	Prior and/or future organizations or individual(s) who will be observing the initiative	Third parties who are either directly or indirectly associated with the initiative
Regulatory Agencies	**Competitors**	**Affected Individuals or Organizations**
Government body or authority exercising automous oversight	Group who may be interested in benchmarking data or may look to expand or modify their offerings based on the inititive	Those who perceive they may either benefit from or be disadvantaged by the initiative

Memory Game: Stakeholder Salience

INSTRUCTIONS: Make a photocopy of this page, then cut out the pieces along the dotted lines. Without referring to the book, try to match the stakeholder saliences.

Dormant	Dominant	Discretionary
high power, low legitimacy and urgency	high power and legitimacy, low urgency	low power, high legitimacy, low urgency
Core	**Dangerous**	**Dependent**
high power, legitimacy, and urgency	high power, low legitimacy, high urgency	low power, high legitimacy, high urgency
Demanding	**Power**	**Legitimacy**
low power, low legitimacy, high urgency	ability to influence	actions are desirable, proper, or appropriate
	Urgency	
	a call for immediate attention	

Memory Game: Project Charter and Scope Statement

INSTRUCTIONS: Listed below are the key contents of the project charter and project scope statement; practice recalling them using the template on the following page.

Charter	Project Scope Statement
Approval requirements	Acceptance criteria
Assumptions/constraints	Assumptions
Budget (summary)	Constraints
Description (high-level)	Deliverables
Objectives (measurable)	Exclusions
Justification (or purpose)	Scope description
Milestones (summary)	
Requirements (high-level)	
Risks (high-level)	
Stakeholders	
Sponsor	
Project manager (responsibility and authority)	
Success criteria	

Memory Game: Project Charter and Scope Statement

INSTRUCTIONS: Make a photocopy of this page; practice listing the contents of the project charter and project scope statement.

Charter	Project Scope Statement

Memory Game: Business Case

INSTRUCTIONS: Listed below are the key sections and contents of the business case; practice recalling them using the template on the following page.

Business Case		
Opportunity/Problem		
Situation statement	Supporting data	Stakeholders (impacted)
Analysis		
Organizational goals and objectives	Root causes, contributors to success	Supporting data
Capabilities (needed versus existing)	Program/project objectives	Alignment to portfolio
Recommendation		
Feasibility analysis (results)	Assumptions, constraints, risks, and dependencies	Alternatives (rank ordered)
Recommendation (including cost–benefit analysis)	Implementation approach (milestones, dependencies, roles, and responsibilities)	
Evaluation		
Benefits realization plan		

Memory Game: Business Case

INSTRUCTIONS: Make a photocopy of this page; practice recalling the key sections and contents of the business case.

Business Case

Needs Assessment: Practice Questions

INSTRUCTIONS: Make a copy of the answer sheet on page 41. Note the most suitable answer for each multiple-choice question in the appropriate space on the answer sheet.

1. Your sponsor, Rupert, asked for assistance outlining the business case. What are the required components of this important document?
   a. All the details gathered during the Needs Assessment
   b. Rank order alternatives, analysis of the situation, affected stakeholder groups, and the problem or opportunity
   c. Rank order alternatives, analysis of the situation, affected stakeholder groups, the problem or opportunity, detailed valuation analysis
   d. Rank order alternatives, analysis of the situation, affected stakeholder groups, and the problem prompting the need for action

2. Your project manager, Hailey, asked for help creating goals and objectives that are specific and clearly understood. You recall that goals and objectives should be:
   a. Specific, measurable, attainable, relevant, and time bound
   b. Written down, agreed to, and frequently revisited to assess relevance
   c. Specific, meaningful, achievable, realistic, testable
   d. Time-tested, time-bound, meaningful, achievable, realistic

3. Claudia, the VP of the project management office, asked for your assessment as to whether the product approved by the executive sponsor is in alignment with business goals and objectives. You schedule a meeting to review the following:
   a. Problems to be solved and any potential opportunities; an assessment of the organization's current capabilities; an outline of desired future state; capability gaps; the outline of the business case that will enable the organization to achieve its vision, mission, goals, and overall business objectives.
   b. Problems to be solved; an assessment of the organization's current capabilities; an outline of desired future state; capability gaps; the outline of the business case that will enable the organization to achieve its vision, mission, goals, and overall business objectives.
   c. Problems to be solved and any potential opportunities; an assessment of the organization's current capabilities; an outline of desired future state; capability gaps; the outline of the problem statement that will outline how the organization should consider potential alternatives.
   d. The traceability matrix detailing how each business requirement is solidly anchored to the organization's vision, mission, goals, and overall business objectives.

4. You're working with Amanda and Laura—two SMEs from the Animal Life Sciences Division of your company who are experts in veterinary sciences. As their business analyst, you've begun the process of identifying stakeholders based on their power, legitimacy, and urgency. You're creating a:
   a. Power and interest diagram
   b. Salience diagram
   c. Stakeholder matrix
   d. Power and interest matrix

5. Your executive sponsor, Evan, asked for assistance outlining the benefits associated with building XYZ's new headquarters. As you work on the business case, which of the below statements would be preferable?
   a. The new building will be environment friendly; by reducing XYZ's carbon footprint and using natural light, the building will be more comfortable and improve morale.
   b. The new building will be environment friendly; by installing solar panels, geothermal heat pumps, and energy-efficient windows, it's expected the project will reduce XYZ's carbon footprint by 10%.
   c. The new building will be environment friendly; by installing solar panels, geothermal heat pumps, and energy-efficient windows, it's expected the project will reduce XYZ's carbon footprint by 10% year one and incorporate the use of 25% more natural light.
   d. The new building will be environment friendly, incorporating the use of more natural light, thus improving comfort, morale, and employee productivity.

6. You work for a government subcontractor and have prepared a valuation analysis of two projects for the executive sponsor, General Robert Garth. Because there are insufficient resources (people and financial), the general can only select one. Each project will take 18 months and cost $2,500,000; based on the below which project should the general select:
   A. A process optimization that will result in cost reductions of $1,250,000 per year, with benefit realization starting six months after go-live, running with a discount rate of 3%.
   B. A mandated initiative that will have a negative NPV, forecasted to be –$2,500,000 over 10 years.
   a. Option B, the mandated government initiative with a negative NPV.
   b. Option A, because it has the higher NPV.
   c. Option A, then two years after go-live, continue with Option B, which could be funded by the cost reductions achieved from Option A.
   d. There is not enough information to answer the question.

7. You are facilitating a session with the intent of visualizing complex problems. Because there are complex relationships, you decide to create an interrelationship diagram. One of your SMEs, Heather, has noted that there are several instances in which two factors influence each other; how should this be best addressed?
   a. The team needs to note both factors, otherwise something may be overlooked.
   b. Cause and effect factors of significant value should be depicted as squares.
   c. In cases where there is more than one influencing factor, the team needs to determine which factor is stronger, and note only one.
   d. Cause and effect factors of significant value should be depicted as circles.

8. You are the business analyst for snack food company Crunch & Chips. As far back as you can remember, you've enjoyed all their products. Now you're facilitating a study to assess the feasibility of exchanging the main ingredients with ones that will save the company nearly 25% per package. In taste tests, only a small portion (< 15%) can tell the difference between the premium ingredients, which have been used for over 100 years, and the newer synthetic ingredients. During these facilitated sessions, you should:
   a. While facilitating the sessions, help guide the participants to a selection that is aligned with the organization's vision, mission, goals, and objectives.
   b. Defer to product management and marketing; they are best skilled to participate in these sessions.

    c. As the business analyst, you should both facilitate and lead the feasibility study.

    d. Only facilitate the sessions; as a business analyst, you need to remain impartial.

9. As a business analyst for a company that manufactures pencils, you have just presented a plan that includes metrics to assess how well the proposed solution aligns to the organization's overall goals and objectives. You have just presented:

    a. The strategic plan

    b. The vision and mission plan

    c. The business case

    d. The business plan

10. You work as business analyst for a government subcontractor that manufactures combat helicopters. While conducting tests for one product, an engineer discovered a new compound that will make the helicopters 25% lighter and 50% more durable. Becoming aware of this discovery, you promptly commenced a needs assessment and are now assessing the feasibility of using this compound. Which factor is most important to consider?

    a. Time to market

    b. Technology/systemic feasibility

    c. Operational

    d. Effectiveness feasibility

11. The CEO of Celeste Marie Bakeries, Valverde, asked for your help in determining the value proposition for expanding and modernizing the kitchen that bakes their chocolatey, gooey, delicious brownies. You suggest:

    a. There are a few hurdles we'll need to overcome before we can consider expanding and modernizing the kitchen.

    b. We should first determine our stakeholders, then construct a stakeholder matrix, which will serve as inputs to our value proposition.

    c. A first step should be to clearly outline the project goals and objectives.

    d. The solution scope statement will serve as an input to the baseline for prioritizing requirements, which will be data points in the value proposition.

12. As an expert in business analysis, you've been hired by Maggie, a process-mapping consultant for a pet food company. During the initial meeting, she starts to outline the solution scope statement. To this point, Maggie has already:

    a. Determined the alignment of the proposed solution with the organization's goals and objectives.

    b. Collaborated with the project manager in the development of project goals and objectives.

    c. Asked a series of probing questions in an attempt to understand the opportunity.

    d. Determined the NPV, IRR, and ROI for the new line of cat food.

13. Tucker, the chief science officer for your company's division of DDR Memory Chips, asked for your help mapping the division's people, locations, processes, applications, data, and technology. This process method is best known as:

    a. Value stream mapping

    b. Capability mapping

c. Process requirements mapping

d. Enterprise and business architectures

14. Prior to studying business analysis, you majored in creative writing at your community college. Although you don't directly support them, almost all executive sponsors ask for your assistance drafting situation statements. This is because:

a. They are linked to the organization's mission, vision, goals, and objectives.

b. Your situation statements ensure a complete understanding of the opportunity or problem, the contributing effects, and the overall impacts.

c. Your situation statements ensure a complete understanding of the opportunity or problem and the associated benefits, both tangible and intangible.

d. Your situation statements are brief and concise, but most importantly, they are linked to the organization's mission, vision, goals, and objectives.

15. Your project manager, Gwen, requested your assistance with the framework for a tool that will be used during a stakeholder facilitation session to articulate and understand the high-level views surrounding a business proposition.

a. You draw quadrants for strengths, weaknesses, opportunities, and threats on the white board.

b. You begin by outlining a capability table on the easel.

c. You commence by outlining an affinity table, which will be distributed to each participant.

d. You start by outlining the relevant criteria on the SMART board.

16. Elaine and Burke, both senior VPs of recreation for a retirement community, asked for your assistance developing an interrelationship diagram. As you facilitate this session, you find instances in which some factors have many incoming arrows. As the day progresses, another session produces factors with a large number of outgoing arrows. What do these factors represent?

a. Factors with a large number of outgoing arrows are the effects or key outcomes of other factors, whereas factors with a large number of incoming arrows are the causes or source of the concern.

b. The arrows represent the direction of the cause, starting with the stronger influences.

c. The arrows represent the bi-directionality of the factors.

d. Factors with a large number of outgoing arrows are the causes or source of the concern, whereas factors with a large number of incoming arrows are the effects or key outcomes of other factors.

17. Lucy, the chief quality officer for Lucky Organic Vegetables, asked for assistance creating a key project document that will outline the factors that constitute success, who determines success, and ultimately who'll sign off that the project was successful. Lucy is asking for assistance creating:

a. The solution scope statement

b. The business case

c. The charter

d. The solution requirements document

18. Your sponsor, Autumn, has asked you to conduct a study to determine if new opportunities exist within your market. You commission a:

a. Benchmarking study

b. Capability assessment

    c. Market-based affinity study

    d. Monte Carlo study

19. Although there may be some perceived redundancy with the project charter, the PM assigned to your team, Bobby, asked for your assistance drafting a document that outlines the scope, deliverables, exclusions, assumptions, and constraints associated with your project. You're outlining:

    a. A project scope statement

    b. A project business case

    c. A Preliminary project scope statement

    d. A project requirements document

20. You are working with Ellie, the product manager for hair-care products at Tortoise, Inc., a global manufacturer. You're facilitating a session for which you've been asked to draw a diagram to map out the high-level causes of the problems.

    a. You start to draw an affinity table with categories for ROI, NPV, IRR, and BCR.

    b. You start by drawing a few lines and labeling them: methods, measurements, regulations, procedures, manufacturing line.

    c. You create a capability table that outlines the benefits of the new line, including increased market share and high levels of customer satisfaction.

    d. You create a process flow diagram, which outlines the enterprise and business architectures.

Answer Sheet

	Answer Choice				Correct	Incorrect	Predicted Answer			
							90%	75%	50%	25%
1	a	b	c	d						
2	a	b	c	d						
3	a	b	c	d						
4	a	b	c	d						
5	a	b	c	d						
6	a	b	c	d						
7	a	b	c	d						
8	a	b	c	d						
9	a	b	c	d						
10	a	b	c	d						
11	a	b	c	d						
12	a	b	c	d						
13	a	b	c	d						
14	a	b	c	d						
15	a	b	c	d						
16	a	b	c	d						
17	a	b	c	d						
18	a	b	c	d						
19	a	b	c	d						
20	a	b	c	d						

Needs Assessment: Answer Key

1. Your sponsor, Rupert, asked for assistance outlining the business case. What are the required components of this important document?
   a. All the details gathered during the Needs Assessment.
   b. Rank order alternatives, analysis of the situation, affected stakeholder groups, and the problem or opportunity.
   c. Rank order alternatives, analysis of the situation, affected stakeholder groups, the problem or opportunity, detailed valuation analysis.
   d. Rank order alternatives, analysis of the situation, affected stakeholder groups, and the problem prompting the need for action.

   The correct answer is: **B**

   Although the format and templates may vary among organizations, the common elements of a well-formed business case are a description of the problem to be addressed or the opportunity to be pursued; a thorough analysis linking the proposed solution to organizational goals and objectives, noting the affected stakeholder groups; and a comprehensive recommendation providing the rationale for the decision along with rank-ordered alternatives. Once created, this artifact will be updated over the course of the project and used to support effective decision making, and it serves as a reminder of why the initiative was approved.

   Answer Choice A: This answer choice is nonspecific.

   Answer Choice C: This answer choice adds a detailed valuation analysis, which would not be a core element of the business case.

   Answer Choice D: This answer choice only mentions problems—business cases are foundational documents for both problems and opportunities.

   Business Analysis for Practitioners: A Practice Guide, Section 2.6, "Assemble the Business Case."

   PMI Professional in Business Analysis (PMI-PBA)® Examination Content Outline, 2013, "Needs Assessment," Task 5.

2. Your project manager, Hailey, asked for help creating goals and objectives that are specific and clearly understood. You recall that goals and objectives should be:
   a. Specific, measurable, attainable, relevant, and time bound
   b. Written down, agreed to, and frequently revisited to assess relevance
   c. Specific, meaningful, achievable, realistic, testable
   d. Time-tested, time-bound, meaningful, achievable, realistic

   The correct answer is: **A**

   Projects exist to enable change in organizations; to be successful, the resulting product must be linked to the organization's goals and objectives. This question centers on collaboration with the project manager and ensuring alignment between the project's goals and objectives and those of the proposed solution. In this task, it is important to note that if goals and objectives do not exist, it is the business analyst's responsibility to work with the sponsor to create them. Answer choice A is aligned with the SMART acronym: specific, measurable, attainable, realistic, and time bound. Please note, other substitutions can also include the achievable or assignable, and the relevant or reasonable.

Answer Choice B: Does not properly address the question.
Answer Choice C: Meaningful is the distractor.
Answer Choice D: Time-tested is not valid.
Business Analysis for Practitioners: A Practice Guide, Section 2.4.1.2, "SMART Goals and Objectives."
PMI Professional in Business Analysis (PMI-PBA)® Examination Content Outline, 2013, "Needs Assessment," Task 3.

3. Claudia, the VP of the project management office, asked for your assessment as to whether the product approved by the executive sponsor is in alignment with business goals and objectives. You schedule a meeting to review the following:
   a. Problems to be solved and any potential opportunities; an assessment of the organization's current capabilities; an outline of the desired future state; capability gaps and an outline of the business case that will enable the organization to achieve its vision, mission, goals, and overall business objectives.
   b. Problems to be solved; an assessment of the organization's current capabilities; an outline of desired future state; capability gaps; and an outline of the business case that will enable the organization to achieve its vision, mission, goals, and overall business objectives.
   c. Problems to be solved and any potential opportunities, an assessment of the organization's current capabilities, an outline of desired future state capability gaps, and an outline of the problem statement that will show how the organization should consider potential alternatives.
   d. The traceability matrix detailing how each business requirement is solidly anchored to the organization's vision, mission, goals, and overall business objectives.

The correct answer is: **A**

This question is from the perspective of the PMO and inquires if the product approved by the executive sponsor was aligned with the goals and objectives of the organization. Unfortunately, some organizations fall victim to either "pet projects" or "zombie projects," neither of which deliver any value. The Needs Assessment domain includes the activities necessary to understand the opportunities or problems of the business and the tools and techniques to propose viable solutions. It concludes with an approved business case that will address the problems to be solved and any potential opportunities, an assessment of the organization's current capabilities, an outline of the desired future state, and capability gaps—all of which will be evaluated to enable the organization to achieve its vision, mission, goals, and overall business objectives.
Answer Choice B: This answer choice omitted potential opportunities.
Answer Choice C: Provided the *problem statement* as the artifact, whereas answer choice A provides the *business case* as the artifact.
Answer Choice D: This answer choice is incomplete; although the traceability matrix is used for tracing and tracking, answer choice A addresses the question in a more complete manner.
Business Analysis for Practitioners: A Practice Guide, Section 1.8.1 to review the Business Analysts Deliverables as Part of the Needs Assessments.
PMI Professional in Business Analysis (PMI-PBA)® Examination Content Outline, 2013, "Needs Assessment," Task 3.

4. You are working with two SMEs from the Animal Life Sciences Division of your company—Amanda and Laura—who are both experts in veterinary sciences. As their business analyst, you've begun the process of identifying stakeholders based on their power, legitimacy, and urgency. You're creating a:

a. Power and interest diagram
b. Salience diagram
c. Stakeholder matrix
d. Power and interest matrix

The correct answer is: **B**

The stakeholder salience diagramming method is used to assist team members when the need arises to prioritize competing stakeholder requests (Wood, 1997). The purpose of the tool is to categorize stakeholders based on the factors of power, urgency, and legitimacy. As with the stakeholder register and the power/interest grid, the project salience model should be reevaluated over the life of a project.

Answer Choices A & D: The power and interest diagram or matrices are typically a four-quadrant analysis, in which stakeholders are plotted in the quadrants of: monitor, keep informed, actively manage, and keep satisfied. The question is asking for the tool to plot power, legitimacy, and urgency.

Answer Choice C: The stakeholder matrix or stakeholder register would simply list all the stakeholders with a few defining characteristics.

A Guide to the Project Management Body of Knowledge (PMBOK)®, Section 13.1.2, "Tools and Techniques for Stakeholder Analysis."

PMI Professional in Business Analysis (PMI-PBA)® Examination Content Outline, 2013, "Needs Assessment," Task 4.

5. Your executive sponsor, Evan, asked for assistance outlining the benefits associated with building XYZ's new headquarters. As you work on the business case, which of the statements below would be preferable?

a. The new building will be environment friendly; by reducing XYZ's carbon footprint and using natural light, the building will be more comfortable and improve morale.
b. The new building will be environment friendly; by installing solar panels, geothermal heat pumps, and energy efficient windows, it's expected the project will reduce XYZ's carbon footprint by 10%.
c. The new building will be environment friendly; by installing solar panels, geothermal heat pumps, and energy efficient windows, it's expected the project will reduce XYZ's carbon footprint by 10% year one and incorporate the use of 25% more natural light.
d. The new building will be environment friendly and incorporate the use of more natural light, thus improving comfort, morale, and employee productivity.

The correct answer is: **C**

Within the business case, proposed initiatives are linked to the organization's goals and objectives. Properly formed statements follow the SMART principle, in which product benefits are described in a specific manner, they are measurable, agreed to by all parties, realistic to achieve in the allotted schedule, and time bound. Answer choice C completely addresses all the principles of SMART goals and objectives.

Answer Choices A & D: While they are well intentioned, these statements are not measurable.

Answer Choice B: As with answer choices A and D, the statement is well intentioned, but not time bound.

Business Analysis for Practitioners: A Practice Guide, Section 2.4.1.2 for the components of SMART Goals and Objectives.

PMI Professional in Business Analysis (PMI-PBA)® Examination Content Outline, 2013, "Needs Assessment," Task 3.

6. You work for a government subcontractor and prepared a valuation analysis of two projects for the executive sponsor, General Robert Garth. Because there are insufficient resources (people and financial), the general can only select one. Each project will take 18 months and cost $2,500,000, based on the below; which project should the general select?
   A. A process optimization that will result in cost reductions of $1,250,000 per year, with bene-fit realization starting six months after go-live, running with a discount rate of 3%.
   B. A mandated initiative that will have a negative NPV, forecasted to be –$2,500,000 over 10 years.
   a. Project B, the mandated government initiative with a negative NPV.
   b. Project A, because it has the higher NPV.
   c. Project A, then two years after go-live, continue with Option B, which could be funded by the cost reductions achieved from Option A.
   d. There is not enough information to answer the question.

The correct answer is: **A**

In this scenario, the business analyst is documenting the expected benefits of a government proj-ect. In the private sector, projects with a positive net present value (NPV) are considered a good investment; however, it's not uncommon for government mandates and initiatives to be approved with a negative NPV. On the exam, although you will not be expected to calculate NPV, you may need to select the preferred project based on the organization.

Answer Choice B: Generates a positive return; however, project B is a government mandate, where NPV is less of a concern than in the private sector.

Answer Choice C: This may seem like a logical choice, but the question is singular, looking for only one project.

Answer Choice D: This is not the best choice, as there is sufficient information to answer the question.

Business Analysis for Practitioners: A Practice Guide, Section 2.5.6.4, "NPV."

PMI Professional in Business Analysis (PMI-PBA)® Examination Content Outline, 2013, "Needs Assessment," Task 2.

7. You are facilitating a session with the intent of visualizing complex problems. Because there are complex relationships, you decide to create an interrelationship diagram. One of your SME's, Heather, has noted that there are several instances in which two factors influence each other; how should this be best addressed?
   a. The team needs to note both factors, otherwise something may be overlooked.
   b. Cause and effect factors of significant value should be depicted as squares.
   c. In cases in which there is more than one influencing factor, the team needs to determine which factor is stronger, and note only one.
   d. Cause and effect factors of significant value should be depicted as circles.

The correct answer is: **C**

Interrelationship diagrams are used for visualizing complex problems and relationships. When using the tool, it's quite common to find factors that influence each other. In this case, the business analyst, working with the subject matter experts, needs to determine which factor is stronger and note only one. As an outcome of this exercise, the business analyst and subject matter experts will be positioned to identify which factors are the leading causes of the problems.

> **Answer Choice A:** While this seems like a logical answer, with interrelationship diagrams there is one factor that will always have slight significance over another.
> **Answer Choices B & D:** These answer choices test your knowledge of the tool.
> *Business Analysis for Practitioners: A Practice Guide,* Section 2.4.4.2 to review Cause-and-Effect Diagrams.
> *PMI Professional in Business Analysis (PMI-PBA) Examination Content Outline,* 2013, "Needs Assessment," Task 2.

8. You are the business analyst for snack food company, Crunch & Chips. As far back as you can remember, you've enjoyed all their products. Now you're are facilitating a study to assess the feasibility of exchanging the main ingredients with ones that will save the company nearly 25% per package. In taste tests, only a small portion (< 15%) can tell the difference between the premium ingredients, which have been used for over 100 years, and the newer synthetic ingredients. During these facilitated sessions, you should:
    a. While facilitating the sessions, help guide the participants to a selection that is aligned with the organization's vision, mission, goals, and objectives.
    b. Defer to product management and marketing; they are best skilled to participate in these sessions.
    c. As the business analyst, you should both facilitate and lead the feasibility study.
    d. Only facilitate the sessions; as a business analyst you need to remain impartial.

> **The correct answer is: D**
>
> This question introduces statistical noise to distract students. At all times, especially during facilitation sessions, business analysts need to remain impartial. This information will be used within the business case as viable solutions are recommended in rank order. The focus should be on, "What problem are we solving?" or, "What problems do our customers have that this opportunity will address?"
> **Answer Choice A:** While this may seem appropriate, business analysts should not guide the conversation or selection.
> **Answer Choice B:** This is a distractor answer choice, as there may be other stakeholders that need to weigh in.
> **Answer Choice C:** Feasibility studies would most likely be led by product management or marketing; the business analyst could offer some assistance with this effort.
> *Business Analysis for Practitioners: A Practice Guide,* Section 2.3, "Identify Problem or Opportunity."
> *PMI Professional in Business Analysis (PMI-PBA)® Examination Content Outline,* 2013, "Needs Assessment," Task 2.

9. As a business analyst for a company that manufactures pencils, you have just presented a plan which includes metrics to assess how well the proposed solution aligns to the organization's overall goals and objectives. You have just presented:
    a. The strategic plan
    b. The vision and mission plan
    c. The business case
    d. The business plan

> **The correct answer is: C**
>
> When approaching questions on the exam pertaining to the business case, there are three key concepts to remember: (a) the business case is an approved document that contains the costs and

benefits for the product and the project; (b) the business case is a living document, which is constantly referenced and updated over the life of a project; (c) at the conclusion of the project, the business case is validated based on actual measurements. Answer choice C is the only option in which metrics could assess how well the proposed solution aligns to the organization's overall goals and objectives.

Answer Choice A: Strategic plans establish the organization's overall goals and objectives and the steps required to achieve these goals.

Answer Choice B: This is a made-up answer.

Answer Choice D: Business plans describe how a company intends to grow, evolve, and change and what the company hopes to achieve over a defined period of time.

Business Analysis for Practitioners: A Practice Guide, Section 2.6 to review the Business Case.

PMI Professional in Business Analysis (PMI-PBA)® Examination Content Outline, 2013, "Needs Assessment," Task 3.

10. You work as business analyst for a government subcontractor that manufactures combat helicopters. While conducting tests for one product, an engineer discovered a new compound that will make the helicopters 25% lighter and 50% more durable. Becoming aware of this discovery, you promptly commenced a needs assessment and are now assessing the feasibility of using this compound. Which factor is most important to consider?

    a. Time to market

    b. Technology/systemic feasibility

    c. Operational

    d. Effectiveness feasibility

The correct answer is: **C**

It's very common for scientists to make discoveries; unfortunately not all of them are viable. An operational feasibility assessment will evaluate several factors, including (a) how well does the discovery meet the nonfunctional requirements of the business? (b) what is the impact to the manufacture and customer? (c) is there a business need for this new discovery?

Answer Choice A: Time to market would only be considered once the offering was deemed viable; it looks at the cycle time from ideation to sale.

Answer Choice B: This study would evaluate if the organization had the technology to create or the skills to support the product. As the discovery was made, this is not a viable choice.

Answer Choice D: This is a distractor and not a valid feasibly study.

Business Analysis for Practitioners: A Practice Guide, Section 2.5.4.1, "Operational Feasibility."

PMI Professional in Business Analysis (PMI-PBA) Examination Content Outline, 2013, "Needs Assessment," Task 2.

11. The CEO of Celeste Marie Bakeries, Valverde, asked for your help in determining the value proposition for expanding and modernizing the kitchen that bakes their chocolatey, gooey, delicious brownies. You suggest:

    a. There are a few hurdles we'll need to overcome before we can consider expanding and modernizing the kitchen.

    b. We should first determine our stakeholders, then construct a stakeholder matrix that will serve as inputs to our value proposition.

    c. A first step should be to clearly outline the project goals and objectives.

    d. The solution scope statement will serve as an input to the baseline for prioritizing requirements, which will be data points in the value proposition.

The correct answer is: **A**

The value proposition is a key component to consider when preparing the cost–benefit analysis and supporting material for the business case. Typical models to support the benefit/value for the investment are: payback period (PB), return on investment (ROI), internal rate of return (IRR), and net present value (NPV). This question tests your knowledge of the fundamental aspects of ROI, where it's common for organizations to establish hurdle rates before a project can be considered.

Answer Choice B: A stakeholder analysis would not be required to determine financial value proposition.

Answer Choice C: The establishment of project goals and objectives would not be required to determine financial value proposition.

Answer Choice D: This is a made-up answer and a distractor.

Business Analysis for Practitioners: A Practice Guide, Section 2.5.6.2, "Return on Investment (ROI)."

PMI Professional in Business Analysis (PMI-PBA)® Examination Content Outline, 2013, "Needs Assessment," Task 2.

12. As an expert in business analysis, you've been hired by Maggie, a process-mapping consultant for a pet food company. During the initial meeting, she starts to outline the solution scope statement. To this point, Maggie has already:
    a. Determined the alignment of the proposed solution with the organization's goals and objectives.
    b. Collaborated with the project manager in the development of project goals and objectives.
    c. Asked a series of probing questions in an attempt to understand the opportunity.
    d. Determined the NPV, IRR, and ROI for the new line of cat food.

The correct answer is: **C**

This question tests your overall domain knowledge and challenges you to consider predecessor activities and deliverables to enable the drafting of the solution scope statement. From the answer choices, the business analyst would need to have an understanding of the opportunity.

Answer Choice A: Alignment of the proposed solution to the organization's goals and objectives would follow the assessment of the value proposition.

Answer Choice B: The development of project goals and objectives follow the assessment of the value proposition.

Answer Choice D: The solution scope statement is the final deliverable from Task 1 in Needs Assessment; this positions the business analyst to begin assessing the value proposition.

Business Analysis for Practitioners: A Practice Guide, Section 2.4.1.1, "Five Why's Problem & Opportunity Analysis."

PMI Professional in Business Analysis (PMI-PBA) Examination Content Outline, 2013, "Needs Assessment," Task 2.

13. Tucker, the chief science officer for your company's division of DDR Memory Chips, asked for your help mapping the division's people, locations, processes, applications, data, and technology. This process method is best known as:
    a. Value stream mapping
    b. Capability mapping
    c. Process requirements mapping
    d. Enterprise and business architectures

The correct answer is: **D**

This question tests your knowledge and the practical use of the tools listed in the answer choices. Enterprise and business architecture techniques are used to map organizational elements such as applications, data, locations, people, processes, and technology. The output of this analysis establishes the groundwork for a capability assessment.

Answer Choice A: Value stream mapping is a technique used to identify non-value added time, commonly referred to as *waste*, within a process.

Answer Choice B: Capability maps are used to understand the needed capabilities for an organization to achieve its objectives and goals and fulfill its mission.

Answer Choice C: Process requirements mapping or process flow charts are tools for visualizing processes; they are a special type of cause-and-effect diagram.

Business Analysis for Practitioners: A Practice Guide, Section 2.4.6, "Assess Current Capabilities of the Organization."

PMI Professional in Business Analysis (PMI-PBA) Examination Content Outline, 2013, "Needs Assessment," Task 5.

14. Prior to studying business analysis, you majored in creative writing at your community college. Although you don't directly support them, most all executive sponsors ask for your assistance drafting situation statements. This is because:
    a. Your situation statements are linked to the organization's mission, vision, goals, and objectives.
    b. Your situation statements ensure a complete understanding of the opportunity or problem, the contributing effects, and the overall impacts.
    c. Your situation statements ensure a complete understanding of the opportunity or problem and the associated benefits—both tangible and intangible.
    d. Your situation statements are brief, concise, but most importantly linked to the organization's mission, vision, goals, and objectives.

The correct answer is: B

Well-written situation statements present a complete understanding of the problem to be addressed or the opportunity to be pursued, along with the contributing effects and the overall impacts. Once the business analyst has documented the problem or opportunity, the deliverable from Task 1 is an approved situation statement.

Answer Choices A & D: Goals and objectives are linked to the organization's mission, vision, goals, and objectives, not the situation statement.

Answer Choice C: Situation statements provide only enough detail to establish the problem (or opportunity) of "x," the effect of "y," with the impact of "z"; they are too brief to provide both tangible and intangible benefits.

Business Analysis for Practitioners: A Practice Guide, Section 2.3.4 to review Situation Statements.

PMI Professional in Business Analysis (PMI-PBA) Examination Content Outline, 2013, "Needs Assessment," Task 1.

15. Your project manager, Gwen, requested your assistance with the framework for a tool that will be used during a stakeholder facilitation session to articulate and understand the high-level views surrounding a business proposition.
    a. You draw quadrants for strengths, weaknesses, opportunities, and threats on the white board.
    b. You begin by outlining a capability table on the easel.
    c. You commence by outlining an affinity table, which will be distributed to each participant.
    d. You start by outlining the relevant criteria on the SMART board.

The correct answer is: **A**

A SWOT analysis is a form of quadrant analysis used for plotting and understanding data elements—in this case the high-level views surrounding a business proposition. The SWOT analysis assesses an organization's internal strengths and weaknesses and its external opportunities and threats.

Answer Choice B: A capability table would be used in conjunction with a capability map to understand the needed capabilities for an organization to achieve its objectives and goals and fulfill its mission.

Answer Choice C: This is a made-up tool.

Answer Choice D: This is a distractor; introducing the SMART board, an interactive display to capture, present, and share relevant criteria, would be unnecessary to understand the high-level views surrounding a business proposition.

Business Analysis for Practitioners: A Practice Guide, Section 2.4.2, "SWOT Analysis."

PMI Professional in Business Analysis (PMI-PBA)® Examination Content Outline, 2013, "Needs Assessment," Task 3.

16. Elaine and Burke, both senior VPs of recreation for a retirement community, asked for your assistance developing an interrelationship diagram. As you facilitate this session, you find instances in which some factors have many incoming arrows. As the day progresses, another session produces factors with a large number of outgoing arrows. What do these factors represent?
    a. Factors with a large number of outgoing arrows are the effects or key outcomes of other factors, whereas factors with a large number of incoming arrows are the causes or source of the concern.
    b. The arrows represent the direction of the cause, starting with the stronger influences.
    c. The arrows represent the bi-directionality of the factors.
    d. Factors with a large number of outgoing arrows are the causes or source of the concern, whereas factors with a large number of incoming arrows are the effects or key outcomes of other factors.

The correct answer is: **D**

This is an experience-based question and relies on your practical use and understanding of interrelationship diagrams, which can be used in scope modeling. Elements or factors with a large number of outgoing arrows are the causes or sources of concern, which require further investigation, whereas factors with a large number of incoming arrows are the effects or key outcomes of other factors.

Answer Choice A: This is the opposite of the correct answer.

Answer Choice B: This is a plausible answer choice, but incorrect.

Answer Choice C: This is a made-up answer and a distractor.

Business Analysis for Practitioners: A Practice Guide, Section 2.4.4.2, "Interrelationship Diagrams."

PMI Professional in Business Analysis (PMI-PBA)® Examination Content Outline, 2013, "Needs Assessment," Task 2.

17. Lucy, the chief quality officer for Lucky Organic Vegetables, asked for assistance creating a key project document that will outline the factors that constitute success, who determines success, and ultimately who'll sign off that the project was successful. Lucy is asking for assistance creating:
    a. The solution scope statement
    b. The business case
    c. The charter
    d. The solution requirements document

The correct answer is: **C**

The charter is the key artifact, which formally recognizes and authorizes the project and provides the project manager with the authority to execute the project. It outlines approval requirements, assumptions and constraints, summary budget, high-level description of the initiative, measurable objectives, justification/purpose/intent, summary milestones, high-level requirements, high-level known risks (opportunities and threats), and stakeholder's roles and responsibilities. It establishes the project manager's authority and outlines the criteria for success.

Answer Choice A: The solution scope statement establishes the boundary for the initiative covering elements such as acceptance criteria, assumptions and constraints, deliverables, exclusions, and scope description. Although there is some overlap, the charter is the only document that establishes both success criteria and who'll sign off on the final deliverable.

Answer Choice B: The business case will serve as an input to the creation of the charter.

Answer Choice D: Requirements for predictive and iterative lifecycle projects are presented in the form of business requirements documents or solution requirements documents; they describe the purpose/intent of the component and metrics to evaluate its impact on the organization. They don't outline the factors that constitute success

A Guide to the Project Management Body of Knowledge (PMBOK®), Table 5-1, 13.1.2 "Contents of the Project Charter."
Business Analysis for Practitioners: A Practice Guide, Section 2.6, "Business Cases."
PMI Professional in Business Analysis (PMI-PBA)® Examination Content Outline, 2013, "Needs Assessment," Task 1.

18. Your sponsor, Autumn, has asked you to conduct a study to determine if new opportunities exist within your market. You commission a:
    a. Benchmarking study
    b. Capability assessment
    c. Market-based affinity study
    d. Monte Carlo study

The correct answer is: **A**

This question tests your knowledge of the tools and their intended use. Benchmarking studies are very common for organizations to use to compare themselves to competitors and for idea generation. Most often they are performed by independent third parties using publicly available data.

Answer Choice B: A capability assessment is conducted to understand the needed capabilities for an organization to achieve its objectives and goals and fulfill its mission.

Answer Choice C: This is a made-up answer.

Answer Choice D: Monte Carlo simulation is a tool to analyze the probability of various outcomes based on repeated mathematical modeling.

Business Analysis for Practitioners: A Practice Guide, Section 2.4.5.3 Benchmarking.
PMI Professional in Business Analysis (PMI-PBA)® Examination Content Outline, 2013, "Needs Assessment," Task 2.

19. Although there may be some perceived redundancy with the project charter, the PM assigned to your team, Bobby, asked for your assistance drafting a document that outlines the scope, deliverables, exclusions, assumptions, and constraints associated with your project. You're outlining:
    a. A project scope statement
    b. A project business case
    c. A preliminary project scope statement
    d. A project requirements document

The correct answer is: **A**

The project scope statement establishes the boundary for the initiative covering elements such as acceptance criteria, assumptions and constraints, deliverables, exclusions, and scope description.
Answer Choice B: The business case will contain sections which outline (a) opportunity/problem, (b) analysis, (c) recommendation, (d) evaluation—all of which are used to frame the project scope statement.
Answer Choice C: A preliminary project scope statement establishes the high-level project objectives.
Answer Choice D: Requirements documents are not going to address overall scope and deliverables.
A Guide to the Project Management Body of Knowledge (PMBOK®), Table 5-1, 13.1.2 to review the contents of the Project Scope Statement.
PMI Professional in Business Analysis (PMI-PBA)® Examination Content Outline, 2013, "Needs Assessment," Task 1.

20. You are working with Ellie, the product manager for hair-care products at Tortoise, Inc., a global manufacturer. You're facilitating a session for which you've been asked to draw a diagram to map out the high-level causes of the problems.
    a. You start to draw an affinity table with categories for ROI, NPV, IRR, and BCR.
    b. You start by drawing a few lines and labeling them: methods, measurements, regulations, procedures, manufacturing line.
    c. You create a capability table, which outlines the benefits of the new line, including increased market share and high levels of customer satisfaction.
    d. You create a process flow diagram, which outlines the enterprise and business architectures.

The correct answer is: **B**

Cause-and-effect diagrams, a variation of which is a Fishbone diagram (aka *Ishikawa diagram*), are used to investigate the high-level causes of problems that are occurring. The problem is placed at the head of the fish (which can face either left or right), and the categories are often used to group causes along the spine.
All other answer choices contain made-up answers.
Business Analysis for Practitioners: A Practice Guide, Section 2.4.4.2, "Cause-and-Effect Diagrams – aka Fishbone or Ishikawa Diagrams."
PMI Professional in Business Analysis (PMI-PBA)® Examination Content Outline, 2013, "Needs Assessment," Task 2.

Chapter 3

Planning

Study Hints

The Planning domain contains the elements that focus on the preparation required to manage the business analysis activities throughout the lifecycle of the project.

If you fail to plan, you are planning to fail!

— Benjamin Franklin

Similar to the planning activities associated with project management, the activities for business analysis planning are critical to ensure that all stakeholders clearly understand the approach and intent for the effective management of product requirements. Concepts introduced in this domain will be referenced and further elaborated over the life of the project.

! *Hint:* On the exam, there are six specific aspects associated with planning (see Figure 3.1).

The initial focus of the business analyst is on understanding the rationale and driving factors supporting the project. The key artifact to support this activity is the business case. As you'll recall from the Needs Assessment domain, the business case contains four sections: (a) an outline of the opportunity or problem that the project is addressing; (b) an analysis of the project's objectives and success criteria; (c) the business analyst's recommendation; and (d) plans for benefits realization.

Over the life of the project, one of the most important strategies and tools for business analysis pertains to requirements traceability. The strategy establishes the approach to traceability, and the tool enables the tracing and tracking to monitor and validate the requirements. The resulting tool, which should be established before the start of the project, with team members properly trained prior to analysis, is the requirements traceability matrix. This tool can be as simple as a spreadsheet or fully integrated into a project management information system (PMIS). It enables all requirements to be documented in detail, traced to the project objectives, and tracked from ideation to evaluation.

! *Hint:* For the exam, please remember that the project team and stakeholders must agree on the format and its use over the duration of the project.

The business analysis plan and the requirements management plan provide the overarching direction throughout the project and formally establish the *how* and *when* for all solution development activities. In organizations with dual project manager/business analyst roles, these documents are produced by the same person, whereas in larger organizations, these documents will be prepared by different individuals who are collaborating. The requirements management plan will cover elements of both the project and product, identifying stakeholders and their roles, establishing the framework for communications, and articulating the guidelines for managing requirements. In contrast, the business analysis plan focuses solely on the activities and deliverables related to the efforts of business analysis. PMI is a proponent of plans: exam questions will focus on the contents, purpose, and intent of these project artifacts. During the plan development, it's not uncommon for the business analyst to refine the stakeholder analysis started during Needs Assessment. The stakeholder activities are iterative throughout the project, and as these are refined, stakeholders should be grouped together based on similar attributes or characteristics.

Another critical artifact created in this domain is the change management plan. Depending on the nature and structure of the project, the requirements change process can be documented in either the business analysis plan or a separate change management plan. This is a collaborative decision between the project manager and the business analyst; the ultimate goal is to have a single source be the definitive reference. As a key project artifact, this plan establishes the ground rules, roles, and responsibilities for how stakeholders will propose changes, how these changes will be documented, and when proposed changes will be reviewed and decisions communicated.

> *! Hint:* For the exam, please remember that any deviation from the agreed-upon scope and requirements baseline, no matter how minute or from whom, must be brought to the Change Control Board for approval.

Other central-planning activities in this domain pertain to document versioning and control and determining metrics and success criteria for use during the Evaluation domain.

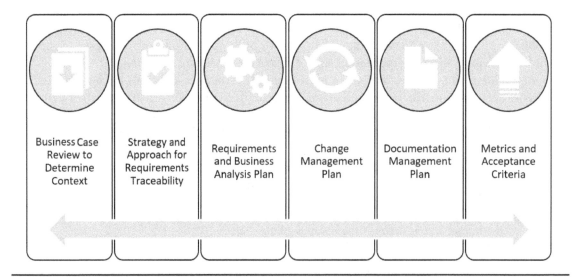

Figure 3.1 Lifecycle of planning.

Figure 3.2 Planning artifacts.

! Hint: On the exam, please remember, enablement tools can be as simple as a spreadsheet or shared drive to store documents; they don't need to be complicated or sophisticated project management information systems (PMIS).

It is paramount that once all plans are established and agreed upon, stakeholders are trained on the project lifecycle, delivery methodology, and the tools and techniques that will be used over the course of the project.

As outlined in the *Examination Content Outline*, Figure 3.1 highlights the six essential business analyst tasks associated with the Planning domain. Please refer to the practice guides and the *PMBOK® Guide* for additional context related to all the items outlined in Figure 3.1.

The key artifacts associated with the planning domain are shown in Figure 3.2.

At the end of this chapter are exercises to reinforce the concepts from this domain, the answers to which can be found throughout this exam prep book. This domain represents 22% (or 44) of the questions that will appear on the PMI-PBA® Exam.

Major Topics

3.1 Business Analysis Activity Context

➢ **Additional Concepts**
  o Critical thinking
  o Political and cultural awareness

3.2 Strategy and Approach for Requirements Traceability

Requirements Traceability Artifact or Tool
  • Backlog management
  • Issue, risk, decision tracking
  • Requirements traceability matrix
  • Dependency analysis

➢ **Additional Concepts**
  o RICE-BW tracker
  o Use cases to acceptance tests
  o Models and diagrams
  o Epics or user stories to features and acceptance tests
  o Kanban board
  o Traceability trees
  o Requirements baseline

3.3 Requirements and Business Analysis Plan

  • Business analysis plan
  • Requirements management plan
  • Stakeholder analysis
  • Work plan
    o Development methodologies
      ▪ Agile, iterative, incremental, lean, waterfall
    o Estimating tools and techniques
      ▪ Analogous, averaging, bottom-up, estimation poker, quadrant analysis, three point
    o Planning tools and techniques
      ▪ Strategic, tactile
    o Scheduling tools and techniques
      ▪ Alternative analysis, decomposition, dependency determination, rolling wave, critical chain method (CCM), critical path method (CPM), precedence diagramming method (PDM), leads and lags
    o Collaboration tools and techniques
      ▪ Brainstorming, social media, web/video conferencing

- Knowledge management systems, project management information systems
  - ○ Facilitation tools and techniques
    - Brainstorming, focus groups, ice breakers, MultiVoting (NGT), scatter diagrams, flow charting, affinity diagramming, consensus decision making, didactic interaction

- ➢ **Additional Concepts**
  - ○ Political and cultural awareness
  - ○ Leadership principles and skills
  - ○ Lessons learned, retrospectives
  - ○ Contingency planning

3.4 Change Management Plan

Change Control Tools and Techniques
- Change management plan
- Project management information system
- Configuration management systems

Conflict Management Techniques
- Withdraw/avoid
- Smooth/accommodate
- Compromise/reconcile
- Force/direct
- Collaborate/problem solve

3.5 Documentation Management Plan

Document Management Tools and Techniques
- Documentation management plan
- Project management information system
- Version control

3.6 Metrics and Acceptance Criteria

Measurement Tools and Techniques
- Acceptance criteria plan
- Planguage
- Service-level agreements

➢ **Additional Concepts**
- ○ Acceptance criteria
- ○ Burndown charts
- ○ Key performance indicators
- ○ Operational-level agreements
- ○ Requirements documentation

Planning
Chapter Exercises

Test Your Knowledge: Planning

INSTRUCTIONS: On the exam, it's not enough to simply recall that an artifact was created in a particular domain; you'll need to understand the context and how it relates to business analysis. Below are some of the key concepts related to this chapter; take the time to research each and fill in the blanks.

Q	Practice Area	Example	Explanation
	These diagrams are used to show the causes of specific events.	Ishikawa or fishbone diagram	High-level causes of a problem. Categories can include machines, materials, measurements, stakeholders, and locations.
1	What project artifact covers requirements management, spanning both the project and product?		
2	What project artifact is focused on the activities and deliverables associated with business analysis?		
3	What process outlines how changes can be proposed, reviewed, documented, and approved?		
4	What are the three most common project delivery methodologies?		

Q	Practice Area	Example	Explanation
5	What are seven characteristics of stakeholders?		
6	With the stakeholder characteristics defined, how can groupings be structured?		
7	List the techniques that can be used to group or analyze stakeholders.		
8	When prioritizing requirements, what factors are influenced by the project delivery methodology?		
9	What other criteria can be used to determine the priority of requirements?		

Q	Practice Area	Example	Explanation
10	What are three processes related to solution requirements that require stakeholder support and buy-in?		
11	What are four techniques that can be used to estimate project or activity durations?		
12	What are some characteristics of projects?		

Memory Game: Requirements Management Plan

INSTRUCTIONS: Listed below are the key contents of the requirements management plan and the overall guidelines for managing requirements. Practice recalling them using the template on the following page.

Requirements Management Plan	
How Requirements Are:	*In Addition:*
Analyzed	Acceptance criteria
Approved	Approach to traceability and metrics
Communicated/reported	Authorization levels
Documented	Decision-making process
Elicited	Framework for communication
Maintained	Roles and responsibilities matrix
Managed	
Prioritized	
Tracked	
Validated	

Memory Game: Requirements Management Plan

INSTRUCTIONS: Make a photocopy of this page; practice listing the contents of the requirements management plan.

Requirements Management Plan	

Memory Game: Business Analysis Plan

INSTRUCTIONS: Listed below are the key contents of the business analysis plan. Practice recalling them using the template on the following page.

Business Analysis Plan	
Plan Components	**Requirements Process Decisions**
Activities to be conducted	Analysis
Approval levels and authorities	Change management
Deliverables produced	Communication
Requirements maintenance	Decision making
Requirements status	Elicitation
Roles and responsibilities	Evaluation and acceptance criteria
Version control	Prioritization
	Traceability
	Validation and verification

Memory Game: Business Analysis Plan

INSTRUCTIONS: Make a photocopy of this page; practice listing the contents of the business analysis plan.

Business Analysis Plan	

Planning: Practice Questions

INSTRUCTIONS: Note the most suitable answer for each multiple-choice question in the appropriate space on the answer sheet on page 70.

1. Following the completion of the registry, Zachary added further details for each stakeholder, noting their office location, how many years they've been with company, and their availability. What did Zachary perform?
   a. A stakeholder grouping analysis
   b. A stakeholder interest analysis
   c. A stakeholder salience analysis
   d. A stakeholder characteristics analysis

2. Margo was recently recruited from a Big 4 consulting company to fill the lead business analyst role on a major software implementation project. Now three weeks into her job, the company has just released its budget for the following year, and the funds previously allocated for this project have been redistributed to other projects. Margo should:
   a. Contact her former employer and inquire about returning to her prior position
   b. Continue working on the requirements management plan until directed otherwise by her sponsor
   c. Consult with her manager on how her talents could be applied to other projects
   d. Talk with other team members about the budget to learn of the implications

3. You're speaking with Ella, the engagement manager for the project, and she voices concern that the client project team doesn't have a clear understanding of the delivery methodology. Your project manager, Rose, suggests:
   a. As the project started, Ella should have better engaged stakeholders across the organization.
   b. At the start of the project, there wasn't sufficient interest in the business analysis processes.
   c. At the start of the project, the stakeholders were slightly overwhelmed and couldn't allocate the time to participate in jointly planned delivery methodology sessions.
   d. As the project started, Ella should have engaged stakeholders across the organization and trained them on all the concepts and methodologies, defining metrics and acceptance criteria that would be used during testing.

4. Your project sponsor, Julian, has expressed concern with the proposed project planning methodologies. Reflecting on the nature of the project and organizational culture, he'd like to hold off on defining the scope for future phases until the current scope of approved work is complete. What can you suggest as the preferred project delivery methodology?
   a. Predictive
   b. Adaptive
   c. Iterative/incremental
   d. Planned

5. Michael is relatively new to the organization, but not to project management. In an effort to drive consistency, minimize rework, and establish the level of traceability to monitor requirements, he suggests:

a. Creating templates unique to the projects to improve work processes
b. Organizational and project assets be reviewed to determine applicability
c. Building a robust project management plan and communication management plan
d. Not using a work plan, as the project requirements will influence the timeline and delivery methodologies

6. You work for an organic whole-food supplement company, and your sponsor, Astrid, has asked that you take the lead in determining the best approach to business analysis for the project.
   a. You start by understanding the characteristics of the project.
   b. Your first task is to review the business case, then the project goals and objectives.
   c. You review the project size, the complexity, and how requirements will be validated and authorized.
   d. You review the stakeholder register, RACI, and the requirements management plan.

7. Your project manager, Ambrose, is on vacation for three weeks as the project enters the planning stage. Upon return from holiday, he'll begin working on the project management work plan and schedule. While he's on holiday, you decide to finalize your work plan and overall approach to business analysis activities. Will Ambrose support your efforts as you define your business analysis activities?
   a. Yes, project management and business analysis activities are independent efforts.
   b. No, on the RACI, Ambrose was identified as the one accountable for these deliverables.
   c. No, Ambrose has some insights he'd like to offer as you build the work plan.
   d. Yes; although there is some overlap, Ambrose will integrate all your planning and scheduling activities into his plan.

8. Due to the complex nature of the proposed solution, you are planning a half-day session to discuss acceptance criteria and metrics, which will be closely monitored both during the project and post go-live to ensure that the solution is aligned with business objectives. Based on the approved stakeholder register, you find a date on which all designated attendees are available. Just before the start of the session, you receive a number of emails requesting the conference line information. What could be the cause of the confusion?
   a. Stakeholder characteristics were not thoroughly refined.
   b. The invitation didn't clearly state the meeting was to be conducted in person.
   c. The refinement of complexity level was limited to the view point of the sponsor and business analyst and didn't take into consideration collaboration with the project manager.
   d. The attendees were not located in the same geography as the meeting.

9. Malcolm is relatively new to business analysis and looks to you as the senior practitioner to help him define the best approach for the enterprise software product that is planned to be deployed across the organization. You describe:
   a. Processes have stakeholder support and buy-in, change management is communicated and rationale understood, processes for approving requirements and solutions are clearly defined and supported, and the organization supports your activities.
   b. Processes for approving requirements and solutions are clearly defined, processes for change management are communicated, and processes for identifying stakeholders are established.

c. The importance of the activities is communicated to the organization and teams during project kickoff.

d. Processes for validating, verifying, and approving requirements and solutions are aligned with stakeholder needs and expectations.

10. You are working on a classified project called "The Lamp Post." Due to the sensitive nature of the solution, your project manager, Eloise, is providing detailed plans for deliverables due within the next 45 days and an outline of what is anticipated from days 46 to 90. As you collaborate with Eloise and review the project with the stakeholders, you:

a. Provide all the details of the requirements management plan through go-live

b. Provide all the details related to the delivery methodology and planning decisions

c. Work with Eloise and the stakeholder to elaborate the plan for days 46 to 90

d. Provide only the details of the business analysis plan that correspond with Eloise's plan

11. Josiah spent considerable time working on the requirements management plan. Prior to distributing it to the team, he had the requirements expert at the local PMI chapter review the plan for any omissions or inconsistencies. Unfortunately, he's having difficulty understanding why the team is deviating, despite having his sponsor also sign off on the document.

a. The requirements management plan was aligned to PMI standards, which were slightly different than the organization's.

b. He was unable to attend the team's weekly "Lunch and Learn," at which key project tasks and artifacts are discussed and reviewed.

c. The project manager added additional core components to the plan, for which the team was unaware.

d. The requirements management plan was a subsidiary plan of integrated change control, which had not yet been released to the team.

12. Having worked with Phyllis, the VP of HR, for a number of years, you've come to appreciate her approach that all requirements are the number-one priority. However, on this highly complex project, you would like to establish a roadmap for delivering the proposed solution that will address:

a. Frequency and timing of prioritization, with a focus on alignment to the vision, mission, goals, and objectives of the organization

b. Frequency of prioritization activities, addressing compliance and regulatory first, then FIFO (first in, first out) to maintain a steady state of development

c. How the product owner, Jennifer, will manage the backlog

d. Frequency and timing of prioritization, with consideration for uncovering issues early and addressing compliance deadlines, opportunity cost, and goodwill benefits

13. As Chloe prepares for a focus group, she is having slight difficulty because the invitee list contains stakeholders with varying levels of interest, influence, and importance. What could Chloe have done to improve the process?

a. Sent the invitations based on the categories established in the stakeholder register

b. Prepared for the focus group by conducting an adaptive stakeholder classification session

c. Had a better understanding of the sociological impacts to the focus group

d. Sent invitations based on stakeholders' common interests

14. Collaborating with your project manager, Layton, you are trying to determine the best lifecycle approach for a relatively small project. As you consider the business analysis processes, what are a few key decisions you'll need to consider?
    a. Managerial preferences, project characteristics
    b. Project context, organizational characteristics, solution requirements
    c. Stakeholder and solution requirements
    d. Requirements management, change management, business process management

15. Mariella, the new business analyst assigned to your team, inquired as to the dependencies before the team could begin assembling the business analysis work plan. You offer:
    a. We'll need approval of the business analysis deliverables, tasks, and activities.
    b. It's based on approval of the business analysis management plan.
    c. It's contingent on approval of the business analysis deliverables and artifacts.
    d. We'll need approval of the requirements and business analysis management plans.

16. Your project has just commenced, and earlier in the week Ryker was assigned as your project manager. It's now Friday, and you're outlining the initial scope and discussing the building of the stakeholder register. Ryker asked that on Monday, you begin drafting the outline of the authorization and key decision-making process. You respond:
    a. This document should also include success factors and planning activities, but we should wait until we have approval of the project office.
    b. This document should also include how requirements will be developed, tracked, managed, and validated.
    c. This document should also include how requirements will developed, tracked, managed, and validated, along with a communication plan.
    d. Are there other core components that should be included?

17. You are debating with Gannon, your senior project manager, about the value of expanding the stakeholder register to include additional characteristics. You contend that by adding just one additional field, the team will be more effective in delivery of business analysis processes. What field would help you to better run requirements walkthroughs and the change management process?
    a. Difficulty
    b. Approach
    c. Culture
    d. Experience

18. Your project manager, Ippy, has asked for your help drafting a key project document that will outline how scope is defined, validated, and controlled. You're collaborating on the creation of the:
    a. Project scope definition plan
    b. Project scope management plan
    c. Requirements management plan
    d. Work breakdown structure and WBS dictionary

19. Having worked together on a previous client engagement, you've been paired with Hamish, who will serve as project manager for the recovery of a highly complex initiative. As the business

analyst, you suggest that one of the first tasks is to interview stakeholders to understand completed activities, in an attempt to learn from their past experiences. Ideally:

a. These sessions are held independent of each other, so as not to influence or confuse project management and business analysis activities.

b. These sessions will be facilitated jointly by the project manager and business analyst.

c. These sessions will focus on lessons learned pertaining to completed project activities.

d. These sessions will focus on lessons learned pertaining to completed requirements.

20. Collaborating with Anna, your project manager, you've developed a fairly comprehensive list of stakeholders, which includes nearly the entire organization. Before presenting to the project Steering Committee for review, you do all of the following except:

a. Suggest adding the guy who runs the food truck outside the office complex

b. Add a government agency, with whom your organization has had no prior affiliation

c. Recommend adding the Stock Exchange on which your company is listed

d. Remove a small external entity, with whom you have no formal affiliation, who is merging with a competitor

Answer Sheet

	Answer Choice				Correct	Incorrect	Predicted Answer			
							90%	75%	50%	25%
1	a	b	c	d						
2	a	b	c	d						
3	a	b	c	d						
4	a	b	c	d						
5	a	b	c	d						
6	a	b	c	d						
7	a	b	c	d						
8	a	b	c	d						
9	a	b	c	d						
10	a	b	c	d						
11	a	b	c	d						
12	a	b	c	d						
13	a	b	c	d						
14	a	b	c	d						
15	a	b	c	d						
16	a	b	c	d						
17	a	b	c	d						
18	a	b	c	d						
19	a	b	c	d						
20	a	b	c	d						

Planning: Answer Key

1. Following the completion of the registry, Zachary added further details for each stakeholder, noting their office location, how many years they've been with company, and their availability. What did Zachary perform?
    a. A stakeholder grouping analysis
    b. A stakeholder interest analysis
    c. A stakeholder salience analysis
    d. A stakeholder characteristics analysis

    The correct answer is: **D**

    Business analysts define stakeholder characteristics such as attitude, complexity, culture, experience, location, and availability to properly plan and manage requirements throughout the project lifecycle. Roles and responsibilities are a component of the requirements management plan, which is the roadmap for delivering the planned solution.

    Answer Choice A: Stakeholders can be grouped based on interest, influence, or power, but a stakeholder grouping analysis would not be used to capture data such as office location, tenure, or availability.

    Answer Choice B: A stakeholder interest table (power/interest grid) maps categories for engagement such as unknown, blocker, neutral, supporter, and champion.

    Answer Choice C: The stakeholder salience diagramming method is used to assist team members, when the need arises, to prioritize competing stakeholder requests (Wood, 1997). The purpose of the tool is to categorize stakeholders based on the factors of power, urgency, and legitimacy. As with the stakeholder register and the power/interest grid, the project salience model should be reevaluated over the life of a project.

    Requirements Management: A Practice Guide, Section 4.2.1.2, "Group and Characterize Stakeholders."
    PMI Professional in Business Analysis (PMI-PBA)® Examination Content Outline, 2013, "Planning," Task 3.

2. Margo was recently recruited from a Big 4 consulting company to fill the lead business analyst role on a major software implementation project. Now three weeks into her job, the company has just released its budget for the following year, and the funds previously allocated for this project have been redistributed to other projects. Margo should:
    a. Contact her former employer and inquire about returning to her prior position
    b. Continue working on the requirements management plan until directed otherwise by her sponsor
    c. Consult with her manager on how her talents could be applied to other projects
    d. Talk with other team members about the budget to learn of the implications

    The correct answer is: **C**

    Shifts in organizational priorities are very common, especially in technology-based and fast-paced organizations. This question focuses on cultural awareness skills. The organization's budget is a very strong indication of the intent and direction of the company. As the lead business analyst, Margo should be looking to best align her talents and strengths with the needs of the organization.

    Answer Choice A: Retreating may seem reasonable; however, the software implementation project should not be the only reason for joining the firm. Having recently left a Big 4 firm, Margo has a lot to offer.

Answer Choice B: Continuing to work on the requirements management plan may seem reasonable; however, she should first talk with her manager, who may very well ask that the requirements management plan be completed, should funds be reappropriated at a later date.

Answer Choice D: While gossip and watercooler talk may be a common practice and often natural, it is frowned upon because it often leads to a distortion of the truth. The best person to consult is her manager.

Examination Content Outline, Knowledge and Skills, #24, "Political and Cultural Awareness."

PMI Professional in Business Analysis (PMI-PBA)® Examination Content Outline, 2013, "Planning," Task 3.

3. You're speaking with Ella, the engagement manager for the project, and she voices concern that the client project team doesn't have a clear understanding of the delivery methodology. Your project manager, Rose, suggests:
   a. As the project started, Ella should have better engaged stakeholders across the organization.
   b. At the start of the project, there wasn't sufficient interest in the business analysis processes.
   c. At the start of the project, the stakeholders were slightly overwhelmed and couldn't allocate the time to participate in jointly planned delivery methodology sessions.
   d. As the project started, Ella should have engaged stakeholders across the organization and trained them on all the concepts and methodologies, defining metrics and acceptance criteria that would be used during testing.

The correct answer is: D

At the start of the project, it's the business analyst's responsibility, in collaboration with the project manager, to train the core team on the project lifecycle, delivery methodology, artifacts, tools, and techniques. Furthermore, the business analyst, working with stakeholders, will define metrics and acceptance criteria, which will be used both during testing and at solution signoff. These steps are essential to ensure there is both buy-in and understanding of the project. All business analyst-led sessions should be highly collaborative, and in some cases conducted jointly with the project manager.

Answer Choice A: This answer choice is incomplete.

Answer Choices B & C: While these may often be true in practice, PMI is a proponent of taking action. Simply accepting that stakeholders were overwhelmed and didn't have time is not an acceptable answer.

Business Analysis for Practitioners: A Practice Guide, Section 3.4.5, "Ensure the Team Is Trained on the Project Life Cycle."

PMI Professional in Business Analysis (PMI-PBA)® Examination Content Outline, 2013, "Planning," Task 6.

4. Your project sponsor, Julian, has expressed concern with the proposed project planning methodologies. Reflecting on the nature of the project and organizational culture, he'd like to hold off on defining the scope for future phases until the current scope of approved work is complete. What can you suggest as the preferred project delivery methodology?
   a. Predictive
   b. Adaptive
   c. Iterative/incremental
   d. Planned

The correct answer is: **C**

Iterative planning methodologies are a hybrid of fully plan-driven and adaptive planning. These planning methodologies are often preferred when there is significant overall scope or when the organization would like to separate the work into predefined phases. Organizations aren't often ready to accept dramatic change, and the iterative planning approach allows the complete solution to be delivered in smaller increments.

Answer Choice A: The predictive planning approach is often associated with the waterfall delivery methodology. The focus is structured on delivering a predefined solution within a specific time-frame, and the scope is entirely planned up front.

Answer Choice B: Adaptive approaches are often associated with agile or change-driven methods, in which it's anticipated that the end product is going to evolve over the duration of the project.

Answer Choice D: This is a distractor. The three main structure types are predicative (fully plan driven), iterative/incremental, and adaptive.

Business Analysis for Practitioners: A Practice Guide, Section 3.4.4, "Understand How the Project Life Cycle Influences Planning Decisions."

PMI Professional in Business Analysis (PMI-PBA)® Examination Content Outline, 2013, "Planning," Task 3.

5. Michael is relatively new to the organization, but not to project management. In an effort to drive consistency, minimize rework, and establish the level of traceability to monitor requirements, he suggests:
    a. Creating templates unique to the projects to improve work processes
    b. Organizational and project assets be reviewed to determine applicability
    c. Building a robust project management plan and communication management plan
    d. Not using a work plan, as the project requirements will influence the timeline and delivery methodologies

The correct answer is: **B**

Whenever possible, business analysts should leverage existing organizational standards, which can either be organizational process assets (OPAs) or project process assets (PPAs). This will minimize rework and lead to quicker adoption. In cases where they may not exist or may need to be refined, the project should look to the organization's project management office (PMO) for guidance and assistance—many have vast libraries or contacts to obtain templates.

Answer Choice A: While sometimes it may be necessary to create templates, the first place the business analyst should consult is the PMO and consider using existing templates.

Answer Choice C: While the project management plan and communication management plan will aid in some respects, they will not fully address the question of establishing the level of traceability to monitor requirements.

Answer Choice D: Work plans are never optional.

Requirements Management: A Practice Guide, Section 4.2.2.2, "Identify Organizational Standards and Guidance."

PMI Professional in Business Analysis (PMI-PBA)® Examination Content Outline, 2013, "Planning," Task 2.

6. You work for an organic whole-food supplement company, and your sponsor, Astrid, has asked that you take the lead in determining the best approach to business analysis for the project.
    a. You start by understanding the characteristics of the project.
    b. Your first task is to review the business case, then the project goals and objectives.

c. You review the project size, the complexity, and how requirements will be validated and authorized.

d. You review the stakeholder register, RACI, and the requirements management plan.

The correct answer is: **B**

There are many factors and characteristics to consider when planning the approach to the business analysis activities on a project. For a newly assigned business analyst, the first step is understanding the rationale and justification for the project; this can be gleaned from the business case. If the business analyst had been involved with the creation of the business case, the next step would be to develop a clear understanding of the project and to define the strategy for requirements traceability.

Answer Choices A & C: Project characteristics are attributes such as time, scope, budget, size, and complexity; although they are certainly aspects to consider, they would only provide limited context.

Answer Choice D: The stakeholder register, responsibility assignment matrix (RACI), and requirements management plan would only provide limited information to help with determining the best approach to business analysis.

Business Analysis for Practitioners: A Practice Guide, Section 3.4.3, "Understand the Project Context."

PMI Professional in Business Analysis (PMI-PBA)® Examination Content Outline, 2013, "Planning," Task 1.

---

7. Your project manager, Ambrose, is on vacation for three weeks as the project enters the planning stage. Upon return from holiday, he'll begin working on the project management work plan and schedule. While he's on holiday, you decide to finalize your work plan and overall approach to business analysis activities. Will Ambrose support your efforts as you define your business analysis activities?

a. Yes, project management and business analysis activities are independent efforts.

b. No, on the RACI, Ambrose was identified as the one accountable for these deliverables.

c. No, Ambrose has some insights he'd like to offer as you build the work plan.

d. Yes; although there is some overlap, Ambrose will integrate all your planning and scheduling activities into his plan.

The correct answer is: **D**

The success of initiatives is highly dependent on the activities associated with business analysis. Furthermore, success is increased when project managers and business analysts collaborate on the development of work plans and schedules for both threads.

Answer Choice A: Project management and business analyst activities are not independent efforts.

Answer Choice B: In some organizations, the project manager may also be accountable for the business analyst's activities. However, in this scenario, there is an assigned business analyst, who would receive support from the project manager.

Answer Choice C: This is a partially correct answer; the project manager will provide insights and support to the business analyst during the development of the planning and scheduling activities.

Business Analysis for Practitioners: A Practice Guide, Section 3.2.2, "Business Analyst Planning and Project Management Planning."

PMI Professional in Business Analysis (PMI-PBA)® Examination Content Outline, 2013, "Planning," Task 1.

---

8. Due to the complex nature of the proposed solution, you are planning a half-day session to discuss acceptance criteria and metrics, which will be closely monitored both during the project and post go-live to ensure that the solution is aligned with business objectives. Based on the approved stakeholder register, you find a date on which all designated attendees are available. Just before the start

of the session, you receive a number of emails requesting the conference line information. What could be the cause of the confusion?

a. Stakeholder characteristics were not thoroughly refined.

b. The invitation didn't clearly state the meeting was to be conducted in person.

c. The refinement of complexity level was limited to the view point of the sponsor and business analyst and didn't take into consideration collaboration with the project manager.

d. The attendees were not located in the same geography as the meeting.

The correct answer is: A

Once the business analyst and project manager have identified the project stakeholders, leading practice is to analyze their characteristics based on significance or relevance to the project. This will help with future elicitation sessions, managing the balance of the initiative, and ensuring there is buy-in, support, and a common/consistent understanding of the undertaking. Common characteristics include attitude, complexity, culture, experience, location, level of influence, and level of authority.

Answer Choice B & D: These are distractors; although possible reasons, they don't get to the root cause of the confusion.

Answer Choice C: This is a fabricated answer; although complexity is a characteristic, it does not address the question.

Business Analysis for Practitioners: A Practice Guide, Section 3.3.2, "Stakeholder Characteristics."

PMI Professional in Business Analysis (PMI-PBA)® Examination Content Outline, 2013, "Planning," Task 3.

9. Malcolm is relatively new to business analysis and looks to you as the senior practitioner to help him define the best approach for the enterprise software product that is planned to be deployed across the organization. You describe:

a. Processes have stakeholder support and buy-in, change management is communicated and rationale understood, processes for approving requirements and solutions are clearly defined and supported, and the organization supports your activities.

b. Processes for approving requirements and solutions are clearly defined, processes for change management are communicated, and processes for identifying stakeholders are established.

c. The importance of the activities is communicated to the organization and teams during project kickoff.

d. Processes for validating, verifying, and approving requirements and solutions are aligned with stakeholder needs and expectations.

The correct answer is: A

Business analysis planning activities establish the framework to warrant that the optimal approach is selected and followed for the project. Although all four answer choices represent portions of a complete answer, choice A is slightly more complete. Within the business analysis plan, the approach will ensure that stakeholders are thoroughly identified; business analysis activities and deliverables are defined and agreed to; processes are established for validating, verifying, and approving requirements and approving solutions; the change management process is defined, communicated, and agreed to; and stakeholders are aware of, and support, the activities and time commitments required to complete the work associated with establishing the solution requirements.

Answer Choice B: The optimal business analysis approach for change management is one that ensures that the process for proposing changes to requirements is both defined and understood by stakeholders. This answer choice limits the process to only communication.

Answer Choice C: While communicating the importance of the activities during project kickoff is important, there is a fundamental step that must first occur in business analysis planning.

Answer Choice D: The process for validating, verifying, and approving requirements and approving solutions needs to be acceptable to stakeholders.

Business Analysis for Practitioners: A Practice Guide, Section 3.1, "Overview."

PMI Professional in Business Analysis (PMI-PBA)® Examination Content Outline, 2013, "Planning."

10. You are working on a classified project called "The Lamp Post." Due to the sensitive nature of the solution, your project manager, Eloise, is providing detailed plans for deliverables due within the next 45 days and an outline of what is anticipated from days 46 to 90. As you collaborate with Eloise and review the project with the stakeholders, you:
    a. Provide all the details of the requirements management plan through go-live
    b. Provide all the details related to the delivery methodology and planning decisions
    c. Work with Eloise and the stakeholder to elaborate the plan for days 46 to 90
    d. Provide only the details of the business analysis plan that correspond with Eloise's plan

The correct answer is: D

Project planning is often an iterative process; when a project starts, some of the details may not be known or are intentionally withheld. As the project evolves, the planning can become more specific and detailed; this is known as *progressive elaboration*. The question is based on rolling wave planning, a type of progressive elaboration, in which planning details should only be shared for a comparable time horizon.

Answer Choices A & B: These could be appropriate answer choices, had the question been based on a predictive planning approach or the waterfall delivery methodology.

Answer Choice C: The question is asking about reviewing and the answer choice suggests elaborating on the plan, inferring that it is not yet complete.

Business Analysis for Practitioners: A Practice Guide, Section 3.4.2.1, "Determining the Proper Level of Detail."

A Guide to the Project Management Body of Knowledge (PMBOK®), Section 6.2.2.2, "Rolling Wave Planning."

PMI Professional in Business Analysis (PMI-PBA)® Examination Content Outline, 2013, "Planning," Task 3.

11. Josiah spent considerable time working on the requirements management plan. Prior to distributing it to the team, he had the requirements expert at the local PMI chapter review the plan for any omissions or inconsistencies. Unfortunately, he's having difficulty understanding why the team is deviating, despite having his sponsor also signoff on the document.
    a. The requirements management plan was aligned to PMI standards, which were slightly different than the organizations.
    b. He was unable to attend the team's weekly "Lunch and Learn," during which key project tasks and artifacts are discussed and reviewed.
    c. The project manager added additional core components to the plan, for which the team was unaware.
    d. The requirements management plan was a subsidiary plan of integrated change control, which had not yet been released to the team.

The correct answer is: **B**

Once the requirements management plan is complete, it is the business analyst's responsibility to secure sponsor and governance approval. Depending on both the initiative and the organization, approval may be formal or informal; it's best to seek guidance from the project management office (PMO) or sponsor for the preferred approach. Once approved, the business analyst must train key stakeholders on this document to engender support and adoption. An optimal time for training is during "Lunch and Learns." Over the course of the project, should it become necessary to update the plan, the business analyst should ensure that these updates are appropriately communicated.

Answer Choice A: This is a distractor; PMI standards are highly flexible and robust and can be easily adaptable to an organization.

Answer Choice C: This is a potential answer choice but doesn't fully address the question.

Answer Choice D: This is a fabricated answer choice; the requirements management plan is not a subsidiary plan of integrated change control.

Requirements Management: A Practice Guide, Section 4.2.4, "Launch the Requirements Management Plan."

PMI Professional in Business Analysis (PMI-PBA)® Examination Content Outline, 2013, "Planning," Task 3.

12. Having worked with Phyllis, the VP of HR, for a number of years, you've come to appreciate her approach that all requirements are the number-one priority. However, on this highly complex project, you would like to establish a roadmap for delivering the proposed solution that will address:
    a. Frequency and timing of prioritization, with a focus on alignment to the vision, mission, goals, and objectives of the organization
    b. Frequency of prioritization activities, addressing compliance and regulatory first, then FIFO (first in, first out) to maintain a steady state of development
    c. How the product owner, Jennifer, will manage the backlog
    d. Frequency and timing of prioritization, with consideration for uncovering issues early and addressing compliance deadlines, opportunity cost, and goodwill benefits

The correct answer is: **D**

It's quite common for stakeholders to communicate that all requirements are the top priority. Although in reality this may be true, to appropriately manage requirements, they must be prioritized. Whereas the lifecycle will influence frequency and timing of prioritization activities, complementary assessments include value, cost, difficulty, regulatory requirements, and risk.

Answer Choice A: This is a partially correct answer; however, the focus is not on alignment to the vision, mission, goals, and objectives of the organization.

Answer Choice B: This is a partially correct answer; however, FIFO (first in, first out) to maintain a steady state of development would not be applicable in this situation.

Answer Choice C: This is a distractor and doesn't address the question.

Business Analysis for Practitioners: A Practice Guide, Section 3.4.9, "Define the Requirements Prioritization Process."

PMI Professional in Business Analysis (PMI-PBA)® Examination Content Outline, 2013, "Planning," Task 3.

13. As Chloe prepares for a focus group, she is having slight difficulty because the invitee list contains stakeholders with varying levels of interest, influence, and importance. What could Chloe have done to improve the process?
    a. Sent the invitations based on the categories established in the stakeholder register

b. Prepared for the focus group by conducting an adaptive stakeholder classification session
c. Had a better understanding of the sociological impacts to the focus group
d. Sent invitations based on stakeholders' common interests

The correct answer is: **D**

Once the business analyst and project manager have completed building the stakeholder register, they can begin to analyze their characteristics. This can often consist of analyzing their attitude, complexity, culture, experience, level of influence, location, and availability. Following this analysis, stakeholders can then be grouped, which is leading practice in preparation for requirements elicitation sessions. Stakeholders can be grouped based on common needs, interests, levels of importance, influence within the organization, and a variety of other characteristics considered important by the project team. This can manage attendance during focus groups, ensuring that an appropriate population is represented.

Answer Choice A: This is a partially correct answer; in preparation for eliciting requirements, stakeholders also need to be grouped.

Answer Choice B: Adaptive stakeholder classification is a made-up analysis.

Answer Choice C: This is a partially correct answer; although the business analyst should have had a better understanding of the sociological impacts to the focus group, invitations should have been sent based on stakeholders' common interests.

Requirements Management: A Practice Guide, Section 4.2.1.2, "Group and Characterize Stakeholders."

PMI Professional in Business Analysis (PMI-PBA)® Examination Content Outline, 2013, "Planning," Task 3.

---

14. Collaborating with your project manager, Layton, you are trying to determine the optimal lifecycle approach for a relatively small project. As you consider the business analysis processes, what are a few key decisions you'll need to consider?
    a. Managerial preferences, project characteristics
    b. Project context, organizational characteristics, solution requirements
    c. Stakeholder and solution requirements.
    d. Requirements management, change management, business process management.

The correct answer is: **A**

The project lifecycle establishes the framework for managing the project and is determined by factors such as managerial preferences, processes established by the project management offices for maintaining and controlling projects, and project characteristics. Depending on the selected project lifecycle (predictive, iterative/incremental, or adaptive), there are a number of process and planning decisions that the business analyst must take into account. For example: (a) what activities will be undertaken; (b) activity order; (c) activity timing; (d) associated deliverables; (e) level of formality required; (f) prioritization approach for requirements; and (g) change control process. Accordingly, the business analyst should factor decisions based on the selected project lifecycle approach.

Answer Choice B: This is a partially correct answer, as project context, organizational characteristics, and solution requirements could influence the lifecycle approach, but they are not the questions you'll need to consider.

Answer Choice C: This is also a partially correct answer. Managerial preferences and project characteristics are a more complete representation of the decisions to consider.

Answer Choice D: These answer choices are representative of processes and are not the questions you'll need to consider.

Business Analysis for Practitioners: A Practice Guide, Section 3.4.4, "Understand How the Project Life Cycle Influences Planning Decisions."

PMI Professional in Business Analysis (PMI-PBA)® Examination Content Outline, 2013, "Planning," Task 3.

15. Mariella, the new business analysts assigned to your team, inquired as to the dependencies before the team could begin assembling the business analysis work plan. You offer:
    a. We'll need approval of the business analysis deliverables, tasks, and activities.
    b. It's based on approval of the business analysis management plan.
    c. It's contingent on approval of the business analysis deliverables and artifacts.
    d. We'll need approval of the requirements and business analysis management plans.

The correct answer is: **A**

The deliverables, tasks, activities, and required resources are the primary inputs to creating a robust business analysis work plan. This is a critical integration and collaboration point with project management. It is essential that the business analyst and project manager agree on the level of detail that will be maintained in the *business analysis* work plan versus the *project* work plan.

Answer Choice B: This a distractor; assembling the work plan is not contingent on approval of the business analysis management plan.

Answer Choice C: This is a partially complete answer. In addition to the deliverables, tasks, activities, and required resources are also elements of the work plan.

Answer Choice D: This a distractor; as assembling the work plan is not contingent on the approval of either requirements or business analysis management plans.

Business Analysis for Practitioners: A Practice Guide, Section 3.5.3, "Assemble the Business Analysis Work Plan."

PMI Professional in Business Analysis (PMI-PBA)® Examination Content Outline, 2013, "Planning," Task 3.

16. Your project has just commenced, and earlier in the week Ryker was assigned as your project manager. It's now Friday, and you're outlining the initial scope and discussing the building of the stakeholder register. Ryker asked that on Monday, you begin drafting the outline of the authorization and key-decision making process. You respond:
    a. This document should also include success factors and planning activities, but we should wait until we have approval of the project office.
    b. This document should also include how requirements will be developed, tracked, managed, and validated.
    c. This document should also include how requirements will developed, tracked, managed, and validated, along with a communication plan.
    d. Are there other core components that should be included?

The correct answer is: **A**

As the business analyst and project manager collaborate on the requirements management plan, there are several factors to consider. The purpose of the requirements management plan is to outline the detailed activities in each of the domains starting with Elicitation, then Analysis, Monitoring and Controlling, concluding with Evaluation. The exercise in this chapter will help to strengthen your understanding of this key artifact. The plan, however, isn't built until the project management planning process group—the scenario in the question—describes activities within the project management initiating process group. Therefore, the business analyst and project manager should wait until they have approval to proceed to the next project phase.

Answer Choice B: The requirements management plan will cover tracking, management, and validation but will not address how requirements are developed.

Answer Choice C: The requirements management plan will not cover how requirements are developed; all other elements are components of the plan.

Answer Choice D: Answer choices that end with a question should be avoided and ignored.

Requirements Management: A Practice Guide, page 19.

PMI Professional in Business Analysis (PMI-PBA)® Examination Content Outline, 2013, "Planning," Task 1.

17. You are debating with Gannon, your senior project manager, about the value of expanding the stakeholder register to include additional characteristics. You contend that by adding just one additional field, the team will be more effective in delivery of business analysis processes. What field would help you to better run requirements walkthroughs and the change management process?
    a. Difficulty
    b. Approach
    c. Culture
    d. Experience

The correct answer is: C

Working with colleagues of diverse cultural backgrounds can significantly strengthen a team and ultimately lead to a better product or solution. In the processes, having a good understanding of the cultural nuances can improve communication and the requirements prioritizations, approval, change control, and signoff processes.

Answer choices A, B, D: While these may be factors to consider, of the answer choices culture is the best characteristic to consider to effectively improve the delivery of business analysis processes.

Business Analysis for Practitioners: A Practice Guide, Section 3.3.2.3, "Culture."

PMI Professional in Business Analysis (PMI-PBA)® Examination Content Outline, 2013, "Planning," Task 3.

18. Your project manager, Ippy, has asked for your help drafting a key project document that will outline how scope is defined, validated, and controlled. You are collaborating on the creation of the:
    a. Project scope definition plan
    b. Project scope management plan
    c. Requirements management plan
    d. Work breakdown structure and WBS dictionary

The correct answer is: B

Business analysts and project managers will collaborate on the creation of the scope management plan, a project artifact that provides the structure as to how scope will be managed throughout the project. It would establish how it is defined, validated, and controlled.

Answer Choice A: This is a made-up artifact.

Answer Choice C: The requirements management plan is focused on how requirements are analyzed, approved, communicated/reported, documented, elicited, maintained, managed, prioritized, tracked, and validated. The question is inquiring about scope.

Answer Choice D: The work breakdown structure is a logical decomposition of the work to be accomplished over the project, and the dictionary is an accompanying document to provide clarification to the WBS. They would be the recipient of how scope is defined, validated, and controlled.

A Guide to the Project Management Body of Knowledge (PMBOK®), 5.1, "Plan Scope Management."

PMI Professional in Business Analysis (PMI-PBA)® Examination Content Outline, 2013, "Planning," Task 1.

19. Having worked together on a previous client engagement, you've been paired with Hamish, who will serve as project manager for the recovery of a highly complex initiative. As the business analyst, you suggest that one of the first tasks is to interview stakeholders to understand completed activities, in an attempt to learn from their past experiences. Ideally:
    a. These sessions are held independent of each other, so as not to influence or confuse project management and business analysis activities.
    b. These sessions will be facilitated jointly by the project manager and business analyst.
    c. These sessions will focus on lessons learned pertaining to completed project activities.
    d. These sessions will focus on lessons learned pertaining to completed requirements.

The correct answer is: **B**

Open sessions, focused on completed project activities and business analysis work, should be conducted jointly with the project manager and business analyst as collaborative efforts. Leading practice in project delivery is to hold these sessions at the conclusion of each phase and at significant milestones in the project. This technique will help the project team to continually learn, improve, and share this valuable information with their colleagues.

Answer Choice A: To facilitate continual learning and improve communication and engagement, these sessions should be conducted collaboratively.

Answer Choices C & D: The sessions will not be limited only to lessons learned.

Requirements Management: A Practice Guide, Section 3.4.6.1, "Lessons Learned."

PMI Professional in Business Analysis (PMI-PBA)® Examination Content Outline, 2013, "Planning," Task 3.

20. Collaborating with Anna, your project manager, you've developed a fairly comprehensive list of stakeholders, which includes nearly the entire organization. Before presenting to the project Steering Committee for review, you do all of the following except:
    a. Suggest adding the guy who runs the food truck outside the office complex
    b. Add a government agency, with whom your organization has had no prior affiliation
    c. Recommend adding the Stock Exchange on which your company is listed
    d. Remove a small external entity, with whom you have no formal affiliation, who is merging with a competitor

The correct answer is: **D**

Stakeholders include all parties who either are affected or who perceive they are affected by an initiative. Depending on the initiative, the list of stakeholders can be fairly comprehensive. In some circumstances, competitors are stakeholders who could be impacted by your project. Perhaps they resell your product via a transfer agreement, or they may be competing head to head.

Answer Choices A–C: These all represent stakeholders who could be affected by your project and should appear on the stakeholder register.

Requirements Management: A Practice Guide, Section 4.2.1, "Stakeholder Analysis and Engagement."

PMI Professional in Business Analysis (PMI-PBA)® Examination Content Outline, 2013, "Planning," Task 3.

Chapter 4

Analysis

Study Hints

The Analysis domain is the single most important domain on the exam. It contains the elements that focus solely on the tasks and activities related to the identification, detailed specification, approval, and validation of product requirements.

With public sentiment, nothing can fail.
Without it, nothing can succeed.

— Abraham Lincoln

The starting point for business analysts is the formal identification of requirements with supporting documentation. To do so, business analysts will facilitate sessions with stakeholders to document the product or solution requirements. The material produced during the Needs Assessment must be referenced and validated, because this provides the basis and the justification for the initiative.

As the subject matter experts outline their requirements, business analysts work diligently to clarify and decompose the RICE-BW objects (hereafter referred to as *objects*) to a reasonable level of granularity. This is a collaborative process with the subject matter experts, as the business analyst also works to uncover objects that are related or that may have dependencies with other objects.

Knowing that capability and capacity are two valuable commodities, business analysts must collaborate with stakeholders, subject matter experts, and the sponsor to formally agree on which product requirements will be accepted, deferred, or rejected.

With the product scope defined, the business analyst can then establish the requirements baseline, a key domain artifact for the project team that is a measurement for future comparison. To do so requires project team collaboration to balance scope, schedule, and budget along with the organization's capability and capacity—all while ensuring that the planned initiative remains solidly aligned to the business case.

To use the requirements baseline as the framework for effective decision making, the business analyst must collaborate with key stakeholders to obtain approval on the artifact. This represents the known universe of work to deliver the product or solution. Once baselined, any changes must follow the change management plan, created during the Planning domain. Leading practice suggests that change control is fully integrated with the entire project and holistically evaluates the implications of change requests.

With a formally approved product requirements baseline, the project team can then begin the task of writing the specification documents. As you'll recall from the introductory chapter (Table 1.6: Examples of Requirement Types, page 13), there are six specific categories of business requirements applicable for business analysts. In predicative and iterative software delivery methodologies, it's common for requirements to be decomposed to functional and technical specifications. The adaptive delivery methodology tends to be more aligned with the Agile Manifesto, which favors working software over comprehensive documentation.

The business case created during the needs assessment provided the justification for the initiative and established the link to the organizational pillars. The business analyst is accountable to ensure that the requirements, subsequent demonstrations, and prototypes remain aligned with the needs of the organization. Considering the evolving nature of business, it's not uncommon for products to either lose relevance or require modification to maintain alignment with the needs of the organization.

With the project well underway, the business analyst will begin to prepare for Evaluation. Attention will be focused on the development of evaluation metrics and acceptance criteria, which are central to ensuring that the delivered solution fulfills the value proposition.

As outlined in the *Examination Content Outline,* Figure 4.1 highlights the eight essential business analyst tasks associated with the Analysis domain. Please refer to the practice guides and the *PMBOK® Guide* for additional context related to all the items outlined below.

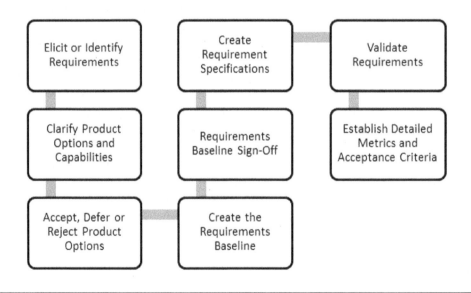

Figure 4.1 Lifecycle of Analysis.

The key artifacts associated with the Analysis domain are shown in Figure 4.2.

Figure 4.2 Analysis artifacts.

At the end of this chapter are exercises to reinforce the concepts from this domain, the answers to which can be found throughout this exam prep book. This domain represents 35% (or 70) of the questions that will appear on the PMI-PBA® Exam.

Major Topics

4.1 Elicit or Identify Requirements

Individual or Group Elicitation Tools and Techniques

- Brainstorming
- Document analysis
- Facilitated workshops
- Focus groups
- Interviews
- Observation
- Prototypes
- Questionnaires
- Surveys

4.2 Clarify Product Options and Capabilities

- Data modeling
- Dependency analysis
- Interface analysis
- Process modeling

➤ Additional Concepts

- Scope models: Goal and business objectives model, ecosystem map, context diagram, feature model, organizational charts, use case diagram, decomposition model, fishbone diagram, interrelationship diagram, SWOT diagram
- Process models: Process flow, use case, user story
- Rule models: Business rules catalogue, decision tree, decision table
- Data models: Entity relationship diagram, data flow diagram, data dictionary, state table, state diagram
- Interface models: Report table, system interface table, user interface flow, wireframes, display-action-response
- Modeling Languages: Business Process Modeling Notation (BPMN), Requirement Modeling Language (RML), System Modeling Language (SysML), Unified Modeling Language (UML)

Problem-Solving and Opportunity Identification Tools and Techniques

- Benchmarking
- Gap analysis
- Operations research
- Scenario analysis
- Systems thinking
- User journey maps
- Value engineering

4.3 Accept, Defer, or Reject Product Options

Decision-Making Tools and Techniques
- Consensus building
- Didactic interaction
- Delphi method
- MultiVoting (nominal group technique)
- Options analysis

Valuation Tools and Techniques
- Cost–benefit analysis
- Net promoter score
- Value stream maps

4.4 Create the Requirements Baseline

Prioritization Tools and Techniques

- Weighted ranking and scoring
- Multi-criteria scoring
- Pair-wise analysis
- MoSCoW
- MultiVoting (nominal group technique)
- Timeboxing

Decision-Making Tools and Techniques
- Consensus building
- Didactic interaction
- Delphi method
- MultiVoting (nominal group technique)
- Options analysis
- Dependency analysis

4.5 Requirements Baseline Sign-Off

Decision-Making Tools and Techniques
- Consensus building
- Delphi method
- Didactic interaction
- MultiVoting (nominal group technique)
- Options analysis

Facilitation Tools and Techniques
- Affinity diagramming
- Brainstorming
- Consensus decision making

- Didactic interaction
- Focus groups
- Flow charting
- Ice breakers
- Scatter diagrams

4.6 Create Requirement Specifications

Business Analysis Requirement Types
- Business requirements
- Stakeholder requirements
- Solution requirements
  ○ Functional requirements
  ○ Nonfunctional requirements
- Transition requirements

Project Management Requirement Types
- Project requirements
- Program requirements
- Quality requirements

Process Analysis Tools and Techniques
- Data flow diagrams
- Dependency graphs
- Events
- Process models
- Use cases
- User stories

4.7 Validate Requirements

Validation Tools and Techniques
- Acceptance criteria
- Demos (conference room pilot – CRP)
- Documentation review
- Inspections
- Peer review
- Prototypes
- User acceptance testing

Conflict Management Tools and Techniques

- Withdraw/avoid
- Smooth/accommodate
- Compromise/reconcile
- Force/direct
- Collaborate/problem solve

4.8 Establish Detailed Metrics and Acceptance Criteria

Validation Tools and Techniques

- Acceptance criteria
- User acceptance testing

Measurement Tools and Techniques

- Service-level agreements
- Operational-level agreements
- Planguage

Analysis
Chapter Exercises

Test Your Knowledge: Analysis

INSTRUCTIONS: On the exam, it's not enough to simply recall that an artifact was created in a particular domain; you'll need to understand the context and how it relates to business analysis. Below are some of the key concepts related to this chapter; take the time to research each and fill in the blanks.

Q	Practice Area	Example	Explanation
	These diagrams are used to show the causes of specific events.	Ishikawa or Fishbone diagram	High-level causes of a problem. Categories can include machines, materials, measurements, stakeholders, and locations.
1	What are five factors to consider when preforming elicitation?		
2	As you are planning elicitation activities, you recall that there are four stages. List each stage sequentially.		

Q	Practice Area	Example	Explanation
3	While interviewing stakeholders, you know that a variety of questions can lead to more thought-out and comprehensive responses. What are four categories of questions?		
4	What are nine types of elicitation activities?		
5	List the two categories of interview techniques and the two methods.		
6	Provide four methods of observation.		

Q	Practice Area	Example	Explanation
7	What are two categories of prototype and two examples?		
8	What are two forms of high-fidelity prototype?		
9	What type of models can be used to analyze the context of the environment?		
10	What are three models that describe elements of a solution, process, or project?		
11	What are three models that are used to administer policies?		

Q	Practice Area	Example	Explanation
12	What types of data models can be used to illustrate information within the process?		
13	What types of models can be used to illustrate process interfaces?		
14	What are four types of feasibility assessment?		
15	What are four methods of prioritizing requirements?		

Q	Practice Area	Example	Explanation
16	List three techniques to build consensus.		
17	What three requirements types are generally considered outside the requirements process?		
18	How does the INVEST acronym relate to user stories?		

Memory Game: Requirements Creation

INSTRUCTIONS: As outlined in *Business Analysis for Practitioners: A Practice Guide*,[1] Section 4.11.5, "Guidelines for Writing Requirements," listed below are the key aspects to consider when creating requirements documents. Practice recalling them using the template on the following page.

Requirements Documentation
Complete
Consistent
Correct
Feasible
Measurable
Precise
Testable
Traceable
Unambiguous

---

[1] PMI. *Business Analysis for Practitioners: A Practice Guide*.

Memory Game: Requirements Creation

INSTRUCTIONS: Make a photocopy of this page; practice listing the key aspects to consider when creating requirements documents.

Requirements Documentation

Analysis: Practice Questions

INSTRUCTIONS: Note the most suitable answer for each multiple-choice question in the appropriate space on the answer sheet on page 102.

1. To show a basic outline, flow of information, general operations, and framework for a proposed customer engagement portal, the project team has asked Electra, an experienced web developer, to present a few options to the VP of customer success. Which option should Electra select?
    a. Prototype
    b. Wireframe
    c. Options analysis
    d. Purpose alignment model

2. One of your lead subject matter experts, Eowyn, is constantly traveling and visiting with clients. It has been impossible to find a time slot on her calendar for a critical interview, which must be completed in 48 hours. What might you try to elicit information regarding her critical requirements for a new teleconferencing system?
    a. Draft an email with your questions, and send Eowyn an invite via the company's video collaboration system to record her responses.
    b. Escalate the scheduling concern to your sponsor, as you know a real-time interview is preferred, based on the business analysis plan.
    c. Flex your schedule and book a time that is mutually convenient for both calendars the following week.
    d. Draft an email with your questions to Eowyn and request that she respond at her convenience.

3. Nikolai has a focus group scheduled next week with the VP of Finance and her team to discuss the accounting system project. Prior to the focus group, what can you suggest to Nikolai so that he can gain insight to the environment?
    a. Collaborate with his project manager to refine the interview technique and questions.
    b. Analyze existing documentation in the configuration management system.
    c. Review the strategy and approach of the asynchronous focus group with his sponsor.
    d. Use the time wisely before the focus group to refine his work plan.

4. During an interview, your subject matter expert, Watson, is providing very detailed responses but does not seem to be answering the questions in a refined, direct manner. To coach Watson, you are going to ask a series of Yes/No questions. Is this approach acceptable?
    a. Yes, these would be considered definitive questions.
    b. Yes, these would be considered context-free questions.
    c. Yes, these would be considered closed-ended questions.
    d. No, it is not appropriate to ask questions with forced answer choices.

5. Your project manager, Cassiopeia, asked for your assistance in collaborating with Marius, the subject matter expert from the local farm, with a document that will outline the compliance requirements for a new line of organic soup. What document will you be assisting with?

    a. Business requirements management plan

    b. Quality requirements

    c. Functional requirements

    d. Stakeholder requirements

6. The subject matter experts from manufacturing, Tyler, Severus, and Braden, seem to be at an impasse and cannot agree on the solution to address the robotic control arm in the plant. All are highly skilled in robotics and considered experts in their field. What can you suggest to mediate the conversation?

    a. Create a grid listing options, so they can be ranked and voted upon.

    b. Facilitate a conversion with the team and have them list the strengths and weaknesses of each proposed solution.

    c. Create a chart and list the forces for and against each option.

    d. Build an options analysis chart and facilitate a conversion to derive a resolution.

7. Following approval of your business plan, your sponsor, Mallory, has provided the team with 45 days to complete the requirements Evaluation process and 90 days for solution development. As the business analyst, what can you do to work within this constraint?

    a. Conduct a feasibility assessment based on technology, system, and cost effectiveness.

    b. Preform a MoSCoW analysis to determine the *must have* features for the release window.

    c. Work within the 45-day period to complete the requirements process, then determine if 90 days is sufficient to complete all the development.

    d. Establish the team's capability within the defined time period, as aligned to the results from a MoSCoW analysis.

8. Your subject matter expert, Ryland, is attempting to describe how a user will interact with a complex website that sells organic natural foods and the functional requirements of the sales order web page that connects farmers globally to consumers. In an effort to minimize rework, Ryland would also like to leverage the material produced for use during testing. How can you approach documenting the requirements of the website and describe a set of scenarios?

    a. Document the requirements from the point of view of the consumer, noting the buttons on the web page and the resulting workflows, thereby representing the functional aspects of the organic natural-food website.

    b. Work with Ryland to create an interoperability diagram, noting the features and function calls on the website and the resulting workflows.

    c. Create a workflow process diagram for use during development and testing.

    d. Build an event analysis to establish the requirements from the point of view of the consumer and the resulting workflows.

9. Estella is an experienced business analyst, and she suspects there may be gaps in the Business Objective Model (BOM). She would like to create a model detailing the lifecycle of an object through various conditions. What tool would be most appropriate to help Estella?

    a. A process flow diagram, aligned to industry leading practice.

    b. A pair-wise table that evaluates the initial and target states for objects.

    c. An entity relationship diagram, further detailing the workflows from the BOM.

    d. A high-fidelity prototype to walk stakeholders through the lifecycle of an object.

10. Your subject matter expert, Malaki from your legal department, has recently submitted over a dozen functional requirement documents for the new contracting application. They are all very well written and succinct. However, they seem to be missing one key element:
    a. They are missing the actors.
    b. There were no preconditions to enable testing.
    c. They should have been from the user's point of view.
    d. Some requirements were labeled TBD, because the team was still addressing some elements.

11. Your sponsor, Calista, is concerned that the product team cannot agree on a proposed solution for a critical business requirement. She has suggested that the team enlist the help of a panel of experts, who can voice their options anonymously. What technique is your sponsor suggesting?
    a. Calista is recommending the team use the nominal group technique.
    b. Calista is suggesting the team use didactic interaction.
    c. Calista is suggesting the team consider a moderated discussion.
    d. Calista is suggesting the team use the Delphi technique.

12. Peregrine, your subject matter expert from the Sustainable Fish Farm, is finding it difficult to explain the artificial breeding process. You've been assigned as his business analyst, as the business team is exploring new artificial ecosystems. Lacking formal training in marine science, how do you gain insight into the project?
    a. Onboard an expert to assist you with the project.
    b. Ask Peregrine to carefully document the situation statement and requirements documents.
    c. Spend a few days with Peregrine at the fish farm.
    d. Suggest to your sponsor, Calix, that another business analyst be assigned who is familiar with sustainable fish farming.

13. Ignatius has requested participants from the new product elucidation team to confirm that the requirements as validated by the senior stakeholders and management meet or exceed the organization's standards of excellence. The organization's project management office has established a series of guidelines to oversee this process. What is one element that is critical for this inspection process?
    a. That all reviewers are provided with advance copies of the material and guidelines.
    b. It is recommended that the participants also include business management.
    c. Participants from the new product elucidation team have strong knowledge of the inspection process.
    d. A checklist.

14. You're leading a session with Katniss, the senior VP of human resources, to establish a prioritization framework. She's leaning toward MoSCow but wants further clarification of the acronym.
    a. W, won't have; S, significant to have; M, might not have; Co, could have
    b. Co, could have; S, should have; W, won't have; Mo, might have
    c. Mo, must have; S, shouldn't have; Co, could have; W, won't have
    d. S, should have; Mo, must have; W, won't have; Co, could have

15. Kellan has submitted a number of non-functional specifications. They are very well written; however, they are lacking one key component.
    a. The specifications are missing version control.

b. The specifications were not loaded into the collaboration portal prior to submission.

c. Test scenarios cannot be created to verify that the condition has been satisfied.

d. Feasibility can only be determined after a number of factors have been analyzed.

16. Veronica is preparing for peer review of functional requirements for her solution. When looking at the resource calendar, several individuals have availability. From the options below, whom should she select to participate in the peer review?

a. The department director of the team that developed the functional requirements; this will provide for final verification.

b. Colleagues who have not been involved with the project thus far; this will ensure that they are impartial and objective.

c. Colleagues from the organizational change management team who are tasked with developing the training material for the solution.

d. Members of the organization's project management office (PMO), who established the guidelines of the peer review.

17. Samara just conducted her first didactic interaction, which went very well. As soon as participants arrived, she separated them into two groups based on their position. Following a few rounds of discussion lasting 30 minutes each, the participants arrived at a mutual consensus. The participants left the 90-minute didactic interaction session with a very good understanding of the rationale and the final decision. In hindsight, were there any opportunities for improvement?

a. Yes, didactic interaction sessions should be time boxed to 45 minutes.

b. No, the session is outlined per the guidelines in the *PMI Business Analysis Practice Guide*.

c. Yes, prior to separating the participants, Samara should have set the stage.

d. Yes, Samara did not conduct a didactic interaction session, but rather an NGT session.

18. Your project is scheduled to last for six sprints, with allocated times between two and three weeks for each sprint. Your scrum master has suggested each user story should be:

a. Independent, negotiable, valuable, estimable, small, and testable

b. Independent, non-negotiable, valuable, estimable, small, and testable

c. Independent, negotiable, valuable, estimable, significant, and testable

d. Independent, navigable, valued, estimable, small, and testable

19. You are a business analyst for Serenity Aerospace, supporting a component of an overall program whose ultimate goal is to colonize the Moon, then eventually Mars. It has been determined that Serenity must replace one of its main computer systems. Before approving the work, the governing body for the program requested an illustration showing all the systems that may be affected by this upgrade. What tool would you recommend?

a. A context diagram detailing all the direct and human boundaries within a system.

b. A visual representation highlighting all the features of the solution, arranged in hierarchical edifice, identifying occasional interface and data requirements with the partner organizations.

c. A map showing all the boundaries within Serenity Aerospace's proposed system, so that partner organizations and the governing body can get a better understanding of the proposed initiative.

d. An ecosystem map showing all the systems within both Serenity and the partner companies, identifying occasional interface and data requirements.

20. At Alastair Co., you've been assigned the daunting task of interviewing Dr. Llewellyn Cason, the 2017 winner of the Nobel Prize in Physics. Although retired from Alastair for several years, it's perceived he may be impacted by your division's antimatter project. How would you best prepare for this 45-minute interview?

  a. Brainstorm with your colleagues and come prepared with a list of questions, but ask only one from the list, using your skill as a business analyst to facilitate an unbiased, unstructured conversation.

  b. Brainstorm with your colleagues and come prepared with a list of questions.

  c. Brainstorm with your colleagues and come prepared with a list of questions, with a goal of asking all the questions on the list, even if it exceeds the allotted time.

  d. Brainstorm with your colleagues, come prepared with a list of questions, but let Dr. Llewellyn Cason guide the conversation.

Answer Sheet

	Answer Choice				Correct	Incorrect	Predicted Answer			
							90%	75%	50%	25%
1	a	b	c	d						
2	a	b	c	d						
3	a	b	c	d						
4	a	b	c	d						
5	a	b	c	d						
6	a	b	c	d						
7	a	b	c	d						
8	a	b	c	d						
9	a	b	c	d						
10	a	b	c	d						
11	a	b	c	d						
12	a	b	c	d						
13	a	b	c	d						
14	a	b	c	d						
15	a	b	c	d						
16	a	b	c	d						
17	a	b	c	d						
18	a	b	c	d						
19	a	b	c	d						
20	a	b	c	d						

Analysis: Answer Key

1. To show a basic outline, flow of information, general operations, and framework for a proposed customer engagement portal, the project team has asked Electra, an experienced web developer, to present a few options to the VP of Customer Success. Which option should Electra select?
   a. Prototype
   b. Wireframe
   c. Options analysis
   d. Purpose alignment model

   The correct answer is: **B**

   Wireframes are outlines, blueprints, or schematics that illustrate the general look and feel of a proposed solution. They are often very flexible and highly adaptable for use in various situations. Commonly used for newsletters and website design, wireframes show the flow of specific logic and business functions, depicting all user touchpoints.
   Answer Choice A: Prototypes are working models of the final product, completed before being sent to production.
   Answer Choice C: An options analysis is a technique to evaluate an investment decision, often used when creating a business case.
   Answer Choice D: A purpose alignment model is used for prioritizing requirements and ensuring alignment with the organization's vision, goals, and objectives.
   Business Analysis for Practitioners: A Practice Guide, Section 4.5.5.7, "Prototyping."
   PMI Professional in Business Analysis (PMI-PBA)® Examination Content Outline, 2013, "Analysis," Task 2.

2. One of your lead subject matter experts, Eowyn, is constantly traveling and visiting with clients. It has been impossible to find a time slot on her calendar for a critical interview, which must be completed in 48 hours. What might you try to elicit information regarding her critical requirements for a new teleconferencing system?
   a. Draft an email with your questions, and send Eowyn an invite via the company's video collaboration system to record her responses.
   b. Escalate the scheduling concern to your sponsor, as you know a real-time interview is preferred, based on the business analysis plan.
   c. Flex your schedule and book a time that is mutually convenient for both calendars the following week.
   d. Draft an email with your questions to Eowyn and request that she respond at her convenience.

   The correct answer is: **A**

   The asynchronous interviewing technique is ideal when interviews cannot be conducted in real time (synchronous). When using the asynchronous interviewing technique, the interview questions are scripted, allowing the interviewee to note their response in either an email or a post to a wiki, or to record in a video collaboration system. In some instances, asynchronous interviews are preferred by project teams, so that they can revisit interviewee responses, which are often used to accommodate a team member's schedules and availability.

Answer Choice B: It's expected the business analyst would know how to address this situation in a reasonable manner, without escalating to his or her sponsor.

Answer Choice C: Had there not been a time constraint of 48 hours, this answer choice would have had some merit.

Answer Choice D: While partially correct, this response is not time bound, and the scenario stated that the critical interview must be completed in 48 hours.

Business Analysis for Practitioners: A Practice Guide, Section 4.5.5.5, "Interviews."

PMI Professional in Business Analysis (PMI-PBA)® Examination Content Outline, 2013, "Analysis," Task 1.

3. Nikolai has a focus group scheduled next week with the VP of finance and her team to discuss the accounting system project. Prior to the focus group, what can you suggest to Nikolai so that he can gain insight to the environment?
   a. Collaborate with his project manager to refine the interview technique and questions.
   b. Analyze existing documentation in the configuration management system.
   c. Review the strategy and approach of the asynchronous focus group with his sponsor.
   d. Use the time wisely before the focus group to refine his work plan.

The correct answer is: **B**

Prior to meeting with stakeholders, it's leading practice to conduct a document analysis to gain understanding and insight into the environment and situation. Information gathered can form the basis for productive conversations. Configuration management systems or project management information systems will often provide relevant and useful information to better understand the current state.

Answer Choice A & D: While these may seem like great ideas and a good use of time, neither will provide insight to the environment.

Answer Choice C: This is a made-up answer choice.

Business Analysis for Practitioners: A Practice Guide, Section 4.5.5.2, "Documentation Analysis."

PMI Professional in Business Analysis (PMI-PBA)® Examination Content Outline, 2013, "Analysis," Task 1

4. During an interview, your subject matter expert, Watson, is providing very detailed responses but does not seem to be answering the questions in a refined, direct manner. To coach Watson, you are going to ask a series of Yes/No questions. Is this approach acceptable?
   a. Yes, these would be considered definitive questions.
   b. Yes, these would be considered context-free questions.
   c. Yes, these would be considered closed-ended questions.
   d. No, it is not appropriate to ask questions with forced answer choices.

The correct answer is: **C**

Business analysts can select from a number of techniques when conducting an elicitation session. Types of questions include open-ended questions, closed-ended questions, contextual questions, and context-free questions. Closed-ended questions elicit responses based on limited answer choices and are often used for confirmation: Yes or No.

Answer Choice A: Definitive is not a type of question.

Answer Choice B: Context-free questions are often used as opening or introductory questions.
Answer Choice D: This is incorrect; closed-ended questions are often used to either force or limit choice and to elicit confirmation.
Business Analysis for Practitioners: A Practice Guide, Section 4.5.2.1, "Types of Questions."
PMI Professional in Business Analysis (PMI-PBA)® Examination Content Outline, 2013, "Analysis," Task 1.

5. Your project manager, Cassiopeia, asked for your assistance in collaborating with Marius, the subject matter expert from the local farm, with a document that will outline the compliance requirements for a new line of organic soup. What document will you be assisting with?
   a. Business requirements management plan
   b. Quality requirements
   c. Functional requirements
   d. Stakeholder requirements

The correct answer is: **B**

When documenting requirements, business analysts must classify them into categories to provide clarity and context. Table 1.7, Examples of Requirement Types (page 14), further elaborates on the categories of business, stakeholder, solution, and transitional requirements. In this scenario, quality requirements establish the measures and metrics to ensure project completion and address compliance requirements and standards.
Answer Choice A: This is a distractor answer choice; this is not an artifact.
Answer Choice C: Functional requirements are based on product attributes; in the case of organic soup, this might include vegetable mix or flavor.
Answer Choice D: Stakeholder requirements are based on critical success factors (CSFs), key performance indicators (KPIs), and objectives and goals.
Requirements Management: A Practice Guide, Section 5.2.2, "Define Types of Requirements."
PMI Professional in Business Analysis (PMI-PBA)® Examination Content Outline, 2013, "Analysis," Task 1.

6. The subject matter experts from manufacturing—Tyler, Severus, and Braden—seem to be at an impasse and cannot agree on the solution to address the robotic control arm in the plant. All are highly skilled in robotics and considered experts in their field. What can you suggest to mediate the conversation?
   a. Create a grid listing options, so they can be ranked and voted upon.
   b. Facilitate a conversion with the team and have them list the strengths and weaknesses of each proposed solution.
   c. Create a chart and list the forces for and against each option.
   d. Build an options analysis chart and facilitate a conversion to derive a resolution.

The correct answer is: **A**

The weighted rankings and scorings technique, also used during the needs assessment to rank solution options, is used during analysis to resolve requirements-related conflicts. The process is identical to what was used during needs assessment, whereby solution options are listed, ranked, and voted upon by team members. The tool is used objectively to compare solution options; the option with the highest score is selected.

Answer Choice B: A strengths and weaknesses analysis is typically used to assess the internal capabilities of an organization.

Answer Choice C: A force-field analysis is a collaborative decision-making tool, in which participants list the restraining forces (obstacles or negatives) and the driving forces (motivators or positives) for the option under consideration.

Answer Choice D: Option analysis charts are used to evaluate an investment decision, often created when building a business case.

Business Analysis for Practitioners: A Practice Guide, Section 4.15.3, "Weighted Ranking."

PMI Professional in Business Analysis (PMI-PBA)® Examination Content Outline, 2013, "Analysis," Task 3.

7. Following approval of your business plan, your sponsor, Mallory, has provided the team with 45 days to complete the requirements Evaluation process and 90 days for solution development. As the business analyst, what can you do to work within this constraint?
   a. Conduct a feasibility assessment based on technology, system, and cost effectiveness.
   b. Preform a MoSCoW analysis to determine the *must have* features for the release window.
   c. Work within the 45-day period to complete the requirements process, then determine if 90 days is sufficient to complete all the development.
   d. Establish the team's capability within the defined time period, as aligned to the results from a MoSCoW analysis.

The correct answer is: **D**

Business analysts can use a variety of methods to conjure requirements prioritization from stakeholders. MoSCoW establishes a set of prioritization rules; MultiVoting elicits active participation from stakeholders; time-boxing analyzes the amount of work that can be completed in a given window of time; and weighted ranking applies where criteria are ranked with a score. Regardless of the lifecycle approach, most initiatives do not have the luxury of infinite time. In this scenario, the sponsor provided two timeframes; it's the business analyst's responsibility to work within the fixed timelines and deliver a solution that is aligned to the stakeholder requirements. The business analyst would start with the time-boxing is a prioritization technique and then conduct a MoSCoW analysis.

Answer Choice A: Feasibility assessments evaluate potential solution options for viability based on key variables or "feasibility factors." This assessment would not be appropriate in this scenario.

Answer Choice B: This is a partially correct answer; the preceding analysis would be to assess what could be completed within the release window.

Answer Choice C: This is the worst possible approach, as it sets the wrong expectations with the stakeholders.

Business Analysis for Practitioners: A Practice Guide, Section 4.11.6.1, "Prioritization Schemes."

PMI Professional in Business Analysis (PMI-PBA)® Examination Content Outline, 2013, "Analysis," Task 2.

8. Your subject matter expert, Ryland, is attempting to describe how a user will interact with a complex website that sells organic natural foods and the functional requirements of the sales order web page that connects farmers globally to consumers. In an effort to minimize rework, Ryland would also like to leverage the material produced for use during testing. How can you approach documenting the requirements of the website and describe a set of scenarios?

a. Document the requirements from the point of view of the consumer, noting the buttons on the web page and the resulting workflows, thereby representing the functional aspects of the organic natural-food website.
b. Work with Ryland to create an interoperability diagram, noting the features and function calls on the website and the resulting workflows.
c. Create a workflow process diagram for use during development and testing.
d. Build an event analysis to establish the requirements from the point of view of the consumer and the resulting workflows.

The correct answer is: A

Use cases are used to describe scenarios and explain complex interactions between users and systems in plain text, articulating how the system should operate, the associated flows, and the expected benefits—all from the vantage point of a lead actor. They represent the functional requirements of a system, operation, or process. They should not be used for non-functional requirements, as they generally apply to an entire platform rather than a single use case.

Answer Choice B: Interoperability diagrams are used in documenting relationships and boundaries of interfacing systems.

Answer Choice C: While helpful in the requirements documentation process, it would not address the need to describe a set of scenarios.

Answer Choice D: This is a made-up answer.

Business Analysis for Practitioners: A Practice Guide, Section 4.10.8.2, "Use Case."

PMI Professional in Business Analysis (PMI-PBA)® Examination Content Outline, 2013, "Analysis," Task 2.

9. Estella is an experienced business analyst, and she suspects there may be gaps in the Business Objective Model. She would like to create a model detailing the lifecycle of an object through various conditions. What tool would be most appropriate to help Estella?
a. A process flow diagram, aligned to industry leading practice.
b. A pair-wise table that evaluates the initial and target states for objects.
c. An entity relationship diagram, further detailing the workflows from the Business Objective Model.
d. A high-fidelity prototype to walk stakeholders through the lifecycle of an object.

The correct answer is: B

Business analysts can use both state tables and state diagrams to model the valid states of an object and any transitions between states over their lifecycle. These techniques allow business analysts to specify the lifecycle of an object in the solution. Furthermore, these techniques can also be used to ensure that nothing has been overlooked in the requirements definition process.

Answer Choice A: Process flow diagrams are powerful tools for visualizing both current (as-is) and future (to-be) states, often referred to as *swim-lane diagrams;* this would not be the best technique for detailing the lifecycle of an object through various conditions.

Answer Choice C & D: These are both made-up answer choices, combining unrelated concepts.

Business Analysis for Practitioners: A Practice Guide, Section 4.10.10.4, "State Table and State Diagrams."

PMI Professional in Business Analysis (PMI-PBA)® Examination Content Outline, 2013, "Analysis," Task 6.

10. Your subject matter expert, Malaki from your legal department, has recently submitted over a dozen functional requirement documents for the new contracting application. They are all very well written and succinct. However, they seem to be missing one key element:

    a. They are missing the actors.
    b. There were no preconditions to enable testing.
    c. They should have been from the user's point of view.
    d. Some requirements were labeled TBD, because the team was still addressing some elements.

The correct answer is: **B**

While drafting requirements, it's acceptable to have some elements marked as "TBD," considering the dynamic nature of projects and the evolution of requirements. The key aspects of requirement documents are that they are complete, consistent, correct, feasible, measurable, precise, testable, traceable, and unambiguous. In this scenario, the functional requirements were not measurable, as they were missing the preconditions to enable testing.

Answer Choice A: Actors are used when creating w, not functional requirement documents.

Answer Choice C: Use cases and user stories are created from the user's point of view.

Answer Choice D: This is a distractor; although aspects of a requirements document could be labeled TBD, a requirement labeled as TBD would only appear in the requirements traceability matrix. Once it's been validated and approved, the team could begin creating the requirements document.

Business Analysis for Practitioners: A Practice Guide, Section 4.11.5, "Guidelines for Writing Requirements."

PMI Professional in Business Analysis (PMI-PBA)® Examination Content Outline, 2013, "Analysis," Task 6.

11. Your sponsor, Calista, is concerned that the product team cannot agree on a proposed solution for a critical business requirement. She has suggested the team enlist the help of a panel of experts, who can voice their options anonymously. What technique is your sponsor suggesting?

    a. Calista is recommending the team use the nominal group technique.
    b. Calista is suggesting the team use didactic interaction.
    c. Calista is suggesting the team consider a moderated discussion.
    d. Calista is suggesting the team use the Delphi technique.

The correct answer is: **D**

The role of the business analyst is to remain impartial and objective over the life of the project. The Delphi technique is one way the business analyst can arbitrate a discussion and help the team arrive an acceptable solution. It's a structured process that is used to facilitate informed decision making, designed to allow participants to comment anonymously on a particular subject matter. When planning to use the Delphi technique, the facilitator establishes clear, predefined completion parameters (for example, number of rounds, mean or median score, etc.). One benefit of using this technique is that it's typically staffed by a panel of experts.

Answer Choice A: The nominal group technique is optimally used so that all attendees can participate equally. The tool is also especially helpful in cases in which the team is having difficulty deriving potential solutions. It's not based on an anonymous panel of experts.

Answer Choice B: Didactic interaction is best used when trying to reach consensus on a "Yes/No" or "Option 1/Option 2" decision. Each team must document the advantages and disadvantages of their position. It's not based on an anonymous panel of experts.

Answer Choice C: A moderated discussion would typically not provide for anonymity, and the answer choice does not address the requirement for a panel of experts.

Business Analysis for Practitioners: A Practice Guide, Section 4.15, "Resolve Requirements-Related Conflicts."

PMI Professional in Business Analysis (PMI-PBA)® Examination Content Outline, 2013, "Analysis," Task 3.

12. Peregrine, your subject matter expert from the Sustainable Fish Farm, is finding it difficult to explain the artificial breeding process. You've been assigned as his business analyst, as the business team is exploring new artificial ecosystems. Lacking formal training in marine science, how do you gain insight into the project?
    a. Onboard an expert to assist you with the project.
    b. Ask Peregrine to carefully document the situation statement and requirements documents.
    c. Spend a few days with Peregrine at the fish farm.
    d. Suggest to your sponsor, Calix, that another business analyst be assigned who is familiar with sustainable fish farming.

The correct answer is: **C**

Observation is an ideal technique for viewing stakeholders in their environment and seeing how jobs are performed. It's especially useful when dealing with complex activities that may be difficult to articulate. Furthermore, observation provides unprejudiced information about the activities and environment. There are four types of observation techniques: passive, active, participatory, and simulation. By spending a few days at the Sustainable Fish Farm, you could gain valuable insight to the project.

Answer Choice A: While this may seem like a logical choice, PMI is a proponent of taking action, and there is nothing to indicate at this point that you need expert assistance.

Answer Choice B: This is a distractor; neither artifact is relevant at this time.

Answer Choice D: This is not an acceptable approach; you've been assigned as the business analyst—it's your responsibly to be assertive and take action.

Business Analysis for Practitioners: A Practice Guide, Section 4.5.5.6, "Observation."

PMI Professional in Business Analysis (PMI-PBA)® Examination Content Outline, 2013, "Analysis," Task 1.

13. Ignatius has requested participants from the new product elucidation team to confirm that the requirements as validated by the senior stakeholders and management meet or exceed the organization's standards of excellence. The organization's project management Office has established a series of guidelines to oversee this process. What is one element that is critical for this inspection process?
    a. All reviewers are provided with advance copies of the material and guidelines.
    b. It is recommended that the participants also include business management.
    c. Participants from the new product elucidation team have strong knowledge of the inspection process.
    d. A checklist.

The correct answer is: **D**

Inspections follow a rigorous process and adhere to certain guidelines that govern how they are performed. They are often performed by colleagues of those who participated in the creation of the requirements documents. Because they are rigorous and comprehensive, for consistency, they are often managed via a checklist of known errors or defects.

Answer Choice A: This is a distractor and doesn't completely address the critical element of the inspection process.

Answer Choice B: Business stakeholders and management are specifically excluded from inspections. Leading practice suggests that inspections should be performed by peers who participated in the creation and documentation of the requirements.

Answer Choice C: While it's preferred that the inspection be performed by those receiving the requirements documentation, their strong knowledge is irrelevant.

Business Analysis for Practitioners: A Practice Guide, Section 4.13.2, "Inspection."
PMI Professional in Business Analysis (PMI-PBA)® Examination Content Outline, 2013, "Analysis," Task 7.

14. You're leading a session with Katniss, the senior VP of human resources, to establish a prioritization framework. She is leaning toward MoSCoW but has asked for further clarification as to the acronym.
    a. W, won't have; S, significant to have; M, might not have; Co, could have
    b. Co, could have; S, should have; W, won't have; Mo, might have
    c. Mo, must have; S, shouldn't have; Co, could have; W, won't have
    d. S, should have; Mo, must have; W, won't have; Co, could have

The correct answer is: **D**

MoSCoW is a prioritization acronym, used to help stakeholders prioritize the established requirements. The method helps to objectively guide the evaluation process, focusing on those items that have priority. Mo: must have, essential to the project; S: should have, central to the project but not critical; Co: could have, easily omitted; W: won't have, outside the scope of the present solution.

Answer Choice A: S, significant to have, and M, might not have, are both incorrect. "S" indicates *should have* requirements, whereas "Mo" represents *must have* requirements.

Answer Choice B: Mo, might have, is incorrect. "Mo" denotes *must have* requirements.

Answer Choice C: S, shouldn't have, is incorrect. "S" indicates *should have* requirements.

Business Analysis for Practitioners: A Practice Guide, Section 4.11.6.1, "Prioritization Schemes."
PMI Professional in Business Analysis (PMI-PBA)® Examination Content Outline, 2013, "Analysis," Task 5.

15. Kellan has submitted a number of non-functional specifications; they are very well written; however, they are lacking one key component.
    a. The specifications are missing version control.
    b. The specifications were not loaded into the collaboration portal prior to submission.
    c. Test scenarios cannot be created to verify that the condition has been satisfied.
    d. Feasibility can only be determined after a number of factors have been analyzed.

The correct answer is: **C**

While drafting requirements, the key aspects are that they are complete, consistent, correct, feasible, measurable, precise, testable, traceable, and unambiguous. In this scenario, the non-functional specifications were missing elements related to measurable and testable, therefore they were not complete.

Answer Choice A: Using version control on requirements documentation is a leading practice; if required, they would be identified in the business analysis plan. Omission would not be considered a key component.

Answer Choice B: While this is good practice, it does not address the omission of one key component.

Answer Choice D: This is a distractor; although feasibility is an aspect of the requirements documentation, it should have previously been established.

Business Analysis for Practitioners: A Practice Guide, Section 4.11.5.1, "Functional Requirements," page 128, Testable.

PMI Professional in Business Analysis (PMI-PBA)® Examination Content Outline, 2013, "Analysis," Task 8.

16. Veronica is preparing for peer review of functional requirements for her solution. When looking at the resource calendar, several individuals have availability. From the options below, whom should she select to participate in the peer review?
    a. The department director of the team that developed the functional requirements; this will provide for final verification.
    b. Colleagues who have not been involved with the project thus far; this will ensure that they are impartial and objective.
    c. Colleagues from the organizational change management team who are tasked with developing the training material for the solution.
    d. Members of the organization's project management office (PMO), who established the guidelines of the peer review.

The correct answer is: **C**

When provided the option, peer reviews are optimally conducted by downstream recipients of the information. In this case, the team developing the training material would be well suited to review the requirements documentation. They will check for consistency, completeness, and testability.

Answer Choice A: Whether formal or informal, reviews of the requirements should be conducted by peers of the business analyst. The department director would not be a peer of the business analyst.

Answer Choice B: This could be a good answer; however, answer choice C is a better option, whereby downstream recipients of the information conduct the peer review.

Answer Choice D: business analysts could have recommended other practitioners—for example, a team member in the business analysis center of excellence (COE) or the project management office (PMO). This could be good answer; however, answer choice C is a better option, whereby downstream recipients of the information conduct the peer review.

Business Analysis for Practitioners: A Practice Guide, Section 4.13.1, "Peer Review."

PMI Professional in Business Analysis (PMI-PBA)® Examination Content Outline, 2013, "Analysis," Task 7.

17. Samara just conducted her first didactic interaction, which went very well. As soon as participants arrived, she separated them into two groups based on their position. Following a few rounds of discussion lasting 30 minutes each, the participants arrived at a mutual consensus. The participants left the 90-minute didactic interaction session with a very good understanding of the rationale and the final decision. In hindsight, were there any opportunities for improvement?
    a. Yes, didactic interaction sessions should be time boxed to 45 minutes.
    b. No, the session is outlined per the guidelines in the *PMI Business Analysis Practice Guide.*
    c. Yes, prior to separating the participants, Samara should have set the stage.
    d. Yes, Samara did not conduct a didactic interaction session, but rather a NGT session.

The correct answer is: **C**

When conducting an elicitation activity, there are four stages in which information is gathered. The first stage is the Introduction, in which the business analyst sets the stage and establishes the overall purpose (this is also known as *setting the table*). The second stage is the question-and-answer section, known as the *body* of the elicitation session. The third stage is the *close*, or the conclusion of the elicitation activity. The last stage is the *follow-up*, whereby the business analyst confirms that the participants understand and agree with the outcome. In this scenario, Samara failed to establish the overall purpose of the session, or "set the stage."

Answer Choice A: The times are a distractor; time boxing is not associated with didactic interaction.
Answer Choice B: This is incorrect, as Samara failed to establish the overall purpose of the elicitation session.
Answer Choice D: The mention of techniques is also a distractor; the question is focused on opportunities for improvement.

Business Analysis for Practitioners: A Practice Guide, Section 4.5, "Conduct Elicitation Activities."
PMI Professional in Business Analysis (PMI-PBA)® Examination Content Outline, 2013, "Analysis," Task 5.

18. Your project is scheduled to last for six sprints, with allocated times between two and three weeks for each sprint. Your scrum master has suggested each user story should be:
    a. Independent, negotiable, valuable, estimable, small, and testable
    b. Independent, non-negotiable, valuable, estimable, small, and testable
    c. Independent, negotiable, valuable, estimable, significant, and testable
    d. Independent, navigable, valued, estimable, small, and testable

The correct answer is: **A**

While typically associated with adaptive lifecycle projects, user stories can be used with virtually any delivery methodology. When properly written, they follow the INVEST acronym: independent, negotiable, valuable, estimable, small, and testable in the format:
As a <role or type of user>, I want <goal/desire>, so that <benefit>.
Large user stories, known as EPICs, are decomposed into manageable user stories and added to the backlog.

Answer Choice B: Non-negotiable is incorrect
Answer Choice C: Significant is incorrect
Answer Choice D: Navigable is incorrect

Business Analysis for Practitioners: A Practice Guide, Section 4.10.8.3, "User Story."
PMI Professional in Business Analysis (PMI-PBA)® Examination Content Outline, 2013, "Analysis," Task 6.

19. You are a business analyst for Serenity Aerospace, supporting a component of an overall program whose ultimate goal is to colonize the Moon, then eventually Mars. It has been determined that Serenity must replace one of its main computer systems. Before approving the work, the governing body for the program requested an illustration showing all the systems that may be affected by this upgrade. What tool would you recommend?
    a. A context diagram detailing all the direct and human boundaries within a system.
    b. A visual representation highlighting all the features of the solution, arranged in hierarchical edifice, identifying occasional interface and data requirements with the partner organizations.

c. A map showing all the boundaries within Serenity Aerospace's proposed system, so that partner organizations and the governing body can get a better understanding of the proposed initiative.

d. An ecosystem map showing all the systems within both Serenity and the partner companies, identifying occasional interface and data requirements.

The correct answer is: D

Ecosystem maps extend beyond the typical boundaries of architectural diagrams. When used holistically, they identify all the organizations and associated systems that are affected, or that may be affected, by the proposed solution.

Answer Choice A: Context diagrams are created to show all the direct system and human interfaces within a solution. This solution could have widespread impacts on partners, which is why the ecosystem map is preferred to the context diagram.

Answer Choice B & C: Both answer choices have merit but are not as specific as answer choice D.

Business Analysis for Practitioners: A Practice Guide, Section 4.10.7.2, "Ecosystem Map."

PMI Professional in Business Analysis (PMI-PBA)® Examination Content Outline, 2013, "Analysis," Task 2.

20. At Alastair Co., you've been assigned the daunting task of interviewing Dr. Llewellyn Cason, the 2017 winner of the Nobel Prize in Physics. Although retired from Alastair for several years, it's perceived he may be impacted by your division's antimatter project. How would you best prepare for this 45-minute interview?

a. Brainstorm with your colleagues and come prepared with a list of questions, but ask only one from the list, using your skill as a business analyst to facilitate an unbiased, unstructured conversation.

b. Brainstorm with your colleagues and come prepared with a list of questions.

c. Brainstorm with your colleagues and come prepared with a list of questions, with a goal of asking all the questions on the list, even if it exceeds the allotted time.

d. Brainstorm with your colleagues, come prepared with a list of questions, but let Dr. Llewellyn Cason guide the conversation.

The correct answer is: A

In this scenario, the business analyst should conduct an unstructured interview. Using this form of interviewing technique, the interviewer comes prepared with a list of questions. However, the interviewee is definitely only asked the first question. The interview progresses on its own, relying on the skill of the business analyst to guide the conversation.

Answer Choice B: This is a partially correct answer.

Answer Choice C: This is a partially correct answer; however, it does not take into account the time constraint.

Answer Choice D: This is a partially correct answer; however, the business analyst should facilitate the conversation, remaining focused on achieving the initial objective.

Business Analysis for Practitioners: A Practice Guide, Section 4.5.5.5, "Interviews."

PMI Professional in Business Analysis (PMI-PBA)® Examination Content Outline, 2013, "Analysis," Task 1.

Chapter 5

Traceability and Monitoring

Study Hints

The Traceability and Monitoring domain contains those tasks and activities that focus on managing the lifecycle of the requirements. This domain focuses on establishing the requirements baseline and continually monitoring, tracking, and communicating the status of requirements. Furthermore, the domain encompasses change control and the management of issues, risks, and decisions.

> *There will always be more good ideas than you and your teams have the capacity to execute.*
>
> — Chris McChesney, *The 4 Disciplines of Execution: Achieving Your Wildly Important Goals*

First identified in the Planning domain, the traceability matrix is built out to include the product requirements, which are traced to the organizational pillars (described in Section 1.2, page 9). For each requirement object, relationships and dependencies are established, owners are agreed upon, and dates are assigned for all associated deliverables. When creating the matrix, Figure 5.1 on the next page elaborates the categories and types of objects that should be considered.

Once requirements objects are agreed to and captured on the matrix, they are governed by the change control plan. Although the project manager or sponsor may have some authority, the plan will outline the thresholds for integrated change control to approve any modifications, additions, or material changes. This is also an ideal time for a baseline report detailing the overall metrics.

As the subject matter experts submit their requirements documentation, the business analyst is focused on monitoring the objects and ensuring that there is appropriate documentation, which has been reviewed and signed off on by designated owners. The requirements traceability matrix is appropriately updated as the requirements progress through the project's lifecycle. Proper maintenance of the requirements traceability matrix can yield vast amounts of statistical information with regard to the requirement objects. It is paramount that the business analyst communicate the status of these objects to key stakeholders, all while keeping them apprised of potential issues, risks, and changes, along with decisions that impact the product. This will provide the project manager and the sponsor with insight into whether development efforts are tracking—either favorably or unfavorably—to plan. If the latter, the project manager can recommend corrective actions.

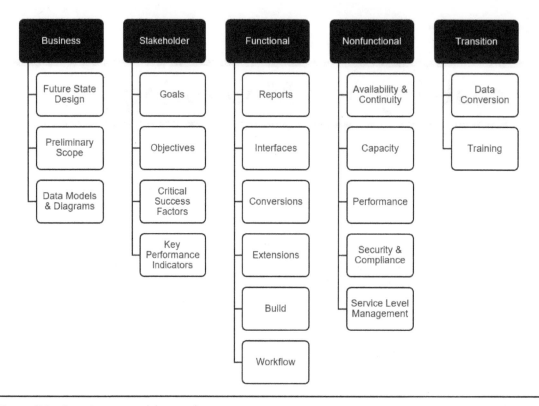

Figure 5.1 Requirements elaborated.

As the subject matter experts and technical resources develop specification documents, build models, and execute test cases, changes are inevitable. It's the business analyst's responsibility to manage requested changes in accordance with the change control plan, which was produced in the Planning domain, all while managing the interrelated aspect of product delivery.

As outlined in the *PMI Professional in Business Analysis (PMI-PBA)® Examination Content Outline,* Figure 5.2 highlights the five essential business analyst tasks associated with the Traceability and

Figure 5.2 Lifecycle of Traceability and Monitoring.

Monitoring domain. Please refer to the practice guides and the *PMBOK® Guide* for additional context related to all the items outlined below.

The key artifacts associated within the Traceability and Monitoring domain are shown in Figure 5.3.

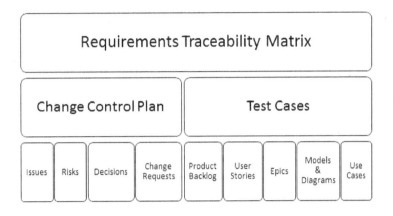

Figure 5.3 Traceability and Monitoring artifacts.

At the end of this chapter are exercises to reinforce the concepts from this domain, the answers to which can be found throughout this exam prep book. This domain represents 15% (or 30) of the questions that will appear on the PMI-PBA® Exam.

Major Topics

5.1 Track Requirements

Traceability Artifact or Tool
- Backlog management
- Dependency analysis
- Issue, risk, decision tracking
- Requirements traceability matrix

➢ Additional Concepts
- RICE-BW tracker
- Epics or user stories to features and acceptance tests
- Kanban board
- Models and diagrams
- Requirements baseline
- Traceability trees
- Use cases to acceptance tests

5.2 Monitor Requirements

Traceability Artifact or Tool
- Backlog management
- Issue, risk, decision tracking
- Requirements traceability matrix

➢ Additional Concepts
- RICE-BW tracker
- Epics or user stories to features and acceptance tests
- Kanban board
- Models and diagrams
- Requirements baseline
- Traceability trees
- Use cases to acceptance tests

5.3 Update Requirements Status

Traceability Artifact or Tool
- Backlog management
- Issue, risk, decision tracking
- Requirements traceability matrix

➢ Additional Concepts
- RICE-BW tracker
- Documentation management
- Epics or user stories to features and acceptance tests

- ○ Kanban board
- ○ Models and diagrams
- ○ Requirements baseline
- ○ Traceability trees
- ○ Use cases to acceptance tests

Reporting Tools and Techniques

➢ **Additional Concepts**
- Documentation management
- Dependency analysis

5.4 Communicate Requirements Status

Traceability Artifact or Tool
- Backlog management
- Issue, risk, decision tracking
- Requirements traceability matrix

➢ **Additional Concepts**
- ○ RICE-BW tracker
- ○ Documentation management
- ○ Epics or user stories to features and acceptance tests
- ○ Kanban board
- ○ Models and diagrams
- ○ Requirements baseline
- ○ Traceability trees
- ○ Use cases to acceptance tests

Reporting Tools and Techniques

➢ **Additional Concepts**
- ○ Documentation management
- ○ Dependency analysis

5.5 Manage Requirements

Change Control Tools and Techniques
- Change Control Board
- Configuration management system
- Version control system

➢ **Additional Concepts**
- ○ Impact analysis
- ○ Dependency analysis
- ○ Conflict management and resolution
- ○ Issue management
- ○ Risk management

Conflict Management Tools and Techniques

- Withdraw/avoid
- Smooth/accommodate
- Compromise/reconcile
- Force/direct
- Collaborate/problem solve

Negotiation Tools and Techniques

Quality Control Tools and Techniques

- Seven basic quality tools
    - Checksheets/checklists
    - Control charts
    - Flow charts/SIPOC
    - Histograms
    - Ishikawa diagrams (aka *fishbone* or *cause-and-effect diagrams*)
    - Pareto diagrams
    - Scatter diagrams

- Statistical sampling
- Inspection
- Approved change request review

Traceability and Monitoring
Chapter Exercises

Test Your Knowledge: Traceability and Monitoring

INSTRUCTIONS: Traceability and Monitoring is a critical function for business analysts. It ensures that the proposed work is solidly linked to the organization's and project's goals and objectives and that all work is traced from conception through delivery validation. Below are a few questions to test your knowledge in this area. Consider each question carefully, and note your answer in the space provided.

Q	Practice Area	Example	Explanation
	These diagrams are used to show the causes of specific events.	Ishikawa or Fishbone diagram	High-level causes of a problem. Categories can include machines, materials, measurements, stakeholders, and locations.
1	When using an adaptive lifecycle, what are three methods to track requirements?		
2	What are three benefits of tracking requirements?		

Q	Practice Area	Example	Explanation
3	For predictive or adaptive projects, name 10 project elements that can be traced or tracked.		
4	What are the attributes for the five stages of product development?		
5	What are two types of change control tools?		

Q	Practice Area	Example	Explanation
6	When conducting an impact analysis, what are some key areas to consider?		
7	What are sequential steps associated with change management?		
8	What tool can be used to monitor and control product scope and link requirements to an organization's goals and objectives?		

Q	Practice Area	Example	Explanation
9	What are the iterative functions associated with requirements traceability?		
10	Your project is considering enlisting the help of an expert; provide a few examples of whom they could retain.		

Q	Practice Area	Example	Explanation
11	What are sequential steps associated with change management?		
12	What tool can be used to monitor and control product scope and link requirements to an organization's goals and objectives?		

Traceability and Monitoring: Practice Questions

INSTRUCTIONS: Note the most suitable answer for each multiple-choice question in the appropriate space on the answer sheet on page 130.

1. Logan, the subject matter expert for Waffles-on-the-Go, has proposed a change to the exterior product packaging, which he believes will result in increased sales in the frozen-food section of grocery stores. Before bringing this request to the Change Control Board, as the business analysts assigned to the project, what should Logan do?
   a. Conduct an assessment to understand the impact on the value proposition and its effect on other requirements.
   b. Present the request to the Change Control Board.
   c. Review with his sponsor, who first evaluates then approves the suggested change.
   d. Conduct an assessment to understand dependent relationships.

2. Your sponsor, Milo, has reservations and is concerned about a recently proposed change. Whom should Milo consult?
   a. Project management office
   b. Regulatory agencies
   c. Suppliers, contractors, and the governance team
   d. All of the above

3. While monitoring the submission of functional requirements, Silas, a recent college graduate, asks you, a veteran business analyst, what the most vital skills for managing this process are? You reply:
   a. Dealing with ambiguity and aligning requirements to the strategic vision for the project
   b. Passive listening, dealing with ambiguity, engaging stakeholders
   c. Dealing with ambiguity, engaging stakeholders, documentation
   d. Keeping the requirements traceability matrix up to date for reporting

4. Ezra has received approval from the Change Control Board (CCB) to descope several requirements because they are no longer linked to the goals and objectives of the organization. As he updates the traceability matrix, he realizes:
   a. The project will complete ahead of schedule and under budget.
   b. The stakeholders will appreciate the removal of non-value–added features.
   c. The traceability trees were not considered by the CCB.
   d. The developers will be able to focus on other value stream improvements.

5. You are working on a very complex project in a highly regulated industry. On your project, there is a significant amount of documentation, which is constantly being revised. Aria, your organizational change management lead, suggests a tool that:
   a. Is well known to all stakeholders and allows for easy updating and minimal training
   b. Traces and tracks requirements from their origin over the lifecycle of the project
   c. Will monitor and evaluate changes for the impact on the organization
   d. Locks documents for editing when they are checked out for updating

6. Sebastian is the product owner for the software development of a new mobile application. The project team has decided to use an adaptive methodology versus an iterative or predictive software delivery approach. Over lunch, he explains that his tool will require grooming over time. What is Sebastian referring to?
   a. The steps to ensure that the backlog contains items that are prioritized and relevant
   b. Backlog prioritization, a foundational agile concept
   c. Agile, scrum, and XP programming, which are tenets of the Agile Manifesto
   d. The process to ensure that the requirements traceability matrix is up to date

7. Your newest team member, Eleanora, creates a poster board with three columns—To Do, Doing, and Done—with development objects posted under each column. Your CIO walks by and approves of the information radiator. What was the true intent of the poster board?
   a. To apprise the team of her current workload
   b. To provide some amount of traceability
   c. For prioritization of requirements
   d. As a reminder of work that has been accomplished and what's scheduled

8. Charlotte, an intern from the local community college, is helping to establish a relationship between functional requirements and deliverables. What tool can you suggest?
   a. Simplex matrix
   b. Use case diagram
   c. CRUD matrix
   d. Traceability matrix

9. At the watercooler, Olivia is discussing the value of the requirements traceability matrix. As you walk back to your desk, you're attempting to recall some of the key items to trace.
   a. Business need, business objectives linked to business validation
   b. Functional requirements, technical requirements linked to business need
   c. Project scope, SDBW deliverables, project objectives
   d. Objectives, development stage, source

10. As a member of what group would Amelia review, evaluate, and consider changes to scope?
    a. Integrated Change Advisory Board (iCAB)
    b. Change Advisory Board (CAB)
    c. Change Control Board (CCB)
    d. Project Steering Committee

11. Asher has just notified key stakeholders and subject matter experts of an approved change. What should he do next?
    a. Evaluate the impact of the change request
    b. Verify the implications of the change request
    c. Ensure that changes are made only after the impact has been assessed and verified
    d. Make note of the change for future reference

12. Having expertise with requirements traceability on predictive projects, Aurora, the senior business analyst assigned to the project, advises:

a. The project sponsor should agree to all items on the matrix prior to development.

b. The team should consider using backlog management to manage change requests.

c. The project team should conduct a dependency analysis prior to beginning the development.

d. She has a worksheet from a prior project that was very useful to trace and track requirements. She suggests trying this tool before creating something from scratch.

13. Jasper is the subject matter expert (SME) for the purchase-to-pay process of your ERP implementation. While reading his submitted requisition approval and invoice approval functional specifications, he sees that they appear to be overly complicated and in some cases confusing. What should you do?

a. Update the statistics to reflect that the functional specifications have been submitted.

b. Request a meeting with Jasper to walk through his submitted documentation.

c. Return the functional specifications for further elaboration, clarification, and simplification.

d. Update the material yourself, as prior to joining the project team you were the SME for the purchase-to-pay workstream.

14. You've requested Adeline, your subject matter expert, to prepare a number of change requests to address solution gaps, because full scope was unknown at the onset. She comments, "The project has just kicked off, what's the need for the formality?" You answer:

a. The Project Steering Committee has committed to keeping costs within the budget and scope aligned with the charter.

b. The Project Steering Committee didn't expect any changes in this project.

c. The Project Steering Committee has already approved the traceability baseline.

d. Predictive and iterative delivery methodologies mandate documentation for changes; however, with agile projects, changes are anticipated and encouraged.

15. Calvin has just baselined the requirements for a project at the Securities and Exchange Commission when his sponsor, Catherine, proposes a change. Following an analysis of the change, Calvin and his sponsor agree to:

a. Bring the change to the Integrated Change Control Board

b. Approve the change without the guidance of the Change Control Board

c. Review the requirements management plan

d. Communicate the proposed change to all stakeholders

16. During a project team meeting, Henry, one of your solution testers, inquired about who had the authority to approve requirements. You suggest:

a. He consult the requirements management plan.

b. Only the sponsor and product owner can approve requirements.

c. The sponsor can approve requirements with insight from the product team.

d. The Project Steering Committee, of which the sponsor is chair.

17. Your sponsor, Cora, has requested a group meeting to review submitted requirements documentation to ensure that the requirements as stated are valid. What type of meeting would be most appropriate?

a. A Delphi-based requirements session

b. A requirements walkthrough session

c. A group focused requirements session

d. A nominal group requirements session

18. Your sponsor, Emma, is concerned about the needed complexity and significant level of detail associated with the requirements traceability matrix and requirements documentation. What might Emma suggest to confirm that the documentation is in compliance with the organizational requirements before a signoff meeting with the project stakeholders?
    a. An informal review by Ethan and Bodhi, two of your peers
    b. The project manager review the documentation
    c. A formal review by the project manager and change lead
    d. A review by the testing and change leads before the meeting

19. You are having a debate with Liam, your project manager, regarding scope management. Which of his following statements are not true regarding requirements?
    a. Product scope management is improved with requirements traceability.
    b. Stakeholder expectations can be managed with requirements traceability.
    c. The requirements traceability matrix is created while eliciting requirements.
    d. Requirements traceability can help determine if there are missing requirements.

20. As Calliope worked to complete the requirements management plan, she included a section that addressed how requirement updates would be managed. What is the primary theme for this section?
    a. Configuration auditing
    b. Configuration control
    c. Configuration identification
    d. Configuration status control

Answer Sheet

	Answer Choice				Correct	Incorrect	Predicted Answer			
							90%	75%	50%	25%
1	a	b	c	d						
2	a	b	c	d						
3	a	b	c	d						
4	a	b	c	d						
5	a	b	c	d						
6	a	b	c	d						
7	a	b	c	d						
8	a	b	c	d						
9	a	b	c	d						
10	a	b	c	d						
11	a	b	c	d						
12	a	b	c	d						
13	a	b	c	d						
14	a	b	c	d						
15	a	b	c	d						
16	a	b	c	d						
17	a	b	c	d						
18	a	b	c	d						
19	a	b	c	d						
20	a	b	c	d						

Traceability and Monitoring: Answer Key

1. Logan, the subject matter expert and business analyst for Waffles-on-the-Go, has proposed a change to the exterior product packaging, which he believes will result in increased sales in the frozen food section of grocery stores. Before bringing this request to the Change Control Board, as the business analyst assigned to the project, what should Logan do?
   a. Conduct an assessment to understand the impact on the value proposition, and its effect on other requirements.
   b. Present the request to the Change Control Board.
   c. Review with his sponsor, who first evaluates then approves the suggested change.
   d. Conduct an assessment to understand dependent relationships.

   The correct answer is: **A**

   Prior to bringing the change request to the Change Control Board, the business analyst should conduct an impact assessment to closely examine the proposed change in relation to approved requirements. This assessment will identify any risks or issues and will serve to clarify the impact to scope, schedule, and the approved budget.
   Answer Choice B: The first step should be to conduct the impact assessment, otherwise the Change Control Board will not have all the information relative to the requested change.
   Answer Choice C: Depending on the level of authority, the sponsor may be able to approve the change. However, before doing so, an impact assessment should be conducted.
   Answer Choice D: A relationship dependency assessment would not be appropriate in this scenario.
   Requirements Management: A Practice Guide, Section 7.3.2, "Impact Analysis."
   PMI Professional in Business Analysis (PMI-PBA)® Examination Content Outline, 2013, "Traceability and Monitoring," Task 5.

2. Your sponsor, Milo, has reservations and is concerned about a recently proposed change. From the choices below, whom should Milo consult?
   a. Project management office
   b. Regulatory agencies
   c. Suppliers, contractors, and the governance team
   d. All of the above

   The correct answer is: **D**

   When considering changes, it may be necessary to consult with industry experts, consultants, or any number of stakeholders who either are impacted or perceive they will be impacted by the project. The groups listed in the answer choices are all categories of stakeholders.
   Answer Choice A: The project management office is the group tasked with defining and managing the standards and providing administrative support for the programs and projects.
   Answer Choice B: Governmental body or authority exercising automous oversight.
   Answer Choice C: Suppliers and contractors are third parties who are either directly or indirectly associated with the initiative. The governance team is the group tasked with ensuring that the components goals are achieved and that the effort is aligned with the organization's overall strategy.

A Guide to the Project Management Body of Knowledge (PMBOK®), Section 4.5.2.1, "Expert Judgment."
PMI Professional in Business Analysis (PMI-PBA)® Examination Content Outline, 2013, "Traceability and Monitoring," Task 5.

3. While monitoring the submission of functional requirements, Silas, a recent college graduate, asks you, a veteran business analyst, what the most vital skills for managing this process are? You reply:
   a. Dealing with ambiguity and aligning requirements to the strategic vision for the project
   b. Passive listening, dealing with ambiguity, engaging stakeholders
   c. Dealing with ambiguity, engaging stakeholders, documentation
   d. Keeping the requirements traceability matrix up to date for reporting

The correct answer is: **A**

Functional requirements establish the vital link between what was envisioned during ideation, what was approved in the business case and charter, and what is delivered. They create a clear canvas and ensure that the appropriate technologies are used to deliver the solution. There are several vital skills for managing this process, including dealing with ambiguity, aligning requirements to the strategic vision for the project, active listening, interpreting requirements, communication, and engaging stakeholders.

Answer Choice B: Passive listening is ineffective and results in misinterpretation. Ways to improve include making eye contact, avoid distracting gestures, asking questions, paraphrasing the speaker, not talking over the speaker, and listening thoughtfully.

Answer Choice C: This is a partially correct answer, as dealing with ambiguity and engaging stakeholders are a vital skill—documentation is not.

Answer Choice D: While grooming the requirements traceability matrix for reporting purposes is important, it's not a vital skill for business analysts.

PMI Pulse of the Profession®, Requirements Management: A Core Competency for Project and Program Success, page 13, "The Most Important Skills for Requirements Management."

PMI Professional in Business Analysis (PMI-PBA)® Examination Content Outline, 2013, "Traceability and Monitoring," Task 2.

4. Ezra has received approval from the Change Control Board (CCB) to descope several requirements because they are no longer linked to the goals and objectives of the organization. As he updates the traceability matrix, he realizes:
   a. The project will complete ahead of schedule and under budget.
   b. The stakeholders will appreciate the removal of non-value–added features.
   c. The traceability trees were not considered by the CCB.
   d. The developers will be able to focus on other value stream improvements.

The correct answer is: **C**

The requirements traceability matrix is the essential link between the baselined scope and organizations pillars (mission, vision, goals, and objectives). In software development, this is often referred to as the RICE-BW tracker (reports, interfaces, conversions, extensions/enhancements, build, and workflow). Prior to meeting with the Change Control Board, a dependency analysis should be conducted to understand the relationships between requirements on the matrix. If there are dependencies, all factors must be considered by the Integrated Change Control Board. A common way to illustrate dependencies is through the use of traceability trees.

Answer Choice A: This a distractor answer; the project could actually take longer to complete if dependencies were not properly considered.

Answer Choice B: What may appear to be non-value–added features could in fact be essential to deliver on core functionality, which could only be noted on traceability trees or through a dependency analysis.

Answer Choice D: This a distractor answer; by failing to properly identify dependencies and improvements to the value stream, the process could inadvertently worsen.

Requirements Management: A Practice Guide, Section 7.3.1, "Dependency Analysis."

PMI Professional in Business Analysis (PMI-PBA)® Examination Content Outline, 2013, "Traceability and Monitoring," Task 1.

5. You are working on a very complex project in a highly regulated industry. On your project, there is a significant amount of documentation, which is constantly being revised. Aria, your organizational change management lead, suggests a tool that:
    a. Is well known to all stakeholders and allows for easy updating and minimal training
    b. Traces and tracks requirements from their origin over the lifecycle of the project
    c. Will monitor and evaluate changes for the impact on the organization
    d. Locks documents for editing when they are checked out for updating

The correct answer is: **D**

Documentation is an essential facet of every project, and the proper management is critical for auditing, compliance, communication, and sharing lessons learned. A properly implemented system takes into consideration features such as document tagging, security, workflow, integration, and archiving. Once a requirements document is signed off on and approved, leading practice is to adopt version control to monitor and control updates. This can be as simple as a table in the beginning of the document or as robust as an automated system. Using a tool that locks checked-out documents will minimize the possibility for overwrites by other team members.

Answer Choice A: Fundamentally, this seems like a reasonable answer, but it doesn't completely answer the question.

Answer Choice B: This describes the requirements traceability matrix.

Answer Choice C: This is a distractor; although it seems reasonable due to the dynamic nature of the organization, it would be impractical to have such a tool.

Business Analysis for Practitioners: A Practice Guide, Section 5.8.2.2, "Version Control Systems."

PMI Professional in Business Analysis (PMI-PBA)® Examination Content Outline, 2013, "Traceability and Monitoring," Task 5.

6. Sebastian is the product owner for the software development of a new mobile application. The project team has decided to use an adaptive methodology versus an iterative or predictive software delivery approach. Over lunch, he explains that his tool will require grooming over time. What is Sebastian referring to?
    a. The steps to ensure that the backlog contains items that are prioritized and relevant

b. Backlog prioritization, a foundational agile concept
c. Agile, scrum, and XP programming, which are tenets of the Agile Manifesto
d. The process to ensure the requirements traceability matrix is up to date

---

The correct answer is: **A**

Adaptive software development, a form of agile, consists of four phases: communication and planning, analysis, design and development, and testing and deployment. As with all agile methods, as requirements are confirmed, they are added to the backlog for prioritization. As these requirements progress through the lifecycle, it's necessary to ensure that the designated matrix is appropriately updated. In the case of adaptive methodologies, a product backlog is used; the method of keeping the tool up to date and relevant is known as *backlog grooming.*

Answer Choice B: The concept is *backlog grooming,* the answer choice provides *backlog prioritization.*

Answer Choice C: The Agile Manifesto is based on the premise of "uncovering better ways of developing software."

Answer Choice D: This is a misleading answer choice; when using the adaptive methodology, the matrix is referred to as the *product backlog.* On the exam, remember to use domain-specific vocabulary.

Requirements Management: A Practice Guide, page 43.

PMI Professional in Business Analysis (PMI-PBA)® Examination Content Outline, 2013, "Traceability and Monitoring," Task 3.

---

7. Your newest team member, Eleanora, creates a poster board with three columns—To Do, Doing, and Done—with development objects posted under each column. Your CIO walks by and approves of the information radiator. What was the true intent of the poster board?
    a. To apprise the team of her current workload
    b. To provide some amount of traceability
    c. For prioritization of requirements
    d. As a reminder of work that has been accomplished and what's scheduled

---

The correct answer is: **B**

Information radiators are large, highly visible displays. When used appropriately, they keep team members and others up to date on the progress of the project. Kanban boards are typically used in adaptive life-cycle projects and show work in progress (WIP) across three columns—To Do, Doing, and Done—providing limited traceability.

Answer Choice A, C, D: These are all partially correct answers; however, they don't fully answer the question.

Business Analysis for Practitioners: A Practice Guide, Section 5.2.1, "What Is Traceability?"

PMI Professional in Business Analysis (PMI-PBA)® Examination Content Outline, 2013, "Traceability and Monitoring," Task 1.

---

8. Charlotte, an intern from the local community college, is helping to establish a relationship between functional requirements and deliverables. What tool can you suggest?
    a. Simplex matrix
    b. Use case diagram
    c. CRUD matrix
    d. Traceability matrix

The correct answer is: **D**

Requirements traceability matrices enable tracking and tracing to monitor and validate the requirements. The matrix is used to link product requirements vertically to the goals and objectives of the organization and horizontally to the deliverables that satisfy the requirements.
Answer Choice A: A simplex matrix is used to solve linear programming equations and would not be used in business analysis.
Answer Choice B: Use case diagrams, also known as *behavior diagrams*, are used to visualize use cases.
Answer Choice C: A CRUD matrix is typically used in database development and is an acronym for create, read, update, or delete.
Requirements Management: A Practice Guide, 7.3.3 "Traceability Matrix."
PMI Professional in Business Analysis (PMI-PBA)® Examination Content Outline, 2013, "Traceability and Monitoring," Task 1.

9. At the watercooler, Olivia is discussing the value of the requirements traceability matrix. As you walk back to your desk, you're attempting to recall some of the key items to trace.
   a. Business need, business objectives linked to business validation
   b. Functional requirements, technical requirements linked to business need
   c. Project scope, SDBW deliverables, project objectives
   d. Objectives, development stage, source

The correct answer is: **A**

The requirements traceability matrix is a critical tool for business analysis; on the exam, you can expect a number of questions regarding the attributes. Although in practice the tool is very flexible, for the exam you'll need to know the primary PMI attributes, which also include business need, opportunities, goals, objectives, approach to testing, and listing of requirements.
Answer Choice B: Requirement documents are self-contained and not linked to business need.
Answer Choice C & D: These are made-up combinations.
Guide to the Project Management Body of Knowledge (PMBOK®), 5.2.3.2 "Requirements Traceability Matrix."
Business Analysis for Practitioners: A Practice Guide, Section 5.2.3.1, "Requirements Attributes."
PMI Professional in Business Analysis (PMI-PBA)® Examination Content Outline, 2013, "Traceability and Monitoring," Task 1.

10. As a member of what group would Amelia review, evaluate, and consider changes to scope?
    a. Integrated Change Advisory Board (iCAB)
    b. Change Advisory Board (CAB)
    c. Change Control Board (CCB)
    d. Project Steering Committee

The correct answer is: **C**

The Change Control Board (CCB) or Integrated Change Control Board is tasked with evaluating and considering changes to scope. Change requests are typically preceded by an impact assessment and include a dependency analysis. As defined in the change management plan, the project's CCB has the authority to consider changes and the responsibility for communicating decisions. Leading practice is for change control to be fully integrated across the entire project, with all aspects thoroughly considered to enable informed decision making.
Answer Choice A: This is a made-up answer choice.

Answer Choice B: Change Advisory Boards (CABs) are typically associated with IT service management, whereas CCBs are associated with project management and business analysis.
Answer Choice D: While the CCB could be a subset of the Project Steering Committee, this is not the best answer.
Business Analysis for Practitioners: A Practice Guide, Section 5.4.2, "Approval Levels."
PMI Professional in Business Analysis (PMI-PBA)® Examination Content Outline, 2013, "Traceability and Monitoring," Task 5.

11. Asher has just notified key stakeholders and subject matter experts of an approved change. What should he do next?
    a. Evaluate the impact of the change request.
    b. Verify the implications of the change request.
    c. Ensure that changes are made only after the impact has been assessed and verified.
    d. Make note of the change for future reference.

    The correct answer is: **D**

    Over the life of project, all decisions must be clearly documented. This material can be used for both project auditing and confirming the rationale and will be used during project close. In terms of changes, the change control plan will typically outline the steps to be taken once changes are either approved or rejected.
    Answer Choice A & B: The business analyst would evaluate the impact and verify the implications prior to submitting the change request.
    Answer Choice C: While this is a true statement, before doing so, the business analyst must make note of the change for future reference.
    Requirements Management: A Practice Guide, Section 7.2.4, "Manage Requirements Change Requests."
    PMI Professional in Business Analysis (PMI-PBA)® Examination Content Outline, 2013, "Traceability and Monitoring," Task 4.

12. Having expertise with requirements traceability on predictive projects, Aurora, the senior business analyst assigned to the project, advises:
    a. The project sponsor should agree to all items on the matrix prior to development.
    b. The team should consider using backlog management to manage change requests.
    c. The project team should conduct a dependency analysis prior to beginning the development.
    d. She has a worksheet from a prior project that was very useful to trace and track requirements. She suggests trying this tool before creating something from scratch.

    The correct answer is: **D**

    It's the business analyst's responsibility to capitalize on artifacts and lessons learned from prior projects. The organization's project management office may also have tools—for instance, templates are artifacts that could be leveraged. The requirements traceability matrix is often a worksheet, but it can also be integrated into a robust project management information system (PMIS). On the exam, remember, never select an answer choice which requires you to buy a tool, and furthermore, the correct answer choice will never dictate the complexity of the solution. However, it is very important that, prior to the start of the project, the matrix is agreed to and the team is trained on its use.

Answer Choice A & C: While these may be true statements, from a sequence perspective, answer D is the better choice.

Answer Choice B: This is a made-up answer choice; backlog management is not used for managing change requests.

Requirements Management: A Practice Guide, Section 7.2.1.1, "Set Up System for Managing Requirements and Traceability."

PMI Professional in Business Analysis (PMI-PBA)® Examination Content Outline, 2013, "Traceability and Monitoring," Task 1.

13. Jasper is the subject matter expert (SME) for the purchase-to-pay process of your ERP implementation. While reading his submitted requisition approval and invoice approval functional specifications, he sees that they appear to be overly complicated and in some cases confusing. What should you do?
    a. Update the statistics to reflect that the functional specifications have been submitted.
    b. Request a meeting with Jasper to walk through his submitted documentation.
    c. Return the functional specifications for further elaboration, clarification, and simplification.
    d. Update the material yourself, as prior to joining the project team you were the SME for the purchase-to-pay workstream.

The correct answer is: **B**

While business analysts should have a fundamental understanding of the areas they support, it's not uncommon for subject matter experts to be better versed and more knowledgeable. As project documentation is submitted, the business analyst will review the material in detail to ensure it is complete, consistent, correct, feasible, measurable, precise, testable, traceable, and unambiguous before being handed off to the next level. This is a case of *dealing with ambiguity*—the best option would be to meet with the SME to review the submitted material.

Answer Choice A: While this is a valid task, it doesn't address the question.

Answer Choice C: This would simply result in more rework, as the SME could return the functional specifications, advising they are not confusing to a knowledgeable resource.

Answer Choice D: Business analysts must remain independent and unbiased to be effective. Updating the material yourself would be a conflict of interest.

PMI Pulse of the Profession®, Requirements Management a Core Competency for Project and Program Success, page 14, Section on Important Skills.

PMI Professional in Business Analysis (PMI-PBA)® Examination Content Outline, 2013, "Traceability and Monitoring," Task 2.

14. You've requested Adeline, your subject matter expert, to prepare a number of change requests to address solution gaps, because full scope was unknown at the onset. She comments, "The project has just kicked off, what's the need for the formality?" You answer:
    a. The Project Steering Committee has committed to keeping costs within the budget and scope aligned with the Charter.
    b. The Project Steering Committee didn't expect any changes in this project.
    c. The Project Steering Committee has already approved the traceability baseline.

d. Predictive and iterative delivery methodologies mandate documentation for changes; however, with agile projects, changes are anticipated and encouraged.

The correct answer is: **C**

The guidelines for managing the traceability matrix were established in the Planning domain. Once objects are agreed to and captured on the matrix, they are governed by the change control plan. Therefore, once the traceability matrix is created and signed off on, it's baselined and monitored throughout the project's duration to evaluation. Any changes to scope would require the traceability matrix to be rebaselined, so that status can be communicated and progress measured.

Answer Choice A: While this statement may be true, it does not address the question.

Answer Choice B: By their very nature, projects go through changes. Over their duration, changes should be anticipated and managed by the change control plan.

Answer Choice D: Although agile projects anticipate changes, there is still a defined process for managing changes to scope once the backlog is created.

Business Analysis for Practitioners: A Practice Guide, Section 5.6, "Monitoring Requirements Using a Traceability Matrix."

PMI Professional in Business Analysis (PMI-PBA)® Examination Content Outline, 2013, "Traceability and Monitoring," Task 5.

---

15. Calvin has just baselined the requirements for a project at the Securities and Exchange Commission when his sponsor, Catherine, proposes a change. Following an analysis of the change, Calvin and his sponsor agree to:
    a. Bring the change to the Integrated Change Control Board
    b. Approve the change without the guidance of the Change Control Board
    c. Review the requirements management plan
    d. Communicate the proposed change to all stakeholders

The correct answer is: **C**

The requirements management plan establishes the ground rules and parameters as to how requirements are analyzed, approved, communicated/reported, documented, elicited, maintained, managed, prioritized, tracked, and validated. In this scenario, they are reviewing the requirements management plan to gain further insight into understanding next steps in the process.

Answer Choice A: Prior to bringing the change to the Change Control Board, the business analyst would first need to conduct an impact assessment and determine any dependencies.

Answer Choice B: Although the sponsor may have the authority to approve the change, the first step would be to review the requirements management plan.

Answer Choice D: The change management plan would outline the process for submitting changes; how the business analyst will verify the change request; the process for evaluating the impact of the change; and how the Integrated Change Control Board will review, decide, and communicate the change. This would not be done without first consulting the requirements management plan.

Requirements Management: A Practice Guide, Section 7.3.4, "Change Control Boards."

PMI Professional in Business Analysis (PMI-PBA)® Examination Content Outline, 2013, "Traceability and Monitoring," Task 5.

16. During a project team meeting, Henry, one of your solution testers, inquired who had the authority to approve requirements. You suggest:
    a. He consult the requirements management plan.
    b. Only the sponsor and product owner can approve requirements.
    c. The sponsor can approve requirements with insight from the product team.
    d. The Project Steering Committee, of which the sponsor is chair.

The correct answer is: **A**

The requirements management plan establishes the ground rules and parameters for how requirements are analyzed, approved, communicated/reported, documented, elicited, maintained, managed, prioritized, tracked, and validated.

In this scenario, they are reviewing the requirements management plan to gain further insight into who has the authority to approve requirements.

Answer Choices B–D: While these may be true statements, the requirements management plan will affirm who has the authority to approve requirements.

Business Analysis for Practitioners: A Practice Guide, Section 5.4, "Approving Requirements."

PMI Professional in Business Analysis (PMI-PBA)® Examination Content Outline, 2013, "Traceability and Monitoring," Task 2.

17. Your sponsor, Cora, has requested a group meeting to review submitted requirements documentation to ensure that the requirements as stated are valid. What type of meeting would be most appropriate?
    a. A Delphi-based requirements session
    b. A requirements walkthrough session
    c. A group focused requirements session
    d. A nominal group requirements session

The correct answer is: **B**

As project documentation is submitted, the business analyst will review the material in detail to ensure they are complete, consistent, correct, feasible, measurable, precise, testable, traceable, and unambiguous before being handed off to the next level. Following this review, leading practice is to conduct a requirements walkthrough or "handshake meeting." This walkthrough ensures that submitted documentation and requirements, as written, are understood by the development team, prior to any development or construction activity commences.

Answer Choice A: The Delphi technique is a structured tool that is used to facilitate informed decision making; it is designed to allow participants to comment anonymously on a particular subject matter.

Answer Choice C: This is made-up answer choice.

Answer Choice D: The nominal group technique is optimally used so that all attendees can participate equally. The tool is also especially helpful in cases in which the team is having difficulty deriving potential solutions.

Business Analysis for Practitioners: A Practice Guide, Section 4.12.2, "Requirements Walkthrough."

PMI Professional in Business Analysis (PMI-PBA)® Examination Content Outline, 2013, "Traceability and Monitoring," Task 2.

18. Your sponsor, Emma, is concerned about the needed complexity and significant level of detail associated with the requirements traceability matrix and requirements documentation. What might Emma suggest to confirm that the documentation is in compliance with the organizational requirements before a signoff meeting with the project stakeholders?
    a. An informal review by Ethan and Bodhi, two of your peers
    b. The project manager review the documentation
    c. A formal review by the project manager and change lead
    d. A review by the testing and change leads before the meeting

The correct answer is: **A**

Peer reviews, which can be either formal or informal requirement reviews, can be conducted at any point during the requirements process. Peer reviews are opportunities for colleagues to provide insight, make recommendations to strengthen the process, and ensure that there are no obvious errors or omissions that could cause non-compliance with organizational documentation standards.
Answer Choice B: This is not the best answer. Although the project manager may be knowledgeable about requirements, the better choice would be direct peers.
Answer Choice C: This is a partially correct answer. Although the change lead would be good choice as a downstream recipient of information, the project manager would not be the best choice as a peer reviewer.
Answer Choice D: The testing and change leads would be a good choice to include in the verification review, providing them an opportunity to gain insight into the project and context at the onset. Both of these roles could focus on requirement details and check for consistency, completeness, and testability. This is not a good answer choice, as the review type is nonspecific.
Business Analysis for Practitioners: A Practice Guide, Section 4.13.1, "Peer Reviews."
PMI Professional in Business Analysis (PMI-PBA)® Examination Content Outline, 2013, "Traceability and Monitoring," Task 3.

19. You are having a debate with Liam, your project manager, regarding scope management. Which of his following statements are not true regarding requirements?
    a. Product scope management is improved with requirements traceability.
    b. Stakeholder expectations can be managed with requirements traceability.
    c. The requirements traceability matrix is created while eliciting requirements.
    d. Requirements traceability can help determine if there are missing requirements.

The correct answer is: **C**

The requirements traceability matrix is created during project planning. It enables the valuation of each requirement by linking it to the business and project objectives. It provides a mechanism for tracking requirements throughout the project lifecycle, starting with ideation through evaluation. The requirements traceability matrix also supports the integrated change management process, and as a result it aids in the management of product scope.
Answer Choices A, B, D: These answer choices are benefits of requirements traceability. Other benefits: scope creep is prevented; requirement documents are created for each object and can be referenced across project phases; the complete baseline contains the needed detail to build the product or features; every detail relates to a requirements document.

Business Analysis for Practitioners: A Practice Guide, Section 5.2, "Traceability."

PMI Professional in Business Analysis (PMI-PBA)® Examination Content Outline, 2013, "Traceability and Monitoring," Task 1.

20. As Calliope worked to complete the requirements management plan, she included a section that addressed how requirement updates would be managed. What is the primary theme for this section?
    a. Configuration auditing
    b. Configuration control
    c. Configuration identification
    d. Configuration status control

The correct answer is: **B**

Configuration control is central to change management; it includes the evaluation of all change requests and the impact to project and product scope. Configuration control ensures that there is consistency and that stakeholders are aware and understand the impact of any proposed changes. Once changes are approved, the requirements traceability matrix baseline can be updated to reflect any changes and the associated dependencies.

Answer Choice A: A configuration audit is an independent evaluation of the delivered product to confirm that the as-built solution is in alignment with the requirements documentation.

Answer Choice C: Configuration identification is the process of identifying attributes for product, solution, or configuration item.

Answer Choice D: This is a made-up answer choice.

Requirements Management: A Practice Guide, Section 7.2.6, "Document and Communicate Requests."

PMI Professional in Business Analysis (PMI-PBA)® Examination Content Outline, 2013, "Traceability and Monitoring," Task 5.

Chapter 6

Evaluation

Study Hints

The Evaluation domain contains the business analyst's tasks and activities that (a) validate test results; (b) analyze and communicate gaps; (c) fulfill the work with stakeholders to obtain signoff; and (d) conclude with an evaluation of the deployed solution.

The most serious mistakes are not being made as a result of wrong answers.
The truly dangerous thing is asking the wrong question.

— Peter Drucker

Many of the concepts, tools, and techniques used in this domain were introduced in prior chapters to initiate an activity. For example, the business case was the key artifact produced in the Needs Assessment domain. The concluding activity in this domain is to validate the cost–benefit analysis or value proposition for the solution. The Evaluation can be based on the full solution or a segment of the solution, using both qualitative and quantitative approaches. As with all the other material reviewed, the concepts and tools are highly flexible and robust; therefore, they can be used with predictive, iterative, and adaptive lifecycles.

Because projects represent an organization's desire to change and involve a commitment of resources (financial, human capital, material, equipment, technology, etc.), the common theme within this domain is "evaluate and measure early" and "evaluate and measure frequently," with an emphasis on quality. One of the biggest challenges with predictive methodologies (e.g., waterfall) is that each phase (i.e., plan, design, build, test, and deploy) must be completed before the next phase begins, and there are generally established exit criteria for each phase. Leading up to the testing phase, organizations have invested a significant amount of resources. The goal is to introduce the concept of Evaluation early on in the process. In the case of adaptive methodologies (e.g., those covered under the agile umbrella), the focus shifts to the principles outlined in the Agile Manifesto, which are intended to deliver value at an increased velocity than that achieved with the predictive development lifecycle. On the exam, PMI will not promote one development lifecycle over another; rather, candidates will be expected to understand the elements of each and how they relate to the domains from which the questions originate.

In an effort to validate whether the delivered solution or product meets the requirements of the business, a sequence of prescribed tests is performed, typically coordinated by a testing lead. During each cycle, additional functionality may be provided to the stakeholders and subject matters experts to validate. Leading practice dictates that there are entry and exit criteria for each test cycle.

As evaluation continues, the business analyst should pay particular attention to quality assurance (QA), working with the project team members and stakeholders to determine any solution gaps or deltas. The intent is to resolve discrepancies referencing project artifacts (e.g., business case, scope statement, charter, and tracker) to what was actually delivered.

At the conclusion of a test cycle or in preparation for go-live, the business analyst will work with the sponsor to obtain signoff—with feedback and input from the stakeholders—on the delivered solution.

Following go-live, the business analyst will commence work on one of the most important aspects of the delivered solution: assessing the value proposition and its performance over time. The business case provided the cost–benefit justification for the solution; the focus now shifts to evaluating the degree to which the solution fulfills the value proposition.

As outlined in the *PMI Professional in Business Analysis (PMI-PBA)® Examination Content Outline*, there are four essential business analyst tasks associated with the Evaluation domain; these are shown in greater detail in Figure 6.1. Refer to the practice guides and the *PMBOK® Guide* for additional context related to all the items outlined below.

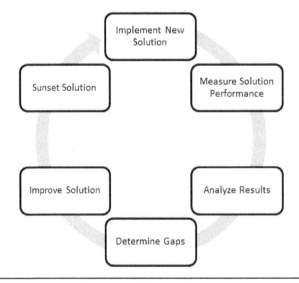

Figure 6.1 Lifecycle of Evaluation.

The key artifacts associated with the Evaluation domain are shown in Figure 6.2.

Figure 6.2 Evaluation artifacts.

At the end of this chapter are exercises to reinforce the concepts from this domain, the answers to which can be found throughout this exam prep book. This domain represents 10% (or 20) of the questions that will appear on the PMI-PBA® Exam.

Major Topics

6.1 Validation of Test Results and Reports

Documentation Management Tools and Techniques

- Lessons learned
- Project management information system (PMIS)
- Reports
- Retrospectives

Verification Tools and Techniques

- Desk checking
- Inspection
- Peer review
- Test
- Walkthrough

➤ Additional Concepts

- ○ Checksheets/checklists

Evaluation Tools and Techniques

- Comparative results
- Day-in-the-life (DITL) testing
- Exploratory testing (ET)
- Surveys and focus groups
- System integrated testing (SIT)
- User acceptance testing (UAT)
- Independent verification and validation (IV&V)
- Outcome measurement and financial calculations

➤ Additional Concepts

- ○ Continuity testing (CT), performance testing (PT), system testing (ST), unit testing (UT)

Validation Tools and Techniques

- Acceptance criteria (given-when-then)
- User acceptance testing (UAT)
- Demonstration

6.2 Quality Assurance (QA), Gap Analysis, Resolution of Discrepancies

Analytic Tools and Techniques

- Decomposition
- Dependency analysis
- Gap analysis
- Impact analysis
- Progressive elaboration
- Risk analysis and assessment

Quality Management and Control Tools
- Affinity diagrams
- Process decision program charts (PDPC)
- Interrelationship digraphs
- Tree diagrams (WBS, RBS, OBS)
- Prioritization matrices
- Activity network diagrams
- Matrix diagrams

6.3 Obtain Signoff

Decision-Making Tools and Techniques
- Consensus building
- Delphi technique
- MultiVoting (nominal group technique)
- Options analysis

➤ Additional Concepts
- Brainstorming
- Didactic interaction
- Organizational readiness

6.4 Solution Evaluation

Evaluation Criteria
- Customer metrics
- Functionality
- Operational metrics and assessments
- Sales and marketing metrics
- SMART goals and objectives

Measurement Tools and Techniques
- Planguage
- Service-level agreements

➤ Additional Concepts
- Operational-level agreements
- Key performance indicators (KPIs)

Valuation Tools and Techniques
- Cost–benefit analysis
  - Internal rate of return
  - Net present value
  - Payback period
  - Return on investment
- Force-field analysis
- Kano model
- Net promotor score
- Purpose alignment model
- SWOT analysis
- Value stream map

Evaluation
Chapter Exercises

Test Your Knowledge: Evaluation

INSTRUCTIONS: When using predictive development methodologies, Evaluation can occur at the end of phase, prior to go-live, or post-implementation. In cases of iterative or adaptive development methodologies, Evaluation occurs following each iteration, sprint, release, or minimum viable product. Below are some of the key concepts related to this chapter; take the time to research each and fill in the blanks.

Q	Practice Area	Example	Explanation
	These diagrams are used to show the causes of specific events.	Ishikawa or fishbone diagram	High-level causes of a problem. Categories can include machines, materials, measurements, stakeholders, and locations.
1	List three development methodologies and provide examples of each.		
2	What are the two most important aspects of testing?		

Q	Practice Area	Example	Explanation
3	Solution evaluation is used to validate a delivered product or increment. What are the two categories of measures?		
4	What are two type of agreements to manage expectations?		
5	What tools can be used during testing to validate processes and roles and also be used for system configuration?		
6	What tool can be used to address ambiguous and incomplete nonfunctional requirements?		

Test Your Knowledge: Group Decision-Making Techniques

INSTRUCTIONS: Securing sign-off from the sponsor, with feedback from stakeholders, on the delivered solution is the objective of the third element discussed in this chapter. In the below table, identify the four methods discussed and provide a brief explanation of each.

Q	Voting Method	Explanation
1		
2		
3		
4		

Check Your Answers: Group Decision-Making Techniques

INSTRUCTIONS: Securing sign-off from the sponsor, with feedback from stakeholders, on the delivered solution is the objective of the third element discussed in this chapter. In the below table, compare your to answers in the previous table.

Q	Voting Method	Explanation
1	Nominal Group Technique (NGT)	Enables all attendees to participate equally. Also helpful in cases in which the team is having difficulty deriving potential solutions.
2	Brainstorming	Participants are encouraged to share unstructured and sometimes radical or "left-field" ideas.
3	Delphi Technique	Participants comment anonymously on a particular topic. At least two rounds of informed decision-making.
4	Didactic Interaction	"Yes/No" or "Option 1/Option 2" decision. Teams must document the advantages and disadvantages of their position.

Evaluation: Practice Questions

INSTRUCTIONS: Note the most suitable answer for each multiple-choice question in the appropriate space on the answer sheet on the "Answer Sheet" on page 155.

1. You are the product manager for a cosmetic company, Endless Youth Ltd., and are responsible for their age-defying skin care products. You've reviewed the results from a benchmarking study and are concerned about the findings. You ask Willow, the business analyst assigned to your team, to initiate a study that will be used to gauge the customer's willingness to recommend your skin care products to their friends. What tool can be used to gauge loyalty to the product line?
   a. Customer segmentation
   b. Perceptions of usability
   c. True intent
   d. Net promoter score

2. Your software project is using an iterative methodology to validate whether the delivered solution is functional, based on a typical day. Elise, the subject matter expert for the accounts payable group, participates in which of the following:
   a. User acceptance testing (UAT)
   b. Day-in-the-life (DITL) testing
   c. Performance testing (PT)
   d. Unit testing (UT)

3. Isaac, the VP of marketing for Vroom Automotive, Inc., an automotive accessories business, has requested your assistance defining a set of metrics that can aid in determining whether the delivered solution will help to increase overall market share. To evaluate if Vroom is positioned to achieve its goals and objectives, you define which of the following:
   a. Critical success measures
   b. Clearly defined goals and objectives
   c. The parameters for a benchmarking study
   d. Key performance indicators

4. Wyatt, the testing lead for your software implementation project, has scheduled a series of focus groups to discuss impressions from a group of industry experts skilled in general ledger software evaluation. In what evaluation approach did the experts partake?
   a. Exploratory testing
   b. Value stream testing
   c. Product feasibility testing
   d. System integrated testing

5. Your project manager, Lucas, and Ivy, the auditor assigned to your project, are debating who is accountable for signoff on the delivered solution. The project has received high praise from all stakeholders, especially your sponsor, Felix, because the project was completed ahead of schedule, was under budget, and meets every defined requirement. You suggest:
   a. Felix, as the sponsor, has accountability for signoff—after all, he provided the resources.
   b. Because the project was such a success, signoff is unnecessary.

    c. Looking in SharePoint, the project teams document repository to review the RACI.

    d. Prior to signoff, the project needs to be formally closed, along with all contracts.

6. During system integrated testing, it's determined that the vendor hosting your cloud solution is not meeting the contractual service level agreement. As the business analyst, you first . . .

    a. Log the result and assign the matter to an appropriate resource for investigation/resolution.

    b. Notify the contract officer assigned to the project, so they can notify the vendor.

    c. Notify the vendor that they've breached the SLA.

    d. Perform the test again to determine if it falls within the Rule of Seconds.

7. Your company's ERP system has been operational for nearly 10 years, with only minor patches applied. During a recent meeting with the CIO, you advise that it's time to upgrade to the cloud-based ERP solution. You schedule a follow-up meeting and provide:

    a. A force-field analysis detailing the factors for and against an upgrade

    b. A value stream and cost–benefit analysis for the upgrade

    c. An opportunity analysis to reduce the carbon footprint of the data center

    d. Cumulative performance metrics, a cost–benefit and force-field analysis

8. During system testing (ST), your subject matter experts (SMEs) have voiced concern that the delivered solution does not address the solution requirements, specifically those outlined in the functional specification documents. As a result, the Executive Steering Committee has voted to cancel the project. Audrey, the project manager, suggests that formal documentation is:

    a. Optional; the slide deck from the Executive Steering Committee Meeting is sufficient.

    b. Required, and should indicate why the project was terminated, along with lessons learned, and be made available for future use.

    c. Required, be made available for future use, and identify why VB Script should no longer be used.

    d. Required, but saved to the secure area of the team's wiki due to the sensitive nature of the content.

9. Elodie, your test lead, and Harlow, your business SME, are satisfied with the test results and have requested a go/no-go meeting to explain their actual versus expected results to all internal stakeholders. They have a disagreement has to how the results should be presented. As the business analyst, what can you recommend?

    a. Advise Elodie to share the details of the solution testing during the meeting.

    b. Advise Harlow to present the results, using graphs and charts, at a summary level.

    c. Advise both Elodie and Harlow that the stakeholders should be provided with summarized details ahead of the meeting, and that during the meeting results should be presented using graphs and charts at a summary level.

    d. Advise both Elodie and Harlow that the individuals who make go/no-go decisions would like an impact analysis to accompany the presentation; either format is acceptable.

10. In planning the go-live for your project, you've determined that a downtime of 48 hours will be required to address all the cutover activities. You ask Belle, the communication lead, to:

    a. Present this impact at the next Change Advisory Board (CAB) meeting.

    b. Confirm with the stakeholders that this is acceptable and clearly communicated.

    c. Coordinate all cutover activities with the program office.

    d. Update and distribute all work assignments prior to go-live.

11. Your test lead, Saylor, is reviewing the results from a recent round of testing with subject matter experts from the supply chain business unit. While defining their nonfunctional requirements, acceptable value ranges were defined and agreed to by the project team. What were the expected value ranges?
    a. Best-case value, wished-for value, worst-case value
    b. Target value, minimum acceptable value, worst-case value
    c. 99.9999% uptime, with clearly defined maintenance windows
    d. Worst-case value, target value, wished-for value

12. Zoelle has just completed system testing and is working to document the solution gaps. As part of her analysis, Zoelle will also be:
    a. Comparing expected results to actual results.
    b. Posting the test results to the team's SharePoint site.
    c. Conducting a didactic interaction session before an opportunity analysis.
    d. Reviewing the value stream map and comparing the results to defined metrics.

13. You work for a hospital and are implementing a complex patient care system. After nearly 36 months, the project has successfully passed all tests. In a hallway conversation with the CIO, Mea, and the sponsor, Gideon, they have approved the go-live for next week and directed the project manager to plan project close events four weeks after go-live. You advise Iris, the project manager:
    a. During the planning of the project, the project management office advised the sponsor and key organizational stakeholders that they would need to provide a "wet signature" indicating their approvals.
    b. Per ITIL, the preproduction review and presentation to the organization's Change Advisory Board (CAB) should be planned for no less than two weeks from now.
    c. Because we are in a heavily regulated industry, we'll need to ensure that we have adequate coverage for the live event—and that means "all hands on deck."
    d. The SMEs for the clinical teams have sufficiently tested the solution, and the go-live will be seamless—not an event at all.

14. You are the sponsor for a small software development project, working for a satellite radio company, Music to the Stars. You are hearing mixed reviews from your subject matter experts regarding the last round of testing. The business analyst, Tenley Sift, suggests:
    a. The team participate in another round of exploratory testing (ET).
    b. She coordinate a hands-on session, demonstrating that the software meets the intended functions.
    c. We schedule a focus group to review all the concerns and use the MultiVoting technique to arrive at a unified decision.
    d. You sign off on the testing round, as the exit criteria clearly stated that the anticipated functionality would not be available until the last round of system integrated testing.

15. In a meeting with the organization's project management office, you are discussing the potential strategies for the phase-out of a highly complex enterprise software application. You are proposing a segmented cutover, others are suggesting either time-boxed coexistence or perhaps even a longer-term coexistence. Your CIO, Molly, listens intently to the conversation and mentions that she is not in favor of coexistence. She proceeds to recommend:
    a. A segmented cutover of the applications being phased out.
    b. Coexistence, with data replicated to both platforms, so users can gradually migrate to the new system.

c. The organization hire a consulting firm to advise on this matter.

d. A massive one-time cutover event, in which all modules are replaced over a long weekend.

16. Your project team has just completed integration testing in an environment that will nearly mirror production. During this test cycle, the subject matter experts referred to test scripts and what other artifact that will help ensure the testers were operating in isolated, production-ready environment?

a. Process flow diagrams

b. Force-field analysis with acceptance criteria

c. Test scenarios

d. User stories and epics

17. During system testing, Teagan, the subject matter expert from compliance, realizes he forgot to submit a very simple functional requirement when he drafted the department's specifications for the new software program. As the department's business analyst, you've called a brief meeting with Teagan, the developer, and the project manager; all agree this is very simple and clearly an oversight. What should you do?

a. Because the change is negligible, ask the project manager to approve and begin development.

b. Work with the project manager to draft a solution statement for the project office to review.

c. Add the change to an isolated environment, then conduct an impact assessment.

d. Work with Teagan to document the requirements, analyze the impact across the system, and then submit a change request to the project office.

18. Felicity, the organizational change management lead for your enterprise software project, is creating a monthly newsletter and planning Town Hall events designed to prepare the stakeholders for eventual launch of the solution. In the planning for each, Felicity . . .

a. Requests a meeting with you, the business analyst, to discuss the existing solution as an input to her communication and rollout activities.

b. Works with the project manager to understand the solution scope.

c. Consults only with the project management office on her activities.

d. Defers to the projects communication lead, based on the communication matrix.

19. During the requirements definition phase of the project, subject matter experts were struggling to clearly define the actual nonfunctional requirements. One of your consultants, Tom Gilb, suggested:

a. Using a template provided by the PMO, which contains lessons learned from prior projects.

b. Adding all the requirements to the traceability matrix, so that development could be monitored.

c. Using a planning language to capture requirements, which can then be used during testing for validation.

d. Performing exploratory testing so the users could validate that the solution meets clearly defined acceptance criteria.

20. Using the nominal group technique, your project team has approved solution signoff. Julia, your sponsor, inquires as to the steps involved with knowledge transfer. You advise they are:

a. Interviews, focus groups, and observations, concluding with a Delphi method to transfer knowledge.

b. Documentation reviews, assessments, sharing for future use in the team's wiki.

c. Identification, capturing, and publishing for future use in the team's wiki.

d. Assessing, sharing, identifying, capturing, and applying.

Answer Sheet

	Answer Choice				Correct	Incorrect	Predicted Answer			
							90%	75%	50%	25%
1	a	b	c	d						
2	a	b	c	d						
3	a	b	c	d						
4	a	b	c	d						
5	a	b	c	d						
6	a	b	c	d						
7	a	b	c	d						
8	a	b	c	d						
9	a	b	c	d						
10	a	b	c	d						
11	a	b	c	d						
12	a	b	c	d						
13	a	b	c	d						
14	a	b	c	d						
15	a	b	c	d						
16	a	b	c	d						
17	a	b	c	d						
18	a	b	c	d						
19	a	b	c	d						
20	a	b	c	d						

Evaluation: Answer Key

1. You are the product manager for a cosmetic company, Endless Youth Ltd., and are responsible for their age-defying skin care products. You've reviewed the results from a benchmarking study and are concerned about the findings. You ask Willow, the business analyst assigned to your team, to initiate a study that will be used to gauge the customer's willingness to recommend your skin care products to their friends. What tool can be used to gauge loyalty to the product line?
   a. Customer segmentation
   b. Perceptions of usability
   c. True intent
   d. Net promoter score

   The correct answer is: D

   The net promotor score (NPS) is a metric system envisioned by Fred Reicheld (Reicheld, 2003, designed to gauge customer loyalty and brand satisfaction. It can be based on a variety of scales; however, the most common is zero to ten, based on a single open question, where zero is "Not at all . . ." and ten is "Extremely . . ." in response to the question; common qualifiers are "Satisfied" and "Likely."

   Answer Choice A: Customer segmentation is the practice of separating customers into groups based on relevant criteria or characteristics for analysis, marketing, or other business need.

   Answer Choices B & C: These are a made-up answer choices.

   Examination Content Outline, Knowledge and Skills, #38, "Valuation Tools and Techniques."
   PMI Professional in Business Analysis (PMI-PBA)® Examination Content Outline, 2013, "Evaluation," Task 4.

2. Your software project is using an iterative methodology to validate whether the delivered solution is functional based on a typical day. Elise, the subject matter expert for the accounts payable group, participates in which of the following:
   a. User acceptance testing (UAT)
   b. Day-in-the-life (DITL) testing
   c. Performance testing (PT)
   d. Unit testing (UT)

   The correct answer is: B

   Day-in-the-life (DITL) testing is a semiformal activity based on typical usage and is performed by a team member with in-depth business knowledge. The results obtained from DITL enable both validation and evaluation, confirming if a product or solution provides the functionality for a typical day of usage for the role, based on the test scenario.

   Answer Choice A: User acceptance testing is one of the last phases of user software testing before go-live, in which users test the solution to ensure that it can handle the required tasks in real test scenarios.

   Answer Choice C: Performance testing determines how a system or product performs under various workloads; it's also used to validate stability.

Answer Choice D: Unit testing is the method whereby the smallest aspect of a product, solution, or application is tested, independent of the larger solution.

Business Analysis for Practitioners: A Practice Guide, Section 6.6.3.

PMI Professional in Business Analysis (PMI-PBA)® Examination Content Outline, 2013, "Evaluation," Task 1.

3. Isaac, the VP of marketing for Vroom Automotive, Inc., an automotive accessories business, has requested your assistance defining a set of metrics that can aid in determining whether the delivered solution will help to increase overall market share. To evaluate if Vroom is positioned to achieve its goals and objectives, you define which of the following?:
   a. Critical success measures
   b. Clearly defined goals and objectives
   c. The parameters for a benchmarking study
   d. Key performance indicators

The correct answer is: **D**

Key performance indicators are quantifiable (measurable) performance indicators tied to a specific critical success factor that help organizations to achieve their goals. KPIs are linked to the goals and objectives of projects; they provide the basis for what should be measured during Evaluation.
Answer Choice A: This is a made-up answer choice.
Answer Choice B: Objectives can be both short-term and long-term *milestones*, designed to evaluate whether the organization is effectively executing its strategic plan. Goals are short-term and long-term *outcomes*, which position an organization to achieve its objectives.
Answer Choice C: Benchmarking parameters would not be relevant in this scenario.

Business Analysis for Practitioners: A Practice Guide, Section 6.5.2.

PMI Professional in Business Analysis (PMI-PBA)® Examination Content Outline, 2013, "Evaluation," Task 1.

4. Wyatt, the testing lead for your software implementation project, has scheduled a series of focus groups to discuss impressions from a group of industry experts skilled in general ledger software evaluation. In what evaluation approach did the experts partake?
   a. Exploratory testing
   b. Value stream testing
   c. Product feasibility testing
   d. System integrated testing

The correct answer is: **A**

Exploratory testing is designed so that participants can learn how the software works.
Answer Choice B: This is a made-up answer choice.
Answer Choice C: Product feasibility testing generally occurs during product development and would not be applicable in this situation.
Answer Choice D: System integrated testing is testing that verifies interoperability and coexistence with other products, applications, and solutions.

Business Analysis for Practitioners: A Practice Guide, Section 6.1.

PMI Professional in Business Analysis (PMI-PBA)® Examination Content Outline, 2013, "Evaluation," Task 1.

5. Your project manager, Lucas, and Ivy, the auditor assigned to your project, are debating who is accountable for signoff on the delivered solution. The project has received high praise from all stakeholders, especially your sponsor, Felix, as the project was completed ahead of schedule, was under budget, and meets every defined requirement. You suggest:
   a. Felix, as the sponsor, has accountability for signoff; after all, he provided the resources.
   b. Because the project was such a success, signoff is unnecessary.
   c. Looking in SharePoint, the project team's document repository, to review the RACI.
   d. Prior to signoff, the project needs to be formally closed, along with all contracts.

The correct answer is: **C**

The responsibility assignment matrix, otherwise known as a RACI, which was developed during business analysis planning, will identify the individual(s) accountable for solution signoff. This will also be documented in the requirements management plan and the business analysis plan. In this scenario, SharePoint serves as the team's project management Information system (PMIS) or document management system and would store all these documents.

Answer Choice A: While this is typically the sponsor, it cannot be assumed, and others may need to sign off as well that the solution has fulfilled its value proposition.

Answer Choice B: In some organizations, signoff is very informal, which can lead to problems later. It's leading practice to have formal signoff on the delivered solution—*this is something you'll want to remember for the exam.*

Answer Choice D: While this may be true in practice, it doesn't answer the question as to who is accountable for signoff.

Business Analysis for Practitioners: A Practice Guide, Section 6.9.

PMI Professional in Business Analysis (PMI-PBA)® Examination Content Outline, 2013, "Evaluation," Task 3.

6. During system integrated testing, it's determined that the vendor hosting your cloud solution is not meeting the contractual service level agreement. As the business analyst, you first:
   a. Log the result and assign the matter to an appropriate resource for investigation/resolution.
   b. Notify the contract officer assigned to the project, so they can notify the vendor.
   c. Notify the vendor that they've breached the SLA.
   d. Perform the test again, to determine if it falls within the Rule of Seconds.

The correct answer is: **A**

Regardless of pass or fail, all results from testing must be logged in the format agreed upon during planning. In the case of defects, once properly logged, they can be assigned for investigation and resolution.

Answer Choice B: In cases in which contract officers manage the relationship with the vendor, they could have the responsibility for notifying them of the breach, after the defect was logged.

Answer Choice C: Service level agreements (SLA) are contracts between service providers (which can be either internal or external to an organization) and a customer. They quantifiably define the level of service expected for a given facet. They can be defined at the customer level, a service level, or a combination (multilevel). This could very well be a subsequent step, but the first step is to log the defect.

Answer Choice D: Although tests can be performed multiple times to confirm and validate the results, the first step is to properly log and document all pass/fail results. *The Rule of Seconds is also made up; be wary of these types of answer choices on the exam.*

Business Analysis for Practitioners: A Practice Guide, Section 6.7.3.

PMI Professional in Business Analysis (PMI-PBA)® Examination Content Outline, 2013, "Evaluation," Task 2.

7. Your company's ERP system has been operational for nearly 10 years, with only minor patches applied. During a recent meeting with the CIO, you advise that it's time to upgrade to the cloud-based ERP solution. You schedule a follow-up meeting and provide which of the following:
   a. A force-field analysis detailing the factors for and against an upgrade
   b. A value stream and cost–benefit analysis for the upgrade
   c. An opportunity analysis to reduce the carbon footprint of the data center
   d. Cumulative performance metrics, a cost–benefit and force field analysis

   The correct answer is: **D**

   As you approach this question, it's important to recognize you're in the Evaluation domain. The life-cycle of Evaluation is: (a) implement new solution; (b) measure solution performance; (c) analyze results; (d) determine gaps; (e) improve solution; (f) sunset solution; start again. When designed into the system, year-over-year and quarter-over-quarter performance metrics provide a very good insight into the operational stability of a solution. Other tools used during the Needs Assessment can also be useful to justify the project. These metrics form the basis for determining gaps and drafting a situation statement.

   Answer Choice A: Originally developed for use in social science by Kurt Lewin as a means for qualitative assessments, force-field analysis has morphed into a collaborative decision-making tool when comparing and contrasting *restraining forces* (obstacles or negatives) and *driving forces* (motivators or positives) for the option under consideration. Although this is a good start, it is not a complete answer.

   Answer Choice B: A cost–benefit analysis would come later in the process, and a value stream map would not be appropriate in this situation.

   Answer Choice C: An opportunity analysis is a forecasting technique, which considers the projections within a potential market for a given solution.

   Business Analysis for Practitioners: A Practice Guide, Section 6.10.2.

   PMI Professional in Business Analysis (PMI-PBA)® Examination Content Outline, 2013, "Evaluation," Task 4.

8. During system testing (ST), your subject matter experts (SMEs) have voiced concern that the delivered solution does not address the solution requirements, specifically those outlined in the functional specification documents. As a result, the Executive Steering Committee has voted to cancel the project. Audrey, the project manager, suggests that formal documentation is:
   a. Optional; the slide deck from the Executive Steering Committee Meeting is sufficient.
   b. Required, and should indicate why the project was terminated, along with lessons learned, and be made available for future use.
   c. Required, be made available for future use, and identify why VB Script should no longer be used.
   d. Required, but saved to the secure area of the team's wiki due to the sensitive nature of the content.

The correct answer is: **B**

Project termination can occur for myriad reasons. In this case, the delivered solution is not properly addressing the requirements as outlined in the functional specifications. To improve future project outcomes and share lessons throughout the organization, the business analyst and project manager should collaborate to properly close out the project. This includes properly documenting the rationale for terminating the project and making the content available internally for future use.

Organizations that document lessons learned significantly improve project outcomes.

Answer Choice A: While the slide deck may accompany the other documentation, it's not sufficient to properly close out and terminate the project.

Answer Choice C: This is a partially correct answer; however, answer choice B offers a slightly more complete rationale.

Answer Choice D: While this is a good answer, it also needs to clearly note why the project was cancelled for future reference.

Requirements Management: A Practice Guide, Chapter 9, page 53.

PMI Professional in Business Analysis (PMI-PBA)® Examination Content Outline, 2013, "Evaluation," Task 3.

9. Elodie, your test lead, and Harlow, your business SME, are satisfied with the test results and have requested a go/no-go meeting to explain their actual versus expected results to all internal stakeholders. They have a disagreement on how the results should be presented. As the business analyst, what can you recommend?
    a. Advise Elodie to share the details of the solution testing during the meeting.
    b. Advise Harlow to present the results, using graphs and charts, at a summary level.
    c. Advise both Elodie and Harlow that the stakeholders should be provided with summarized details ahead of the meeting, and that during the meeting results should be presented using graphs and charts at a summary level.
    d. Advise both Elodie and Harlow that individuals who make go/no-go decisions would like an impact analysis to accompany the presentation; either format is acceptable.

The correct answer is: **C**

This is a complicated question, with focus on running efficient meetings and properly preparing attendees. As with all meetings, there should be a clear agenda with known objectives; ideally, meeting minutes follow within 24 hours. In this scenario, the objective is to solicit a vote on a go/no-go decision. To facilitate a smooth discussion, all the details should be provided in advance of the meeting, whereas summary data is presented during the meeting.

Answer Choice A: In this case, the meeting attendees would not be properly prepared to discuss the test results.

Answer Choice B: While appropriate for during the meeting, answer choice C is a better answer, as participants are provided with details in advance.

Answer Choice D: This is a distractor; impact analyses are generally used when considering change requests.

Business Analysis for Practitioners: A Practice Guide, Section 6.8.

PMI Professional in Business Analysis (PMI-PBA)® Examination Content Outline, 2013, "Evaluation," Task 3.

10. In planning the go-live for your project, you've determined that a downtime of 48 hours will be required to address all the cutover activities. You ask Belle, the communication lead, to:
    a. Present this impact at the next Change Advisory Board (CAB) meeting.
    b. Confirm with the stakeholders that this is acceptable and clearly communicated.
    c. Coordinate all cutover activities with the program office.
    d. Update and distribute all work assignments prior to go-live.

The correct answer is: B

When planning cutover activities and go-live events, communication is critical. In many cases, over-communicating is even preferred, as this will ensure that everyone is knowledgeable and prepared for the live event. Furthermore, key stakeholders must agree to any impact to business operations, and solid contingency plans must be in place.

Answer Choice A: This is a good answer, and a required task before commencing the cutover activities; however, it wouldn't be the most immediate task. We would first need to check with our stakeholders to confirm that this window was acceptable.

Answer Choice C: In most cases, the program office would be aware of the activities and would recommend that you first confirm with the stakeholders that this is acceptable and clearly communicated.

Answer Choice D: This would be a successor task, once all stakeholders were in agreement as to the timeline.

Business Analysis for Practitioners: A Practice Guide, Section 6.11.

PMI Professional in Business Analysis (PMI-PBA)® Examination Content Outline, 2013, "Evaluation," Task 4.

11. Your test lead, Saylor, is reviewing the results from a recent round of testing with subject matter experts from the supply chain business unit. While defining their nonfunctional requirements, acceptable value ranges were defined and agreed to by the project team. What were the expected value ranges?
    a. Best-case value, wished-for value, worst-case value
    b. Target value, minimum acceptable value, worst-case value
    c. 99.9999% uptime, with clearly defined maintenance windows
    d. Worst-case value, target value, wished-for value

The correct answer is: D

Nonfunctional requirements are a subset of solution requirements, which describe the environmental conditions of the component. They are typically characterized as quality or additive requirements and are described as "must haves." Examples can include availability, capacity, continuity, performance, security, compliance, and service-level management. During requirements definition, stakeholders agree on acceptance criteria related to tolerance ranges for nonfunctional requirements, which can include worst-case value (minimum acceptable value); target value (most likely value); and wished-for value (best-case value).

Answer Choice A: This is answer choice is partially correct; however, best-case value and wished-for value are synonymous. Worst-case value (aka *minimum acceptable value*) is a valid response.

Answer Choice B: This answer choice is partially correct; however, minimum acceptable value and worst case value are synonymous. Target value (aka *most likely value*) is a valid response.

Answer Choice C: This is not a valid response for expected value ranges.

Business Analysis for Practitioners: A Practice Guide, Section 6.7.2.
PMI Professional in Business Analysis (PMI-PBA)® Examination Content Outline, 2013, "Evaluation," Task 2.

12. Zoelle has just completed system testing and is working to document the solution gaps. As part of her analysis, Zoelle will also be:
    a. Comparing expected results to actual results.
    b. Posting the test results to the team's SharePoint site.
    c. Conducting a didactic interaction session before an opportunity analysis.
    d. Reviewing the value stream map and comparing the results to defined metrics.

The correct answer is: **A**

System testing is performed on a complete solution to evaluate fit/gap to the specified functional requirements. It's a form of black-box testing, which evaluates the functionality of the solution without requiring knowledge of how the system is built. As part of gap analysis, results from system testing will be contrasted to the expected results. All results, not just defects, would be properly noted in the testing log.
Answer Choice B: Once the results are analyzed, they could be posted to the team's document management system. The first task is to compare the expected to the actual results.
Answer Choice C: This is a made-up answer choice, combining two unrelated tools, neither of which would be appropriate in this scenario.
Answer Choice D: Reviewing the value stream map would not be appropriate in this scenario.
Business Analysis for Practitioners: A Practice Guide, Section 6.7.1.
PMI Professional in Business Analysis (PMI-PBA)® Examination Content Outline, 2013, "Evaluation," Task 2.

13. You work for a hospital and are implementing a complex patient care system. After nearly 36 months, the project has successfully passed all tests. In a hallway conversation with the CIO, Mea, and the sponsor, Gideon, they have approved the go-live for next week and directed the project manager to plan project close events four weeks out after the go-live. You advise Iris, the project manager:
    a. During the planning of the project, the project management office advised the sponsor and key organizational stakeholders that they would need to provide a "wet signature" indicating their approvals.
    b. Per ITIL, the preproduction review and presentation to the organization's Change Advisory Board (CAB) should be planned for no less than two weeks from now.
    c. Because we are in a heavily regulated industry, we'll need to ensure that we have adequate coverage for the live event—that means "all hands on deck."
    d. The SMEs for the clinical teams have sufficiently tested the solution; the go-live will be seamless, not an event at all.

The correct answer is: **A**

During project planning, the responsibility assignment matrix (RACI) was created, which established accountability for solution signoff. This matrix was included as part of the requirements management plan and business analysis plan. Some organizations may require a physical (aka *wet*) signature, whereas others will allow for electronic signatures or voting. Remember, formal signoff was outlined during the planning phase of the project.

Answer Choice B–D: These answer choices do not correctly address the question. The question is focused on planning project close events; these responses focus on the live event itself.

Business Analysis for Practitioners: A Practice Guide, Section 6.9.

PMI Professional in Business Analysis (PMI-PBA)® Examination Content Outline, 2013, "Evaluation," Task 3.

14. You are the sponsor for a small software development project, working for a satellite radio company, Music to the Stars. You are hearing mixed reviews from your subject matter experts regarding the last round of testing. The business analyst, Tenley Sift, suggests:
    a. The team participate in another round of exploratory testing (ET).
    b. She coordinate a hands-on session demonstrating that the software under question meets the intended functions.
    c. We schedule a focus group to review all the concerns and use the MultiVoting technique to arrive at unified decision.
    d. You sign off on the testing round, because the exit criteria clearly stated that the anticipated functionality would not be available until the last round of system integrated testing.

The correct answer is: **B**

The best approach to evaluating a solution is to solicit input from stakeholders. Although there are many techniques, one common class used for evaluation is examination, which includes demonstrations. In this scenario, demonstrations would be an effective means of proving that the delivered solution meets stakeholder's requirements and intended functions.

Answer Choice A: The examination class of techniques for evaluation can also include testing, both exploratory and user acceptance. Testing validates that the delivered solution meets the evaluation and acceptance criteria as outlined in the requirements management plan and business analysis plan.

Answer Choice C: Focus groups and MultiVoting can be used during the evaluation process to solicit stakeholder feedback in regard to a developed solution. However, in this scenario they would not be appropriate for demonstrating that the software meets the intended functions.

Answer Choice D: This answer choice doesn't address the problem.

Requirements Management: A Practice Guide, Section 8.3.1.

PMI Professional in Business Analysis (PMI-PBA)® Examination Content Outline, 2013, "Evaluation," Task 1.

15. In a meeting with the organization's project management office, you are discussing the potential strategies for the phase-out of a highly complex enterprise software application. You are proposing a segmented cutover; others are suggesting either time-boxed coexistence or perhaps even a longer-term coexistence. Your CIO, Molly, listens intently to the conversation and mentions that she is not in favor of coexistence; she proceeds to recommend:
    a. A segmented cutover of the applications being phased out.
    b. Coexistence, with data replicated to both platforms, so users can gradually migrate to the new system.
    c. The organization hire a consulting firm to advise on this matter.
    d. A massive one-time cutover event, in which all modules are replaced on a long weekend.

The correct answer is: **D**

When planning solution replacement/phase out, there are four strategies the business analyst should consider: (a) a massive one-time cutover; (b) a segmented cutover approach; (c) time-boxed coexistence; and (d) permanent coexistence. Although it's very well known that massive cutovers are risky, depending on the situation, the risk may need to be accepted. A cost–benefit analysis and a risk assessment can help the project team with this determination.

Answer Choice A: A segmented cutover is based on a temporary coexistence of the replacement and existing solutions. This answer choice is the opposite of the desired approach.

Answer Choice B: This is not a good answer choice, as the CIO explicitly stated she was not in favor of coexistence.

Answer Choice C: While this answer choice may seem reasonable, it's a distractor, as the business analyst should be able to address the situation.

Business Analysis for Practitioners: A Practice Guide, Section 6.11.

PMI Professional in Business Analysis (PMI-PBA)® Examination Content Outline, 2013, "Evaluation," Task 4.

16. Your project team has just completed integration testing in an environment that will nearly mirror production. During this test cycle, the subject matter experts referred to test scripts and what other artifact that will help ensure that the testers were operating in an isolated, production-ready environment?
    a. Process flow diagrams
    b. Force-field analysis with acceptance criteria
    c. Test scenarios
    d. User stories and epics

The correct answer is: **A**

Integration testing is performed to evaluate the interoperability of the solution, in the broader context of the organization and with third-party systems. Integration tests will often occur in environments that are very similar to production and are more encompassing than systems tests. The process flow diagrams, created during Needs Assessment, will aid testers in following defined steps to uncover any solution gaps and validate that the solution was produced according to the established requirements.

Answer Choices B–D: While these artifacts can be used during testing, they cannot ensure that testers were operating in an isolated, production-ready environment.

Business Analysis for Practitioners: A Practice Guide, Section 6.6.4.

PMI Professional in Business Analysis (PMI-PBA)® Examination Content Outline, 2013, "Evaluation," Task 2.

17. During system testing, Teagan, the subject matter expert from compliance, realizes that he forgot to submit a very simple functional requirement when he drafted the department's specifications for the new software program. As the department's business analyst, you've called a brief meeting with Teagan, the developer, and the project manager; all agree this is very simple and clearly an oversight. What should you do?
    a. Because the change is negligible, ask the project manager to approve and begin development.
    b. Work with the project manager to draft a solution statement for the project office to review.
    c. Add the change to an isolated environment, then conduct an impact assessment.
    d. Work with Teagan to document the requirements, analyze the impact across the system, and then submit a change request to the project office.

The correct answer is: **D**

This scenario focuses on the application of several key tasks: (a) document management; (b) root-cause analysis; and (c) change management. The requirements management plan and business analysis plan created during planning will establish the procedures for logging and assigning the resolution of defects. Prior to proceeding with any additional development efforts, an integrated change request must be presented, accompanied by clearly documented requirements and an impact analysis.

Answer Choice A: This is an easy trap. *Remember that all changes must follow the change management process.*

Answer Choice B: Solution statement is incorrect; they are looking for an impact assessment and change request.

Answer Choice C: While this may seem like a logical approach, adding the change to an isolated environment could be a waste of time if not approved by the project office.

Business Analysis for Practitioners: A Practice Guide, Section 6.7.3.

PMI Professional in Business Analysis (PMI-PBA)® Examination Content Outline, 2013, "Evaluation," Task 2.

18. Felicity, the organizational change management lead for your enterprise software project, is creating a monthly newsletter and planning town hall events designed to prepare the stakeholders for eventual launch of the solution. In the planning for each, Felicity:

    a. Requests a meeting with you, the business analyst, to discuss the existing solution as an input to her communication and rollout activities.
    b. Works with the project manager to understand the solution scope.
    c. Consults only with the project management office on her activities.
    d. Defers to the project's communication lead, based on the communication matrix.

The correct answer is: **A**

As the team prepares for the eventual launch of the solution, proper and systemic communication is critical. Although communication can encompass newsletters and town hall events, other facets of change management to consider include completion of training material, updating standard operating procedures (SOPs), integration with third-party solutions, and organizational coordination. The key elements to consider are planning, communication, training, and support. The people side of change encompasses awareness, desire, knowledge, ability, and reinforcement. It's leading practice for all project leads (project management, communication, training, and organizational change management) to consult and leverage the business analyst's knowledge of the existing system as they prepare for solution launch.

Answer Choice B: This is a partially correct answer, but the understanding must go deeper than solution scope.

Answer Choice C: The consultation should go beyond the project management office.

Answer Choice D: This is also a partially correct answer; there are many others who need to be consulted in addition to the project communication lead.

Business Analysis for Practitioners: A Practice Guide, Section 6.11.

PMI Professional in Business Analysis (PMI-PBA)® Examination Content Outline, 2013, "Evaluation," Task 4.

19. During the requirements definition phase of the project, subject matter experts were struggling to clearly define the actual nonfunctional requirements. One of your consultants, Tom Gilb, suggested:

    a. Using a template provided by the PMO, which contains lessons learned from prior projects.
    b. Adding all the requirements to the traceability matrix, so that development could be monitored.
    c. Using a planning language to capture requirements, which can then be used during testing for validation.
    d. Performing exploratory testing, so the users could validate that the solution meets clearly defined acceptance criteria.

The correct answer is: **C**

This is a difficult question, and one that's likely to appear on the exam. As you'll recall from the May 2014 Pulse of the Profession® Study, PMI found the leading cause of project failure to be inaccurate requirements management. To what cost? The study revealed that, for every US dollar spent, nearly 5.1% is wasted due to poor and inadequate requirements management. Tom Gilb created the concept of a planning language (Planguage) to address ambiguous and incomplete nonfunctional requirements. The tool uses defined identifiers (e.g., tags), to qualify and quantify the quality elements of requirements. On the exam, questions will pertain to the situational and conceptual use of the tool. Remember, Planguage is an informal, structured, keyword-based planning language.

Answer Choice A: This is a partially correct answer; however, answer choice C is slightly more complete.

Answer Choice B: While they will be added to the traceability matrix, doing so will not help the team to define the actual nonfunctional requirements.

Answer Choice D: Exploratory testing would follow once a solution is ready for testing.

Examination Content Outline, Knowledge and Skills, #20, "Measurement Skills and Techniques."

PMI Professional in Business Analysis (PMI-PBA)® Examination Content Outline, 2013, "Evaluation," Task 1.

20. Using the nominal group technique, your project team has approved solution signoff. Julia, your sponsor, inquires as to the steps involved with knowledge transfer. You advise they are:
    a. Interviews, focus groups, observations, concluding with a Delphi method to transfer knowledge.
    b. Documentation reviews, assessments, sharing for future use in the team's wiki.
    c. Identification, capturing, and publishing for future use in the team's wiki.
    d. Assessing, sharing, identifying, capturing, and applying.

The correct answer is: **D**

This question begins with a distractor—mentioning the nominal group technique is completely superfluous information. As you'll recall from the May 2014 Pulse of the Profession® Study, organizations that are effective at knowledge transfer improve project outcomes by nearly 35%. The real question is, what are the steps involved with knowledge transfer: (a) begin by determining what knowledge needs to be transferred (identifying); (b) then assemble the knowledge that must be transferred (capturing); (c) institute methods of transferring knowledge (sharing); (d) stakeholders use the knowledge that was transferred (applying); (e) evaluate the benefits of the shared knowledge (assessing). By following these steps in the knowledge transfer lifecycle, organizations continually improve.

Answer Choice A: This is an incorrect answer choice, which simply lists elicitation techniques (interviews, focus groups, observations) and the unrelated Delphi.

Answer Choice B: This includes assessing and sharing, incorrectly includes documentation reviews, and omits identifying, capturing, and applying.

Answer Choice C: This is a partially correct answer, listing Identification, capturing, and publishing (sharing), but omitting assessing and applying.

Requirements Management: A Practice Guide, Section 9.2.3.

PMI Professional in Business Analysis (PMI-PBA)® Examination Content Outline, 2013, "Evaluation," Task 3.

Chapter 7

About PMI and the PMI-PBA® Credential

7.1 About the Project Management Institute

The Project Management Institute (PMI) is the world's largest not-for-profit project management professional organization. It was founded in 1969 to provide the means for "project managers to associate, share information, and discuss common problems."[1] Now, 48 years later, PMI has over 473,037 members and is helping to advance collaboration, education, and research for over 2.9 million working professionals in 209 countries and territories; globally there are 283 chartered and 12 potential chapters.[2]

> *It began as a dinner in early 1969 among three men at the Three Threes Restaurant, a small, intimate gathering place just a few blocks from City Hall in Philadelphia, Pennsylvania, USA.*
>
> — Project Management Institute

PMI currently has nine credentials/certifications, each developed through a comprehensive process *by* practitioners, *for* practitioners. "The strength of PMI's certifications is that they are portable and not tied to any single method, standard or organization."[3] Eight of the credentials/certifications outlined in Table 7.1 are active, as PMI is preparing to discontinue OPM3 in 2017.

For each credential, PMI has instituted a three-factor criterion for eligibility; this ensures that all certification holders are held to same standards and have ". . . demonstrated their competence through fair and valid measures."[4]

For the PMI-PBA® credential, candidate competency is assessed based on the elements below, each of which are explained in detail throughout this book:

1. Education and experience
2. Exam proficiency
3. Professional development/credential maintenance

---

[1] PMI. "Learn About PMI." Retrieved October 17, 2017, from http://www.pmi.org/about/learn-about-pmi/founders

[2] PMI. "PMI Fact File." *PMI Today®*, 2017, p. 4.

[3] PMI. "About PMI's Certification Program." *PMI-PBA® Handbook,* 2016.

[4] *Ibid.*

Table 7.1, PMI Credential Holders Globally, provides details for each of the active PMI credentials/certifications. For all credentials, the statistics were obtained from the May 2017 issue of *PMI Today®*.

Table 7.1 PMI Credential Holders Globally

Credentials/Certifications		Year Launched	Total Active Holders
CAPM®	Certified Associate in Project Management®	2005	33,383
PMP®	Project Management Professional®	1984	761,905
PfMP®	Portfolio Management Professional®	2013	412
PgMP®	Program Management Professional®	2007	1,880
PMI-RMP®	PMI Risk Management Professional®	2008	4,013
PMI-SP®	PMI Scheduling Professional®	2008	1,652
PMI-PBA®	**PMI Professional in Business Analysis®**	**2014**	**1,282**
PMI-ACP®	PMI Agile Certified Practitioner®	2011	15,115
OPM3®	Organizational Project Management Maturity Model®	2003–2014	< Not Published >

7.2 About the PMI-PBA® Credential

Fast forward to the present day. In 2012, recognizing a gap in the role of project practitioners, PMI commissioned a Business Analyst Role Delineation Study (RDS), which became the basis for the PMI-PBA® credential. Then in May, 2014, PMI conducted a research study on requirements management.[5] The study of over 2,066 professionals found that in cases of unsuccessful and failed projects, the leading cause (47%) was attributed to inaccurate requirements management. Furthermore, when inadequate or poor communication was cited as the primary cause of project failure, 75% of respondents reported that the issue impacted requirements management more than any other area of the initiative.

What happens when organizational initiatives don't meet their original objectives and goals?

The study revealed that organizations are wasting $0.051 for every US dollar spent on programs and projects. That amounts to $51 million USD for every $1 billion USD spent. In the case of low-performing organizations, the study found that more than half of the initiatives were unsuccessful, resulting in waste of $0.10 for every US dollar spent.

As organizations mature and develop robust processes, the correlation rate drops from 47% to 11% stemming from poor or inadequate requirements management, with waste calculated at $0.01 for every US dollar spent.

---

[5] PMI. "Requirements Management: A Core Competency for Project and Program Success." *Pulse of the Profession®*, 2014.

How can organizations improve?

The study found that for organizations to improve the effectiveness of requirements management, they needed to focus on three critical areas: culture, people, and processes. Furthermore, as shown in Table 7.2, the PMI Global Executive Council highlighted the necessary skills that practitioners need to embrace for effective requirements management.

Table 7.2 Effective Skills for Requirements Management

	Effective Skills for Requirements Management
Critical Skills	❖ Active listening
	❖ Interpreting and clearly articulating stakeholder needs and requirements
	❖ Aligning the stakeholder needs and requirements to the strategic vision for the project
Important Skills	➢ Communication
	➢ Dealing with ambiguity
	➢ Stakeholder engagement
Business Acumen	◆ Discover opportunities and problems that stakeholders would not be able to voice on their own
	◆ Understand the complexity of the business processes
	◆ Recognize the implications of change (strategic, tactile, transformative, etc.) on the business
	◆ Clearly communicate the value of the proposed solution to stakeholders

As the global leader in research, education, and collaboration, the Project Management Institute recognizes the critical importance of business analysis and requirements management in the discipline of program and project management. PMI is endeavoring to be the change agent and transformative catalyst to help organizations realize the positive impact business analysts can have on an organization.

7.3 Eligibility Requirements

For each credential, PMI produces their handbook (see Tip below), which outlines important information for applicants. In addition to information on how to apply for the credential, the handbook also contains information about the policies and procedures for credential maintenance, along with the PMI Code of Ethics and Professional Conduct. See Figure 7.1 for eligibility requirements.

☛ *Tip:* PMI requires that all applicants read the entire handbook. For your convenience, following is the link to the *PMI Professional in Business Analysis (PMI-PBA)® Handbook,* updated March 15, 2016: http://bit.ly/2e9h8V0

❗ *Hint:* When determining your business and project management experience, a good rule of thumb is that there are only 2,080 working hours per year. As most professionals are not utilized at 100%, this can be a red flag on your application; consider a maximum utilization of 75% or 1,560 hours per year (this also accounts for vacations and holidays).

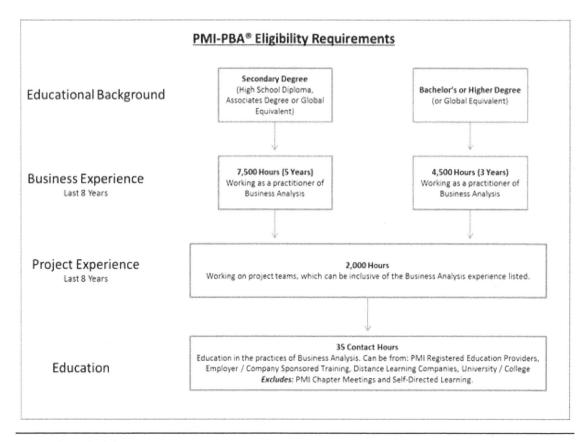

Figure 7.1   Eligibility requirements.

7.4 Role Delineation Study (RDS)

The cornerstone of every PMI credential is a Role Delineation Study (RDS), which is based on industry leading practices and the fundamental responsibilities associated with the profession. The study is conducted when initially developing a credential, then repeated every five to seven years to ensure that the certification exam is still valid for the role. In 2012, PMI commissioned the RDS Study for the PMI-PBA® Credential and validated the results on a world-wide basis.

The Study Outlined:

- The process order for each of the tasks
- The tools and techniques for each process
- The required skills and knowledge of a business analyst

7.5 Examination Content Outline

The *PMI-PBA® Exam Content Outline* (ECO) is created based on the Role Delineation Study. The advisory committee determines the percentage of questions for each domain to reasonably assess a candidate's knowledge.

For each chapter in this exam prep book that corresponds to a domain, we have presented the content in the ECO in a revised manner to highlight the *activity,* the corresponding *tool or technique,* and the *output.* To pass the exam, it is paramount that you understand the ECO and can answer situation-based questions related to the content.

☛ *Tip:* For your convenience, following is the link to the ECO: http://bit.ly/2e9vZiz

7.6 PMI-PBA® Exam

The PMI-PBA® Exam consists of 200 multiple-choice questions, most of which are scenario based and are anchored to the ECO (see Table 7.3). The exam comprises both scored and pretest questions, all of which are randomly placed by domain throughout the exam. The pretest questions are not scored and are used to test the validity of future questions.

Table 7.3 PMI-PBA® Examination Content Outline

Domain	Tasks	Questions		
		Percentage	Scored	Pretest
Needs Assessment	5	18%	32	4
Planning	6	22%	39	5
Analysis	8	35%	61	9
Traceability and Monitoring	5	15%	26	4
Evaluation	4	10%	17	3
Total		100%	175	25

❗ *Hint:* During the exam, you have no way of knowing which questions are scored and which are pretest, so answer each question as if it counts. If you encounter a question that seems unreasonable, it could very well be a pretest question. You may decide to mark the question for later review.

The time permitted each candidate is four (4) hours for the computer-based test (CBT), preceded by a fifteen (15)-minute tutorial and followed by a short survey. The time for both the tutorial and survey are not included within the four-hour allotted time period. Hints and tips for the exam are detailed in Chapter 9: Ready, Set . . . Exam Success!

☛ *Tip:* Although PMI does not publish passing scores, based on my experience we believe the passing score to be between 68% and 70%. When using our exam prep, we recommend candidates score at least 80% on all practice tests before scheduling their exam.

7.6.1 Special Accommodations for the Exam

Although PMI takes the administration of their exams very seriously, they also understand that some candidates may require special accommodations. For candidates with documented medical conditions and/or disabilities, PMI provides reasonable and appropriate test accommodations.

Table 7.4 illustrates the process for requesting special accommodations, followed by a few examples.

Table 7.4 Exam Special Accommodations Process

Special Accommodations Process	
Step	**Description**
1	When submitting payment to PMI for the exam, indicate your request for special accommodations.
2	Once you are approved for the exam, PMI will provide you with an Accommodations Request Form.
3	Return the completed form, along with supporting evidence; **this must be submitted in one package, not mailed separately.**
4	PMI has up to 30 business days to respond to candidates' requests.

Special or personalized accommodations may include:

- Large-font exam
- Additional time
- Separate testing room
- Additional breaks
- Reader
- Scribe

! Hint: Remember to keep a copy of all material submitted to PMI.

Note: Accommodations are individualized on a case-by-case basis, and in some situations, there may be preapproved items (e.g., back cushion).

7.6.2 Exam Administration

The PMI-PBA® Exam is managed in two formats, the most common being the computer-based test (CBT) administered at a Prometric Test Center, the second a paper-based test (PBT), which is only available under the following circumstances:

- Exam candidate resides at least 150 miles (240 km) from the nearest Prometric Test Center.
- A Prometric CBT site is not available within the country of residence, and traveling across country borders is prohibited/unduly burdensome.
- Corporate sponsors and employers request to administer the exam for their employees only.

Because a majority of candidates will be taking the CBT, this section will concentrate on that aspect. Prometric is PMI's partner for exam administration; they have more than 8,000 test centers located across 160 countries. PMI is only one of the close to 400 sponsoring organizations that Prometric serves.

It's not uncommon for there to be slight differences among test centers, and the closest test center may not be the optimal choice based on number of factors (see Table 7.5).

Prometric and PMI take the exam delivery process very seriously; please follow the directions provided on your exam confirmation, and remember to bring with you the required forms of identification.

☛ Tip: In some instances, I've encouraged exam candidates to take advantage of the Prometric "Test Drive." For $30 USD, candidates can familiarize themselves with the test center and the check-in process; to learn more, visit http://bit.ly/1peb1s9

Table 7.5 Prometric Test Center Considerations

Test Center Proximity	What is the relative distance from home or work?
Parking	Does the test center have a parking garage or its own lot, or do you need to locate public or street parking?
Public Transportation	Is the test center close to a bus or subway line?
Exam Seat Availability	Does the closest center offer your preferred time?

7.6.3 Permitted Forms of Identification

As mentioned in the prior section, PMI and Prometric take these examinations very seriously and have strict guidelines regarding candidate ID verification. For admission into the Prometric Exam suite, candidates must bring a valid, non-expired, government-issued identification. The identification needs to be current and include:

- English characters/translation
- Your photograph
- Your signature

The following are acceptable forms of government-issued identification:

✓ Driver's license
✓ Military ID
✓ Passport
✓ National identification card

If for any reason the government-issued identification does not display a photograph or a signature, a secondary form of photo identification may be used.

The following are acceptable forms of secondary photo identification:

✓ Employee ID
✓ Military ID
✓ Bank (ATM) card

The following are NOT acceptable forms of secondary identification:

✘ Social Security cards
✘ Library cards

7.6.4 Exam Fees

Table 7.6, Exam Fees, outlines PMI-PBA® fees based on examination type; please note that computer-based testing (CBT) is the standard delivery method. As outlined in the previous section, paper-based testing (PBT) is only available for candidates who reside 186.5 miles (300 km) from a Prometric Test Center or for employers/corporate sponsors administering the exam to their employees.

Table 7.6 Exam Fees

Exam Type	PMI Member Status			
	Member	Nonmember	Member	Nonmember
Computer-based testing (CBT)	$405	$555	€340	€465
Paper-based testing (PBT)	$250	$400	€205	€335
Re-examination (CBT)	$275	$375	€230	€315
Re-examination (PBT)	$150	$300	€125	€250
CCR Certification Renewal	$60	$150	USD Only	USD Only

The paper-based testing (PBT) examinations have additional limitations:

- On a world-wide basis, there are a limited number of exams offered during the year.
- Exams are only delivered per the published schedule.
- Exam scoring is a manual process and results take up to six weeks.
- Exams require candidates to obtain and use a group testing number (provided by PMI once approved for PBT) prior to scheduling.

Table 7.7 compares the candidate's PMI membership type to the relative cost of the exam.

Table 7.7 Membership Fees

Membership Type	USD	CBT Exam Member	Total Cost	Savings
Individual Member	$139	$405	$544	$11
Student Member	$32	$405	$437	$118
Retiree	$65	$405	$470	$85

! Hint: Prior to submitting your application, it's recommended that you become a PMI member. In addition to saving some money on the exam, you'll get access to all digital editions of the Standards and Practice Guides, professional development, discounts at the PMI Store, and ProjectManagement.com for connecting with colleagues and practitioners globally.

7.6.5 Exam Results

As mentioned in the exam section of this book, the PMI-PBA® Exam consists of 200 multiple-choice questions, and (although not disclosed by PMI) we believe the passing score to be between 68% and 70%. To determine the point at which candidates should pass the exam, and the precise level of difficulty, PMI uses subject matter experts from around the word to establish the proficiency and passing score. Each question answered correctly is worth one point, and the final score is determined by totaling all the questions answered correctly. Upon completion of the CBT Exam, you will immediately receive the results. The overall number of correctly answered questions determines your overall proficiency rating on the exam report.

The PMI-PBA® Exam is scored in two manners:

- Overall Pass/Fail based on the candidate's performance, determined by psychometric analysis
- Level of proficiency, based on the number of correct answers for each domain

The exam is challenging for a number of reasons:

Intensity. Candidates have approximately 72 seconds per question—even less if the goal is to allow for time at the end for a review.

Duration. The exam is four hours, in an environment that is unfamiliar, uncomfortable, and prone to distractions.

Knowledge. Candidates need to understand, assimilate, and prescribe to PMI's methodology and practices. Table 7.8 illustrates the delineation for each of the proficiency ratings.

Table 7.8 PMI-PBA® Exam Results

Exam Performance by Domain	
Above Target	Performance *exceeds* the minimum required for the exam
Target	Performance *meets* the minimum required for the exam
Below Target	Performance is *slightly below target and fails to meet* minimum requirements.
Needs Improvement	Performance is *far below target and fails to meet* the minimum requirements.

! Hint: You do not need a score of *Above Target* or *Target* in every domain to pass the exam. If you score *Below Target* in Traceability and Monitoring (15%) or Evaluation (10%) and score *Above Target* in all others, there is a very good likelihood you can still pass the exam. However, scoring *Below Target or Needs Improvement* in Analysis (35%) and scoring *Above Target* in the other domains could result in less than desirable results.

Tip: Candidates have a one-year eligibility period, during which they are granted three attempts to pass the exam. Due to seat availability at Prometric, should you need to retake the exam, it's our recommendation that you carefully assess your options before the expiration of your eligibility.

7.7 Credential Maintenance

With the exception of Certified Associate in Project Management (CAPM®), each of PMI's certifications have specific professional development requirements. These requirements are captured in the form of professional development units (PDUs) and are a foundational element of PMI's Continuing Certification Requirements (CCR) Program. The CCR Program is designed with an emphasis on continuous professional development, with a concentration on two distinct areas:

Education. Focused on professional development to expand and enhance practitioners' skills and competencies.

Giving Back. Comprises activities that share knowledge related to the credential. In December 2015, PMI introduced the Talent Triangle™, with each side of the triangle representing a core educational competency for each credential. The PMI Talent Triangle™ comprises technical, leadership, and business competencies:

- **Technical.** Expanding knowledge and skills in the field of business analysis.
- **Leadership.** Expanding knowledge and skills to help organizations achieve expected outcomes through requirements management.
- **Business.** Expanding knowledge in your field that can improve business outcomes.

For the PMI-PBA® credential, business analyst professionals must earn a total of 60 PDUs every three years, referred to as the *Certification Cycle.* The cycle starts the day the exam is passed. Table 7.9, PMI-PBA® CCRs, illustrates the PDUs by category.

The CCR Program is designed to be both flexible and robust; Table 7.10 provides examples in the areas of education and giving back.

Table 7.9 PMI-PBA® CCRs—Three-Year Cycle

PMI-PBA® Continuing Certificate Requirements	Minimum	Maximum
Education	**35**	
Technical	8	
Leadership	8	
Strategic and business management	8	
Remaining PDUs can be applied to any of the above educational areas	11 +	
Giving Back		**25**
Volunteering		17
Creating knowledge		
Working as a professional		8

7.8 PMI-PBA® Application

The PMI-PBA® application comprises seven sections (see Table 7.11 on page 178); we'll walk through each area and point out some useful hints and tips along the way.

7.8.1 Application Requirements Explained

As previously discussed in the Eligibility Requirements Section of this chapter, three distinct requirements must be met before candidates can apply for the credential (see Figure 7.1, page 170).

- **Business Analysis Experience.** This section comprises three parts:

    a. Candidates begin by identifying a business initiative, corresponding dates, the sponsoring organization, and a contact (sponsor, client, manager/director, or primary stakeholder).

    ☛ **Tip:** Consider advising your organization contact beforehand. PMI will only contact them in the event of an audit, and providing advance notice will simplify the process.

Table 7.10 Opportunities to Earn PDUs

Opportunities to Earn PDUs		
Education	Course or Training	Instructor-led (in person or online) • Courses offered by a PMI REP • Educational events held by a PMI chapter • Instructor-led courses from PMI SeminarsWorld® • On Demand e-Learning courses • Academic education programs • Courses from other third-party providers
	Self-Paced Learning	Online learning or courses offered by a PMI REP • ProjectManagement.com® • e-Learning on Demand+
	Informal Learning	Structured learning or mentoring
	Organization Meetings	PMI chapter and third-party events
	Reading	Self-paced reading
Giving Back	Creating Content	Authoring material: • Books, blogs, articles • Presentations or webinars
	Hosting a Presentation	Presenting on material relevant to your credential • PMI chapter event, industry conference, at work, etc.
	Sharing Knowledge	Sharing knowledge relevant to your credential • Mentoring, teaching, etc.
	Volunteering	Volunteering with non-client or employer organization
	Working as a Professional	Being actively employed

! Hint: There are many free resources available; remember to visit projectmanagement.com for more information. Prior to the end of your three-year certification cycle, all PDUs must be reported to maintain an active certification status. Once requirements are met for the current cycle, up to 20 PDUs earned in the last year (12 months) of a cycle can be applied to the following cycle.

☛ Tip: Consult PMI's Continuing Certification Requirements Handbook for further information about the CCR Program. For your convenience, following is the link to the handbook, updated August 22, 2016: http://bit.ly/2czTnB4

b. In the next section, candidates document the hours they worked in each of the domain areas, as outlined in the ECO.

☛ Tip: A good rule of thumb is that there are only 2,080 working hours per year. Most professionals are not utilized at 100%, so this can be a red flag on your application. Consider a maximum utilization of 75%, or 1,560/year; this will account for vacations and holidays. Note: the online application will not allow more than 320/month to be recorded.

c. In the last section, credential candidates detail the business analysts tasks for which they were directly responsible. The response must be between 300 and 550 Characters.

Table 7.11 PMI-PBA® Credential Application Overview

PMI Professional in Business Analysis Application	
Section	**Description**
Contact Address	In this section, credential applicants can enter their home and work address, noting their preferred mailing address.
Contact Information	In the fields provided, credential applicants confirm their preferred email address and provide a contact phone number.
Attained Education	This section is used by credential applicants to report their highest level of attained education. Options include: high-school diploma or associates, bachelors, masters, and doctorate degrees. ☛ *Tip:* In the event you are audited, PMI will request transcripts from the schools reported in this section. In the event a school has consolidated or closed, it is very important that you are able to provide proof of completion. ! *Hint:* Before submitting the application, I recommend to applicants that they first secure this documentation to simplify the audit process. Remember, at any point PMI can request an audit of your application, not just at time of submission.
Requirements	In this section you provide the details for your business analysis experience, general project experience, and business analysis education.
Optional Information	Credential referral source (employer suggested or required, professional development). Also note whether you have taken a prep course sponsored by a PMI chapter.
Certificate	The name *exactly* as you would like it to appear on the certificate.
Agreement	Before submitting the application, you need to agree to abide by the terms and conditions set forth by PMI, including, but not limited to, those identified in the *CCR Handbook* and the PMI Code of Ethics.

☛ *Tip:* Consider having five lines, each related to a domain, and heavily abbreviating your response. For example:

Abbreviation	The abbreviated response would be derived from
NA: ID prob & stk hldrs	**Needs Assessment:** Identify problem and stakeholders
PL: Dev BA Plan	**Planning:** Develop business analysis plan
A: Plan & cond elict;	**Analysis:** Plan and conduct elicitation
TM: Bsln req ...	**Traceability and Monitoring:** Baseline requirements
EV: Doc BU goals ...	**Evaluation:** Document business unit goals

• **Project Experience.** This section is similar to the Business Analysis Experience Section, and also comprises three sections:

    a. Candidates begin by identifying project specific information, corresponding dates, the sponsoring organization, and a contact (sponsor, client, manager/director, or primary stakeholder).

    b. In the next section, candidates document their project experience.

c. In the last section, candidates detail their experience working on the project. The response must also be between 300 and 550 Characters. As with the other response, candidates will need to heavily abbreviate.

🖎 **Note:** For those candidates who already hold their PMI Project Management Professional (PMP)® certification, this section is waived.

• **Business Analysis Education.** A minimum of 35 hours, which relate directly to business analysis. One hour of classroom instruction qualifies for one contact hour toward the required 35.

☛ **Tip:** Some employer-sponsored training may qualify, provided it's directly related. If you're audited, PMI may ask for a syllabus or an agenda from the class, in addition to the instructor's name.

7.8.2 Remember the Golden Rule

When preparing your application and writing your summaries, please remember that *at any point PMI can initiate an audit.* If you are audited, PMI will require you to substantiate all information provided, which can also extend to existing credentials. It is the author's opinion that it's much better to understate experience than to have to explain before the Ethics Committee or a listed contact any embellished information.

7.9 Application Stages and Timeline

Table 7.12 details the timeline for the PMI-PBA® credential process, along with some useful hints.

Table 7.12 Credential Process Timeline

Stage	Description	Timeline
1	Application Submission	Once started, you have 90 days to complete and submit your application to PMI. If for any reason you are unable to do so, PMI will close your application. **! Hint:** Should this happen, simply call PMI Customer Care and they will gladly reopen your application, providing you wait another 90 days to submit your application. Also, please note, if you plan on joining PMI, please do so prior to submitting your application.
2	Application Completeness Review	Once submitted, PMI will review your application to ensure that you meet the eligibility requirements, you have completed the necessary education, and your experience summary is consistent with the role of a business analyst. PMI's goal is to process your application within five (5) calendar days.
3	Payment Process	Once payment is remitted to PMI, your application will move on to the next stage in the process. **! Hint:** Remember to note requests for any special accommodations.

(continues on next page)

Table 7.12 Credential Process Timeline *(continued)*

4	Audit	PMI randomly selects applications for audit, following receipt of payment for the exam. Provided you followed the golden rule from the prior section, this process will be very simple. PMI will ask you to submit the following supporting documentation:

PMI randomly selects applications for audit, following receipt of payment for the exam. Provided you followed the golden rule from the prior section, this process will be very simple. PMI will ask you to submit the following supporting documentation:

a) Copies of your diploma or college transcripts

b) Signatures from the individuals listed in the experience verification section of the applications

c) Course completion records from the training institute(s) for each course listed on the application to meet the contact hours

You will have 90 days to submit the requested material. Once it is received, the audit department aims to complete its review in between five and seven days.

> ✎ *Note:* It's very important that all the material be submitted at one time, in a single envelope.

> ❗ *Hint:* One of the most common questions we receive is, "What if I can't get in touch with an individual listed in the experience verification section?" My recommendation is to contact someone from human resources, or another person who can verify your experience, to sign, provided you include a reasonable explanation for the substitution.
> The second most common question is, "The business or entity no longer exists; how can I proceed?" PMI is very reasonable and understands the changing nature of business. My suggestion is to only list people who can attest to your experience; it's also a good idea to contact them before you submit your application so that they know you're providing them as a reference.

> ☛ *Tip:* If for any reason you fail the audit, PMI will only refund $200 USD of your paid application fee.

> ✎ *Note:* PMI can audit your application at any point, even long after you've earned the credential.

5	Exam	Once your application is approved, the one-year eligibility clock begins. PMI provides you with three attempts to pass the exam during this eligibility cycle.

Once your application is approved, the one-year eligibility clock begins. PMI provides you with three attempts to pass the exam during this eligibility cycle.

> ❗ *Hint:* Once approved to take the exam, my recommendation is to visit the Prometric website to:
>
> a) Determine your preferred and backup test centers
> b) Understand seat availability over a three-month period

> ❗ *Hint:* **Under no circumstance** should you book the exam at this time. The timing of when to do so will be covered in a subsequent section.

Chapter 8

Preparing for the Exam

Based on our experience, candidates who are performing the role of business analyst can prepare for and pass the challenging PMI-PBA® Exam in about 90 days. In this section, we'll cover the important topics and share some hints and tips to help you prepare for and pass the exam.

In my opinion, preparing for a PMI-PBA® Exam is similar in nature to an athlete preparing for a game. Athletes cannot simply walk onto the field and expect to compete, much less win, when they're matched against worthy opponents, without some amount of preparation. Athletes who are exceptional spend countless hours training across many disciplines in the hopes of a favorable outcome.

It can be argued that the same is true for the Project Management Institute (PMI) exams. With each role delineation study (RDL) and standard/practice guide update, the exams become a little more challenging. Much like watching game films of your opponents, the following should help to prepare you and, we hope, alleviate some stress and lead to a successful outcome for all your hard work.

8.1 Prerequisite Education

As outlined in Figure 7.1, Eligibility Requirements (page 170), before applying for the PMI-PBA® credential, candidates must complete 35 contact hours of training in the practices of business analysis. This is truly foundational—the difference between passing the exam and a less than desirable score can be traced to the prerequisite education. This *Exam Practice Test and Study Guide* is not endorsing one product over another; rather it is stressing to candidates that they thoroughly evaluate and consider all available options. My personal preference tends to be hands-on, interactive classes.

PMI registered education providers (REPs) are held to very rigorous standards for quality, content, and effectiveness. In preparation for the PMI-PBA® Exam, we strongly encourage candidates to consider training with a PMI REP.

> ☛ *Tip:* For your convenience, following is the link to PMI's registered education providers database (https://www.pmi.org/learning/training-development/reps/find)

8.2 Studying

Just as PMI is a proponent of taking action and planning for business situations, the same philosophy should be incorporated into your studies and preparation for the PMI-PBA® Exam. As outlined in

Step 5 of Stages and Timeline (Chapter 7), once PMI has accepted your application, you will receive an email containing your Prometric confirmation number. My recommendation is to log into Prometric's website to gauge seat availability over the next three months at your preferred or alternate test center.

Our recommended approach to studying and preparing for the exam encompasses eleven aspects. Your overall progress will determine when to schedule the exam. With just a few hours per day, we believe this to be attainable within 90 days. The eleven areas that we'll explore include:

8.2.1   How Best to Use This Exam Practice Test and Study Guide
8.2.2   PMI Reference Material
8.2.3   Code of Ethics
8.2.4   Flash Cards
8.2.5   PMI Lexicon
8.2.6   Memory Games and Course Exercises
8.2.7   Exam Aids
8.2.8   Simulation Tests
8.2.9   Exam Cadence
8.2.10 Optimal Time for Breaks
8.2.11 Sacred Principle

8.2.1 How to Best Use This Exam Practice Test and Study Guide

This *PMI-PBA® Exam Practice Test and Study Guide* is aligned with PMI's *Business Analysis for Practitioners: A Practice Guide, Requirements Management: A Practice Guide,* and the *PMI-PBA® Examination Content Outline;* in addition, it references key content from *A Guide to the Project Management Body of Knowledge (PMBOK®).* Each of the chapters, as outlined below, correspond to a domain as defined in the 2012 Role Delineation Study (RDS) of Business Analysts. For further information on the RDS, please refer to Appendix A: Role Delineation Study (RDS) Process, in the *Examination Content Outline.*

Introductory Chapter

The introductory chapter offers further context to foundational elements, which are carried throughout all the domains, and provides a necessary baseline for the subject material.

Needs Assessment

The Needs Assessment domain includes the tasks necessary to understand the opportunities or problems of the sponsoring organization; the tools and techniques to determine the value proposition; the creation of project goals and objectives linked to the organizational pillars; and a comprehensive stakeholder assessment. The activities in this domain produce deliverables that are used throughout the project, including the situation statement, the solution scope statement, the business case, and various stakeholder assessment artifacts.

Planning

The Planning domain contains the elements that focus on the preparation required to manage the business analysis activities throughout the lifecycle of the project.

Analysis

The Analysis domain is the single most important domain on the exam. It contains those elements that focus solely on the tasks and activities related to the identification, detailed specification, approval, and validation of product requirements.

Traceability and Monitoring

The Traceability and Monitoring domain contains those tasks and activities that focus on managing the lifecycle of the requirements. This domain focuses on establishing the requirements baseline and the continual monitoring, tracking, and communicating the status of requirements. Furthermore, the domain encompasses change control and the management of issues, risks, and decisions.

Evaluation

The Evaluation domain contains those tasks and activities that validate test results, analyze and communicate gaps, and obtain stakeholder signoff, concluding with an evaluation of the deployed solution.

Based on our experience with adult students, we are advocating the following approach to prepare for the exam:

Step 1: Read the *PMI Professional in Business Analysis (PMI-PBA)® Handbook* in its entirety.

Step 2: Read *Requirements Management: A Practice Guide* in its entirety.

Step 3: Read *Business Analysis for Practitioners: A Practice Guide* in its entirety.

Step 4: Read a domain chapter in this *PMI-PBA® Exam Practice Test and Study Guide*.

Step 5: Reread the corresponding domain chapter in *Requirements Management: A Practice Guide* and *Business Analysis for Practitioners: A Practice Guide*.

Step 6: Read the supporting chapter in the *PMBOK® Guide*.

Step 7: Practice the exercises at the end of the relevant chapter of this book, then take the 20-question domain quiz.

Step 8: Review and understand the rationale for correct and incorrect answers.

Step 9: Repeat this process for all the domains.

Step 10: Take the final practice exam on paper.

Step 11: Review and understand the rationale for correct and incorrect answers.

Step 12: Use our online exam bank (see Page *xxii,* Step 12, for instructions) to simulate the actual exam.

8.2.2 65-Day Study Plan

For your convenience, we've created a simple study plan (see Table 8.1 on next page), which is designed to improve your retention as you learn new concepts and theories, along with the practical application of the tools and techniques associated with business analysis. At the same time, this will condition your mind and improve your focus as you prepare for the four-hour, 200-question PMI-PBA® Exam.

There is no substitute for hard work.

— Thomas Edison

✐ Hint: The following study plan can easily be modified, but don't skip this step. Remember you have a higher likelihood of accomplishing a goal if it's written down.

Table 8.1 65-Day Study Plan

Day	Activity	Material
Foundational		
1	Read	*PMI Professional in Business Analysis (PMI-PBA)® Handbook.*
1	Read	The introductory chapters of this book, up to Chapter 2, on Needs Assessment, and complete the introductory exercises.
2–3	Read	*Requirements Management: A Practice Guide.*
4–8	Read	*Business Analysis for Practitioners: A Practice Guide.*
9–10	Create	Your flash cards based on terms, concepts, tools, and techniques that are unfamiliar.
10–12	Paperwork	Complete and submit your application.
Needs Assessment		
12	Read	Chapter 2, Needs Assessment, from this book.
12	Read	Needs Assessment material from *Requirements Management: A Practice Guide.*
12	Read	Needs Assessment material from *Business Analysis for Practitioners: A Practice Guide.*
13–15	Read	Integration, scope, and stakeholder management material from the *PMBOK® Guide.*
16–17	Study	Review all flash cards.
16–17	Game	Memory games from this book for the Needs Assessment domain.
16–17	Exercise	Test Your Knowledge Exercise.
18	Quiz	Take the 20-Question Needs Assessment Quiz.
18–19	Review	Review and understand the rationale for correct and incorrect answers.
20	Study	Review all flash cards.
20	Game	Memory Games for the Needs Assessment domain; review Test Your Knowledge Exercise.
Planning		
21	Read	Chapter 3, Planning, from this book.
21	Read	Requirements management planning from *Requirements Management: A Practice Guide.*
21	Read	Business analysis planning from *Business Analysis for Practitioners: A Practice Guide.*
22–24	Read	Time, communication, and human resource management from the *PMBOK® Guide.*
25–26	Study	Review all flash cards.
25–26	Game	Memory games for the Needs Assessment and Planning domains.
25–26	Exercise	Test Your Knowledge Exercise.
27	Quiz	Take the 20-Question Planning Quiz.
27–28	Review	Review and understand the rationale for correct and incorrect answers.

(continues on next page)

Table 8.1 65-Day Study Plan *(continued)*

29–30	Study	Review all flash cards.
29–30	Game	Memory Games for the first two chapters; review completed Test Your Knowledge Exercises.
Analysis		
31	Read	Chapter 4, Requirements Elicitation and Analysis, from this book.
31	Read	Requirements elicitation material from *Requirements Management: A Practice Guide*.
31	Read	Requirements elicitation material from *Business Analysis for Practitioners: A Practice Guide*.
32	Read	Project cost management from the *PMBOK® Guide*.
33–34	Study	Review all flash cards.
33–34	Game	Memory Games for the first three chapters; review completed Test Your Knowledge Exercises.
33–34	Exercise	Test Your Knowledge Exercise.
35	Quiz	Take the 20-Question Analysis Quiz.
35–37	Review	Review and understand the rationale for correct and incorrect answers.
37–39	Study	Review all flash cards.
37–39	Game	Memory Games for the first three chapters; review completed Test Your Knowledge Exercises.
Traceability and Monitoring		
40	Read	Chapter 5, Traceability and Monitoring, from this book.
40	Read	Requirements monitoring and control material from *Requirements Management: A Practice Guide*.
40	Read	Traceability and Monitoring material from *Business Analysis for Practitioners: A Practice Guide*.
41	Reread	Project scope management from the *PMBOK® Guide*.
42–44	Study	Review all flash cards.
42–44	Game	Memory Games for the first four chapters; review completed Test Your Knowledge Exercises.
42–44	Exercise	Test Your Knowledge Exercise.
45	Quiz	Take the 20-Question Traceability and Monitoring quiz.
45–46	Review	Review and understand the rationale for correct and incorrect answers.
47–49	Study	Review all flash cards.
47–49	Game	Memory Games for the first three chapters; review completed Test Your Knowledge Exercises.
Evaluation		
50	Read	Chapter 6, Evaluation, from this book.
50	Read	Solution evaluation and closure material from *Requirements Management: A Practice Guide*.

(continues on next page)

Table 8.1 65-Day Study Plan *(continued)*

50	Read	Solution evaluation material from *Business Analysis for Practitioners: A Practice Guide.*
51	Read	Project quality management from the *PMBOK® Guide.*
52	Study	Review all flash cards.
52	Game	Memory Games for all five chapters; review completed Test Your Knowledge Exercises.
52	Exercise	Test Your Knowledge Exercise
53	Quiz	Take the 20-Question Evaluation quiz.
53–54	Review	Review and understand the rationale for correct and incorrect answers.
55–56	Study	Review all flash cards.
55–56	Game	Memory Games chapters and review Test Your Knowledge (all chapters).
Practice Exam		
57	Study	Review all flash cards.
57	Game	Memory Games and review Test Your Knowledge (all chapters).
58	Exam	Final Practice Exam from this book.
59	Review	Review and understand the rationale for correct and incorrect answers.
Review		
60–62	Study	Review all flash cards.
60–62	Game	Memory Games and review Test Your Knowledge (all chapters).
60–63	Review	Review and understand the rationale for correct and incorrect answers.
Online Practice Exam		
64	Exam	Final Online Practice Exam (see Page *xxii*, Step 12, for instructions).
65	Review	Review and understand the rationale for correct and incorrect answers.

☛ *Tip:* When both your aggregate and domain scores are above 80% using the Final Online Practice exam, we recommend scheduling your exam for three weeks out. In Chapter 9, Ready, Set . . . Exam Success, we'll cover the preparation leading up to exam day.

8.2.3 PMI Reference Material

Subject matter experts and volunteers from a broad range of industries and backgrounds contribute material, which in aggregate ultimately makes up the question bank for the PMI-PBA® Exam. The exam is not written by any one individual; you'll find it's mostly scenario based to assess candidate's competency, knowledge, and skills from both a hands-on and a classroom perspective.

☛ *Tip:* As a courtesy to assist candidates in their preparation, PMI has compiled Reference Materials for PMI® Professional in Business Analysis (PMI-PBA)® Examination, available at http://bit.ly/ 2e0SS2d

While I've had the opportunity to read many of the publications on this list, this *Exam Practice Test and Study Guide* is designed so that the most important concepts and frameworks are outlined and explained. Aside from the three PMI publications listed below, I'd suggest that the others are optional and worth reading at some point, but not in your preparation for the exam.

> ☛ **Tip:** As a PMI member, all of the foundational standards, practice guides, and frameworks are available for download at no cost.

Recommended Reading

Business Analysis for Practitioners: A Practice Guide
Publisher: Project Management Institute
ISBN13: 9781628250695

Requirements Management: A Practice Guide
Publisher: Project Management Institute
ISBN13: 9781628250893

A Guide to the Project Management Body of Knowledge (PMBOK®) Guide, 5th Ed.
Publisher: Project Management Institute
ISBN13: 9781935589679

8.2.4 PMI Code of Ethics

As the world's leading project, program, and portfolio management professional organization, PMI is guided by a fundamental mission and vision. At the core are its guiding principles, with one of the primary tenets being *professionalism*. PMI prides itself on fostering and encouraging research, diversity, community, and volunteerism. Furthermore, it recognizes the impact its members and credential holders can have both on an organization's results and society on a global basis. The PMI Code of Ethics and Professional Conduct is based on the values of "honesty, responsibility, respect and fairness" and "making the best possible decisions concerning people, resources and the environment" (Project Management Institute). To ensure that its members and credential holders have a clear understanding of the framework, The Code of Ethics and Professional Conduct is included as part of the *PMI-PBA® Handbook,* and candidates must acknowledge they have read and will abide by it as part of the application process. In addition, during each renewal cycle, credential holders must agree to conduct themselves in accordance with the underlying tenets.

> ❗ **Hint:** Questions on the exam will always be focused on "doing the right thing," which may not always be the most practical.

> ☛ **Tip:** As required reading, following is the link to the PMI Code of Ethics and Professional Conduct: http://bit.ly/2e6ycZ7; and the Ethical Decision-Making Framework: http://bit.ly/2eR004O

8.2.5 Flash Cards

When studying new material and concepts, flash cards are a fantastic and often underutilized tool. They engage active recall and metacognition; they allow for self-directed learning; and they help students to assess their progress over time. My recommendation is for students to create flash cards based on both the glossary and important content from each of the practice guide's chapters. Because their effectiveness is based on both repetition and incorporation into one's job, our suggestion is to break them into three groups and use them at different times during the day. This way your mind has time to process the information and they become more than simply words on cards.

☞ *Tip:* When creating flash cards, remember quality over quantity; if you're already familiar with the concept, tool, or technique, consider not creating a flash card. In my experience, students become comfortable with the content on flash cards after about a week of consistent practicing.

8.2.6 PMI Lexicon

The *PMI Lexicon of Project Management Terms* is akin to a secret handshake; developed by volunteer experts, the *Lexicon* enables a common language and understanding among project team members. Considering the critical role of business analysts and their close integration with project teams, my recommendation is that candidates for the PMI-PBA® credential read the *Lexicon* and begin to incorporate the terminology in their work efforts.

☞ *Tip:* For your convenience, following is the link to the PMI Lexicon of Project Management Terms Version 3.1 © 2016: http://bit.ly/2nVgUEP

8.2.7 Memory Games and Course Exercises

The foundation of the PMI-PBA® Exam is the *Professional in Business Analysis Examination Content Outline* (ECO), so it's paramount that candidates can navigate the activities, the corresponding tools and techniques, and the outputs for each domain. Understanding the proper order and sequence will help candidates to comfortably answer a number of the more challenging questions that may appear on the exam.

Memory games are a highly effective tool to help students relate content that at the outset may seem difficult and unrelated. At the conclusion of Chapter 2, Needs Assessment, are cut-outs of important domain concepts (see Figure 8.1).

Concept	Concept	Concept
Definition	Definition	Definition
Definition	Definition	Definition
Definition	Definition	Definition

Cut along the dotted lines

Figure 8.1   Memory game template.

We recommend students make a copy of these pages, cut out the pieces, and then mix them up on the table. Then, without referring to the reference material, try to reassemble the pieces in the correct sequence.

For the exam, it is very important that candidates have a firm grasp of all the domain-specific elements, their situational use, and the practical application of the associated tools and techniques. To that end, following each chapter are comprehensive exercises to reinforce the material from the ECO, the practice guides, and the *PMBOK® Guide*.

8.2.8 Exam Aids

Following the start of the PMI-PBA® Exam, candidates can use the provided scratch paper to write down information that may be of use while answering questions. This process is commonly referred to as *brain dumping*—when candidates create their exam aids or memory maps.

Exam aids and memory maps are very personal; unfortunately, there is not a one-size-fits-all or even fits-most solution when it comes to exam toolkits. Experts suggest that at the start of the exam, candidates should write down information that either will be referenced frequently or will reduce stress during the exam.

At a minimum for the PMI-PBA® Exam, I recommend candidates take two minutes at the start of the exam to write down the elements and concepts in Table 8.2.

> ✍ **Note:** As mentioned in Chapter 9, Exam Success, Prometric and PMI prohibit candidates from using the tutorial time to create brain dumps. While permitted, memory maps and exam aids may only be created once the exam clock starts.

Table 8.2 Exam Aid Suggestions

Exam Aid	Rationale	Exam Study Guide Reference Page
Cadence Chart	It is paramount that you monitor cadence during the exam. This will relieve stress and provide quantitative measures if you need to pick up the pace or if you have time for a break.	Page 192
Contents of Charter	On nearly every PMI Exam, there are questions pertaining to the charter and its contents. These are easy questions if you can memorize each of the sections.	Page 32
Contents of Scope Statement	The scope statement, like the charter, appears on nearly every PMI Exam. The scope statement sets the boundaries for the project, and there are several key aspects candidates must understand when approaching these questions.	Page 32
Contents of Business Case	The business case is the deliverable from the Needs Assessment domain. Exam candidates can expect a number of questions pertaining to this key artifact and its contents.	Page 34
Contents of Requirements Management Plan	The requirements management plan will cover elements of both the project and product, identifying stakeholders and their roles, establishing the framework for communications, and articulating the guidelines for managing requirements.	Page 61
Contents of Business Analysis Plan	The business analysis plan focuses solely on the activities and deliverables related to the efforts of business analysis. PMI is a proponent of plans—exam questions will focus on the contents, purpose, and intent of these project artifacts.	Page 63

8.2.9 Simulation Tests

A key aspect of preparing for a PMI exam is practicing simulation tests. Although these are not the actual questions that will appear on the exam, depending on the quality of the source, they can be somewhat representative and help to assess a knowledge baseline. The question bank in this book is very representative of the questions that will appear on the actual PMI-PBA® Exam; furthermore, the questions and rationales are designed to foster learning. While practicing simulation tests, we suggest using this opportunity to mirror the Prometric experience. As such, simulation tests are the perfect time to:

- Predict your score and track progress.
- Monitor exam cadence.
- Determine the optimal time for breaks.

☛ **Tip:** If you find taking tests to be stressful, you may want to consider playing talk radio in the background to condition yourself to an environment that may not be conducive to focusing. Also, you may want to consider wearing over-sized headphones to mute out ambient noise during the exam.

8.2.9.1 Predicting Your Score

This technique is quite simple, and when used consistently, it's a reliable means to predict your overall exam score. My recommendation is to use this technique during your studies, then consider a modified version for the actual exam.

To start, each domain quiz and final practice exam has an accompanying answer sheet (see Figure 8.2); you will note there are three sections:

➢ **Answer Choice** (a, b, c, d)
➢ **Result** (Correct vs. Incorrect)
➢ **Predicted Score**

Step 1: Thoroughly read each question, and notate your answer.

Step 2: After you answer each question, assign a confidence level, as shown in Figure 8.3.

Figure 8.3 Predicting your score.

Figure 8.2 Sample answer sheet.

90% You're 90%–100% sure of the answer.
75% You were able to narrow the answer to two choices.
50% You eliminated one answer choice, but could not decide among the other three.
25% You guessed, as all four answer choices seemed reasonable.

Step 3: Score your quiz, with a check mark in the appropriate column (Correct/Incorrect) based on the provided answer sheet (Figure 8.4).

In this example, the student answered three questions correctly and one incorrectly, for an overall score of 75%. The predicted score was 76.25%, calculated as follows:

Answer Sheet

	Answer Choice	Correct	Incorrect	Predicted Answer 90%	75%	50%	25%
1.	a b c d	✓			✓		
2.	a b c d	✓		✓			
3.	a b c d		✓			✓	
4.	a b c d	✓		✓			
	Score: 75%	3	1	2	1	1	0

Figure 8.4 Score your quiz

.9	*	2	=	1.8
.75	*	1	=	.75
.5	*	1	=	.5
.25	*	0	=	0
Total				3.05
3.05	/	4	=	76.25%

! Hint: By using this method, the objective during the final review is to increase your confidence by moving answers from a lower predicted percentage to a higher percentage, ideally 90%.

☛ Tip: During the actual exam, maintaining this table can be time consuming. Consider using this method only for the questions that you mark for later review. If you find that time is short, start with those marked at 75%.

8.2.9.2 Tracking Your Progress

Why should the tools and techniques of business analysis be limited to business? I believe that, where applicable, they should be extended to our personal lives as well. In an earlier section, I outlined a study plan with the rationale that you have a higher likelihood of accomplishing a goal if it's written down. In this section, the analogy is based on the traceability matrix, with which we are tracking and measuring our domain quizzes and simulation exam results. Have you ever heard the saying, "You can't improve what you don't measure"?

My recommendation is to start by creating a very simple table, then logging the results from each of the domain quizzes, as shown in Figure 8.5, Tracking quiz results.

Domain		7-Jan Saturday	8-Jan Sunday	9-Jan Monday	10-Jan Tuesday	11-Jan Wednesday	21-Jan Saturday
# Questions	%	20	20	20	20	20	20
Needs Assessment	18%	55.00%					75.00%
Planning	22%		60.00%				
Analysis	35%			50.00%			
Traceability and Monitoring	15%				70.00%		
Evaluation	10%					60.00%	

Figure 8.5 Tracking quiz results.

Following each domain quiz, a key learning w is to understand the rationale behind each correct and incorrect answer. This will help to reinforce the material and strengthen areas of deficiency. After completing all the chapters, my suggestion is to retake the quizzes. As shown in Figure 8.6, after completing 100 practice questions, the student improved significantly in each of the 20-question domain quizzes.

Domain		7-Jan	8-Jan	9-Jan	10-Jan	11-Jan	21-Jan	22-Jan	23-Jan	24-Jan
		Saturday	Sunday	Monday	Tuesday	Wednesday	Saturday	Sunday	Monday	Saturday
# Questions	%	20	20	20	20	20	20	40	40	200
Needs Assessment	18%	55%					75%			69%
Planning	22%		60%					80%		80%
Analysis	35%			50%				85%		74%
Traceability and Monitoring	15%				70%				95%	70%
Evaluation	10%					60%			85%	80%

Figure 8.6   Improvement with quiz results.

Having achieved passing scores in each domain quiz, the column to the far right logs the results from the student's first attempt at the 200-Question Simulation Exam. Based on our algorithm, the overall score is 74.5%, which is considered borderline for scheduling the PMI-PBA® Exam.

☛ *Tip:* As previously mentioned, review the rationale for each correct and incorrect answer from the 200-Question Simulation Test. Then retake the exam and log your results; you should note an overall improvement in your score. While it may appear that you're memorizing the questions and answers, you're also becoming more accustomed to wordy, scenariobased questions.

8.2.10 Exam Cadence

When taking a timed exam, it's highly stressful to believe you won't have sufficient time to finish. For the PMI-PBA® Exam, candidates have four hours (240 minutes) to complete 200 questions. That equates to 1 minute 12 seconds per question with no break; this is challenging even for those individuals who are accustomed to scenario-based examinations.

Tracking your cadence throughout the exam can ease your nerves, especially while you tackle some of the more challenging questions that seem to take several minutes. This technique can also be an early indicator if you need to pick up the pace at any point.

For the PMI-PBA® Exam, I advocate breaking the exam into four sprints, with each sprint consisting of 50 questions with duration of 56 minutes. This equates to 1 minute 7.2 seconds per exam question.

	Sprint 1		Break	Sprint 2		Break	Sprint 3		Break	Sprint 4		Review
Minutes	28	56		85	113		142	170		199	227	13
Target Questions	25	50	1	75	100	1	125	150	1	175	200	
Actual												

Figure 8.7   Cadence chart.

In the example provided in Figure 8.7, after each sprint, we suggest taking a one-minute break at your exam station. With this pace, you'll finish the exam with 13 minutes remaining for a final review.

☛ *Tip:* This cadence chart can easily be adapted to your individual needs, and it can be created and tracked on either the paper or the dry erase sheets during the actual exam.

✎ *Note:* These examples do not include time for creating exam aids or a score prediction table. If you intend to use these tools, please add a buffer prior to Sprint 1.

The following section illustrates the practical application and value of the cadence chart.

8.2.10.1 Exam Cadence Example

At the conclusion of Sprint 1, the candidate's goal was to complete 50 questions. Having answered 52 questions, they are ahead of schedule by two questions (see Figure 8.8).

	Sprint 1		Break
Minutes	28	56	
Target Questions	25	50	1
Actual	27	52	

Figure 8.8   Cadence chart (Sprint 1).

By the conclusion of Sprint 2 (Figure 8.9), 113 minutes have elapsed, and the candidate's goal was to have completed 100 questions. Having answered only 90 questions, they are now behind schedule by 10 questions. At the halfway point, the candidate needs to pick up the pace.

	Sprint 1		Break	Sprint 2		Break
Minutes	28	56		85	113	
Target Questions	25	50	1	75	100	1
Actual	27	52		73	90	

Figure 8.9   Cadence chart (Sprint 2).

At just about 200 minutes into the exam, the candidate's goal was to have completed 175 questions. Based on the cadence chart, they have recovered nicely and are now ahead of schedule by seven questions (Figure 8.10).

	Sprint 1		Break	Sprint 2		Break	Sprint 3		Break	Sprint 4	
Minutes	28	56		85	113		142	170		199	227
Target Questions	25	50	1	75	100	1	125	150	1	175	200
Actual	27	52		73	90		125	152		182	

Figure 8.10   Cadence Chart (Sprint 4).

8.2.11 The Optimal Time for Breaks

After concentrated effort, our minds need time to refocus; sometimes just covering our eyes for 60 seconds will be enough to clear our thoughts. Although we are not proponents of leaving the exam station, you may find it necessary to walk away and stretch; even if you think you don't have the time, this may prove to be crucial if your focus is deteriorating.

> *! Hint:* Your simulation tests, both paper and online, should mirror as closely as possible the actual exam day. If your goal is to walk away for five minutes at the end of each sprint, plan this time into your cadence tracker and actually walk away from your desk for the planned time. Remember to allocate for the time it will take to sign back in with the Prometric proctor—perhaps upwards of two to five minutes if there are other students waiting to sign in.

> *☛ Tip:* Plan to cover your eyes and stretch in your seat for 60 seconds every 50 questions.

8.2.12 Sacred Principle

Most importantly, throughout your studies, remember: *Knowing what to do must be translated into doing what you know.* In other words, if you can incorporate the PMI concepts, Lexicon, and material from this *Exam Practice Test and Study Guide* into your daily routine, the exam will be much easier.

Chapter 9

Ready, Set . . . Exam Success

9.1 Scheduling the Exam

When you're able to score at least 80% on the full-length practice exam using our test simulation software and each of the 20-question domain quizzes, consider scheduling your exam with Prometric about three weeks out. My personal preference is to schedule exams for either Monday or Tuesday, between 9:30 and 10:30 AM, after the first time slot at 8:30; this way there is little wait upon your arrival.

☛ *Tip:* In the event that you need to retake the exam, we recommend that you book it at least three months before the expiration of your eligibility.

9.2 Countdown to Exam Day

Over the next three weeks, concentrate your studies, including practicing your memory games, flash cards, and chapter exercises, if possible reviewing multiple times per day. Focus on the domains and areas in which you are the weakest or have the most doubt. Leading up to the exam, try to complete the full-length simulation test at least two more times, each time charting your scores and reviewing the rationale for incorrect questions.

If for some reason you need to cancel or reschedule your exam within 30 calendar days of your appointment, Prometric will charge you a $70 USD rebooking fee. If you cancel your exam within two days of your appointment, per the http://www.pmi.org/GLOBALS/Handbooks.aspx, you will forfeit the entire exam fee. To reschedule a cancelled exam, you will need to submit a re-exam fee, and you will need to be within your one-year eligibility period.

9.3 Leading Up to Exam Day

Get on a Regular Sleep Schedule. Mental acuity is improved with a great night's sleep. Because PMI exams can be stressful and exhausting, I recommend candidates get a full seven hours of sleep each night the week leading up to the exam. This will allow your brain's suprachiasmatic nucleus (SCN) to synchronize your circadian rhythm. (Located in our hypothalamus, the SCN contains over 20,000 nerve cells in the area of our brain just above the intersection of our optic nerve and our eyes.)

Reduce Stress. The days leading up to the exam can be stressful, so this is sometimes easier said than done, as there may be a tendency to stay up late studying. Try to resist this temptation, instead study and practice taking exams at times that coincide with when your exam is scheduled. The days leading up to the exam should be relaxing and relatively stress free. If your exam is on a Tuesday, I recommend taking Monday off from work, perhaps getting some light exercise.

Visualize Success. Elite athletes, and top performers in business, know the power of visualization. Many rely on mediation, yoga, and positive thinking to establish a sense of well-being, to build confidence, and increase mental acuity, thereby giving themselves a competitive edge. Spend a few minutes each day mentally rehearsing all the events of exam day—it's a positive vision of your success. Before you fall asleep, tell yourself, "I'm going to pass the test"; thinking positive thoughts builds positive energy.

Study Smart. Students often ask, "What should I be studying the week before the exam?" My recommendation is to review the ECO along with the tools and techniques, flash cards, and memory games each day. Although another simulation test can build confidence, it may not be the optimal use of time.

Day 1 – Eve of Exam. The most common question I receive with regard to the day before the exam is, "Should I study or not?" In my opinion, reading new material may actually cause confusion and add doubt. Instead, practice memory aids, perhaps review a few flash cards and the chapter exercises, but don't introduce new material or concepts. To further aid memory recall, focus, concentration, and understanding of the questions, I recommend a brisk walk after dinner to decompress. Don't watch TV or play video games; these passive activities could actually impact your ability to get a restful sleep and contribute to a decrease in your mental acuity during the exam.

> **! Hint:** Don't forget to set your alarm clock; perhaps also set a backup.

Day 0: Morning of Exam Day. Eat a healthy breakfast, not one that's overly heavy; it should be something filling and nourishing, as it will be very hard to concentrate if you're hungry. My recommendation is to not consume an abundance of fluids, as bathroom breaks, although permissible, will count against the exam time.

> **! Hint:** If your exam site is a considerable distance away, perhaps plan for a small snack before entering the exam center.

> **☛ Tip:** Consider wearing two layers of shirts to provide you with some flexibility if the test center is either too warm or cold.

9.4 The Prometric Experience

Plan to arrive at the test center at least 45 minutes before the start of your exam. During one of my visits to Prometric, a bridge was out on the main highway. I couldn't help but think it would make a great anecdote, as I drove miles out of my way on the detour. This was relatively stress free, as I had given myself an abundance of time for travel and parking.

Upon checking in at Prometric, you'll be provided a key for a small locker. Remember to bring your driver's license, any required forms of secondary ID, and confirmation of registration.

> **☛ Tip:** At this point, I would recommend using the restroom, as I don't encourage breaks during which you leave the workstation during the exam.

! Hint: Depending on your test center, you may be seated earlier if there is availability.

When your seat is available, a Prometric proctor will have you sign in and verify that your identification matches the exam registration. They will pull out the dry erase sheet or scratch paper (if they have yet to convert to wipe-off sheets). In my recent experiences, I have been provided with a blank exam book approximately the size of two sheets of legal-size paper folded in half with two pencils.

> ☛ *Tip:* If you're provided with pencils, I recommend rolling the pencil tips to confirm they are not broken and are sharpened to your preference before entering the exam suite.

Once your identity has been confirmed, the Prometric proctor will screen you for any contraband. You'll be asked to pull up your pant legs and shirt sleeves and turn your pockets inside out. This screening also includes the use of a metal detector. They are about as thorough as the US Transportation Security Administration at the airport.

Once the proctor selects your test station, you will be escorted into the exam room. Your identification will be placed on the corner of the desk. This is done so that the proctors can confirm the identity of the person seated throughout the exam; you will be randomly observed by different proctors at random intervals, and your exam session is also recorded.

9.5 Exam Success

In the previous section, I suggested that preparing for the PMI-PBA® Exam is similar in nature to an athlete preparing for a game. In this section, it's GAME DAY—time to put all your hard work and studies into motion and **PASS THE TEST!** The true intent of this section is to now focus on execution.

Success doesn't just come and find you. You have to go out and get it.

— Kushandwizdom

Once the Prometric proctor has escorted you to your workstation and you've confirmed you're seated for the correct test, you'll have one minute to press Start to begin the 15-minute tutorial. During this minute's time, prior to the tutorial, I like to take a few deep breaths and center my focus. In some cases, the proctor will leave your workstation and the one-minute countdown clock will start; in others, they may wait for you to click Start to begin the tutorial.

> *! Hint:* While you should only need a short time to become familiar with the computer-based test, I advise not rushing through this tutorial, as there is no way to return once the exam has started. Use any remaining time to center your focus and relax.

> ✎ *Note:* Prometric and PMI prohibit candidates from using the tutorial time to create brain dumps and exam aids. Although permitted, notes may only be created once the exam clock starts.Upon clicking Start to begin the exam, take a few moments to create your cadence chart and memory maps.

> ☛ *Tip:* Prometric is going green, gradually shifting from paper booklets to dry-erase boards. However, if you've been provided with scratch paper that is stapled together, *under no circumstance should you remove the staples.* The material **must be used** in the same form in which it was distributed.

9.5.1 Start of The Exam . . .

The PMI-PBA® Exam is designed for practitioners, by practitioners and is not directly based on the Practice Standards or the *PMBOK® Guide*. Exam questions can be wordy and contain extraneous information. Following are a few hints, tips, and tricks to keep in mind throughout the exam.

Don't Skip Any Questions	Even if you have to guess, answer every question. Mark those for which you weren't 100% confident for review at the end.
Keep Score	If you're keeping score, mark your confidence for each question, or perhaps note your confidence level for those for which you weren't 100% sure of the answer. Either way, this is a proven technique to predict your score.
Track Your Cadence	This is an invaluable technique to assess if you'll finish with enough time for a review, or even perhaps take a break.
Know Your Role and Look for Key Words	Remember, *roles set boundaries.* In the question scenario, are you the sponsor; project, program, or portfolio manager; etc.?
Don't Over-Analyze	Everything you need to answer the question will be included in the context provided. Don't extrapolate by reading more into the questions or answer choices than is already there.
Eliminate Noise and Clearly Wrong Answers	Focus only on what the question is asking. Sometimes you may need to reread the question, starting with the last sentence, so that you can focus on what's important.
Don't Buy a Tool	Some answer choices may suggest spending money buying a tool or software package to solve the problem; more often than not this is a red flag.
Too Technical	Be very cautious of overly technical answer choices and remember to use domain-specific vocabulary.
What Would PMI Suggest?	A significant number of the exam questions are situational, based on scenarios that business analysts would encounter in the real world. Approach these questions from the practical perspective of a PMI business analyst—which in reality may be different from how you approach the role in your organization.
There Can Be Multiple "Correct" Answers	Read all of the exam question answer choices, and don't stop when you come upon one that seems reasonable. Focus on the best answer, which tends to be: • In the correct process order. • Doing the right thing. • Most relevant to the question.
Take Action	Why put off until tomorrow what you can do today? Consider the long- and short-term impacts of each answer choice.

Be a Leader	If you're calling a meeting to discuss a matter, come prepared with options to be considered.
Think Collaboration	One of the fundamental aspects of business analysis is the collaboration and handoff with the project manager.
Put Yourself in Other Shoes	If the question asks ". . . if the sponsor would like a debrief about a problem, should that occur via email, phone call, or perhaps in person?" As a sponsor, if reasonable, I prefer that to occur in person.
Take Breaks	Stretch at 50 questions, and cover your eyes for 30 seconds to clear your thoughts.
Elimination	In some cases, you may be able to quickly eliminate one or even two answer choices. Be prepared—answer choices have intentionally been designed to be very similar in both wording and length. In these cases, look for key words and distinction between choices.
Grammar	PMI is a global organization, and it's very possible exam questions will contain writing based on British, South American, and Australian English. Words such as *learnt* or *whilst* may occasionally appear.
Avoid Generalizations	*All, always, every, everyone, never, must*
Most Importantly	It's not about how things are done at work . . . it's about the PMI methodology. Thinking the "PMI way" will go a considerable distance toward passing the test.

9.5.2 Final Review

Within the few moments remaining, regardless if marked or not, please go back and review the first 10 questions, in addition to all flagged and low-confidence answers. My recommendation is to start with those marked as 75% confident (meaning you were able to eliminate two answer choices) and convert them to 90%–100%. Then try to convert those marked at 50% to either 75% or 90%–100%. When used consistently, this is a fairly accurate technique to improve and predict your overall score.

Keep in mind that some questions or answer choices may refresh your memory and help you with previous or upcoming questions. Take notes throughout the exam, perhaps writing down question numbers you think may be useful along the way. A word of caution, when debating between two answer choices, I generally don't recommend changing an answer unless you have a solid reason for doing so.

9.5.3 When You Click Submit

Once your review is complete, upon clicking Submit, you'll be prompted to complete a short survey about the test center, the exam, and the PMI credential. Upon completing the survey, a message on screen will indicate Pass/Fail. You can then leave the testing workstation and contact the Prometric proctor.

9.5.4 Celebrate Your Success

In the moments following, the proctor will hand you a printed exam summary, which they'll emboss with the Prometric seal. In two to three weeks, you will receive the certificate from PMI.

Congratulations on earning the prestigious PMI-PBA® credential; your hard work and dedication has paid off, and you've joined the ranks of professionals who are dedicated to improving project outcomes. If you've found this *Exam Practice Test and Study Guide* helpful and would like to continue the conversation, or if you have comments for future editions, please connect with me on LinkedIn, https://www.linkedin.com/in/briandwilliamson/, and drop me a note.

200-Question Practice Exam

Questions

This practice test is designed to simulate the PMI-PBA® Certification Exam. You have four hours to answer all 200 questions.

INSTRUCTIONS: Note the most suitable answer for each multiple-choice question in the appropriate space on the answer sheets beginning on page 241.

1. You are collaborating with Blake on the sequencing of activities for the next generation of air transport that your company is pioneering. As the project manager, she is inquiring about any dependencies that may exist in your product. You describe dependencies that are:
   a. Internal and external to the proposed solution
   b. Mandatory and discretionary to the proposed solution
   c. Mandatory external and discretionary internal to the proposed solution
   d. Represented by nodes and graphically linked to logical relationships

2. As the business analyst for a company that manufactures fire suppression systems, you're collaborating with a project manager on the development of the project management plan for a new state-of-the-art fire suppressor. It's the project manager's responsibility to ensure that requirements-related activities are performed on time and:
   a. Are completed within budget
   b. Are within the boundaries of the charter
   c. Are within budget and deliver value
   d. Are approved by the project management office (PMO)

3. In your role as business analyst for a company that manufactures roller coasters, you are preparing for an elicitation session with your primary stakeholders to discuss changes to an existing model. During these sessions, you will want to cover the following types of requirements:
   a. Functional and business
   b. Functional, service quality, solution, and business

c. Nonfunctional, transition, business, quality, stakeholder, solution, and functional

d. Business, functional, stakeholder, solution, transition, and other requirements that describe conditions or qualities required for the product to be effective

4. Your colleague, Alice, has asked for help defining objectives and goals that define context, provide direction, and are easily understood. From the samples provided below, which would you recommend be included in the business case?

a. To meet the organizational goal of reducing its carbon footprint by 10% next year; solar panels will be installed over all the parking lots and all the office buildings

b. To reduce organizational carbon footprint by 10%

c. To meet the organizational goal of reducing its carbon footprint by 10% next year; solar panels will be installed over all the parking lots by the end of Q2 and all the office buildings by the end of Q4

d. To install solar panels over all the parking lots by the end of Q2 and all the office buildings by the end of Q4

5. Cassius has been designated as your subject matter expert from the poultry processing plant. It's been incredibly difficult to schedule time with him to establish the evaluation metrics and measurement tools for the newly designed solution to euthanize turkeys. Rather than delay this meeting any longer, as the business analyst, you decide to:

a. Review the existing service-level agreement

b. Visit the plant unannounced and request a meeting

c. Conduct a review of the consensus criteria with the designated representatives from the plant

d. Perform an event analysis to first understand any potential concerns with the existing evaluation metrics and acceptance criteria, then schedule a meeting with Cassius

6. The executive sponsor of your wind turbine project asked that you present an analysis comparing potential solution options for the next generation of rotor blades made of carbon fiber, which would be commercially available in five years. Your analysis is based on which of the following?

a. Are the options presented aligned with the organization's requirements for sustainability and reliability; can the organization acquire the required technology; is it financially feasible; and can the solution be delivered in the timeframe outlined?

b. Would the project be based on the project management iron triangle considerations of quality as constrained by time, cost, and scope?

c. Would the analysis be based solely on business requirements, stakeholder requirements, and solution requirements?

d. Can the organization acquire the technology; is it financially feasible; and can the solution be delivered in the timeframe outlined?

7. Your company is working on a redesign of a motorized scooter, marketed for active individuals in their golden years. To better understand the concerns and limited adoption of the current model, you decide to take the scooter to the local supermarket and then to the mall for the afternoon. How would this activity be best characterized?

a. Passive

b. Participatory

c. Simulation

d. Active

8. You are mediating a session with Delaney, a well-respected customer, and Meilani, the head of the solution team, trying to come to terms on the final requirements. As a well-trained business analyst, you know there are several techniques for conflict resolution, some of which include compromising, withdrawing, and accommodating, but there can also be others; which one is crucial and most important?
   a. Accommodating
   b. Compromising
   c. Withdrawing
   d. Collaborating

9. You've requested that Charity draft a document outlining how changes to product elements will be completed and communicated over the life of the project. The document is a:
   a. Requirements management plan
   b. Requirements change management plan
   c. Requirements traceability matrix
   d. Requirements work plan

10. You are leading an approval session and are asking for formal signoff on the requirements documents. At a minimum, from whom should you seek formal signoff?
    a. Sponsor, product owner
    b. Product owner, solution team lead
    c. Business analyst, product owner, head of product development
    d. Sponsor, project owner, solution team lead

11. In a meeting with your sponsor, Ellie, she had recommended that you complete a RACI for the two highly visible projects that you're supporting for the multinational agricultural company. As you complete the RACI, who has accountability for providing input to the current state of the company and identifying any opportunities associated with the project?
    a. Your sponsor, Ellie
    b. Your sponsor, Ellie, and you in your role as business analyst
    c. Your sponsor, Ellie; product management; and you in your role as the business analyst
    d. Your sponsor, Ellie; product management; project management; and you in your role as business analyst

12. What tool or technique can be used to determine if a requirement cannot be satisfied without another requirement being present, and what is a potential outcome?
    a. Dependency impact assessment, product documentation
    b. Requirements traceability matrix, issue management
    c. Dependency analysis, traceability tree
    d. Requirements traceability matrix, product documentation

13. It's been requested that you work with Arabella, the head of zipper manufacturing, to address nonconformance concerns with the product. Upon meeting with Arabella and her team, you recommend:
    a. Preventative measures, focusing primarily on training
    b. Preventative measures, addressing the time to manufacture each zipper

    c. Reviewing the 7QC tools to determine which ones are most appropriate within the context of the Deming Cycle (Plan, Do, Check, Act)

    d. Creating a SIPOC diagram to identify the suppliers, the process inputs and outputs, and the customer's requirements

14. Your sponsor, Wyatt, suggested spending additional money during the project to avoid production failures. What are the elements associated with conformance costs?

    a. Building a quality product and destructive testing loss

    b. Prevention costs and liabilities

    c. Testing and liabilities

    d. Testing, liabilities, and rework

15. Amara has started to build the requirements traceability matrix, beginning with high-level items and filling in the details as they become known. What is this process referred to as?

    a. Work breakdown structure (WBS)

    b. Traceability matrix requirements

    c. Detailed traceability

    d. Progressive elaboration

16. Navi, an experienced programmer, has requested that Mahala, the software quality assurance lead, check her conceptual logic in a very complicated algorithm, asking that she manually record the results in hopes of uncovering any problems before she begins coding, which is expected to take several months. What technique will be Mahala be performing?

    a. Desk checking

    b. Algorithm review

    c. Conceptual code review

    d. Planguage review

17. Scarlett is writing an acceptance test for a user story related to a debit card transaction at a gas station. What can you suggest as the ideal format?

    a. Step-by-step verification, based on the Agile Manifesto.

    b. Verification based on the actors and personas.

    c. As a checklist for actions related to actors, personas, events, and logic.

    d. Given that a bank account has a positive balance, when a user processes a transaction less than the current balance, the transaction should complete without any error conditions.

18. Greta is hosting a video conference to elicit information so that she may thoroughly document the solution requirements. Interviews performed in this manner are:

    a. Asynchronous

    b. Unstructured

    c. Synchronous

    d. Structured

19. Nyla, the business analyst for a hedge fund, has team members located in 12 countries on four continents. Which of the following should not be a concern for her global team?

a. Will team members work at locations other than their primary location, and what is a typical work day?

b. How will team members contribute to the decision-making process and address conflict?

c. What is the approach to documentation analysis?

d. How will requirements be elicited, managed, maintained, and approved?

20. Past projects at Soleil, Aubrey & Hart LLC, a leading management consulting firm, have been tumultuous, resulting in losses in the tens of millions of dollars. In a turnaround effort, the new chief project officer, Garth, recognizes that the organization must focus on key areas to improve the effectiveness of requirements management. As his business analyst, you would suggest:

a. Skills development, formalization and standardization of processes, garnering the support of senior leadership

b. Implementing the principles of business analysis and requirements management

c. Using a business analysis plan, a requirements management plan, and change control

d. Following the processes and using the tools and techniques from PMI publications

21. As you verify that the delivered product is fulfilling the outlined requirements, you make note of them, referring to:

a. A checklist in your team's collaboration portal

b. The project management plan

c. The schedule management plan

d. The time management plan

22. You've been assigned to work with Genevieve, the supervisor for the pipefitters at a high-rise office building in midtown Manhattan, to understand why the water pipes are losing pressure. Following a walkthrough, you create an Ishikawa diagram and determine the root cause of the problem:

a. The journeyman's apprentice assembled the areas of the pipes where defects were found.

b. Will necessitate spending money to avoid project failures.

c. Will result in a claim to the trade union.

d. Will be categorized as an internal project failure.

23. Saoirse is the business analyst working in the corporate accounting department of a Fortune 100 multinational company. She is assisting with the deployment of "Hyacinth P&L," a software package designed to consolidate financial information from accounting systems in 106 countries. As part of a companion project, she'd like to ask her global colleagues for their input on an assorted number of topics. What is the most efficient way to engage her colleagues?

a. Send them an email containing the questions, requesting they respond in kind.

b. Schedule an audio conference at a mutually convenient time for all participants.

c. Have your assistant Luna prepare a Survey Monkey.

d. Plan a focus group with all participants within the next week.

24. Talia has witnessed a dramatic change in her organization since they embraced business analysis. They are continuously monitoring capability and capacity, and they are identifying and implementing changes to improve project delivery. It can be said that the organization, its people, and its processes and tools are:

a. Maturing
b. Static
c. Aligned
d. Focused

25. Everett has drafted documents that accurately describe the solution to be built. It can be said that his requirements documents are:
a. Complete
b. Precise
c. Unambiguous
d. Correct

26. You work for the Great Brick Company, which for the past 150 years has had a reputation for producing the strongest, most durable bricks on the market. Over the last few months, contractors have reported sporadic cases of bricks cracking and crumbling. You have been asked by Oliver, a senior product manager, for assistance determining the cause of the defect. To begin the analysis, you decide to start with a:
a. Monte Carlo analysis
b. Tornado diagram
c. Pareto chart
d. Ishikawa diagram

27. Construction has just completed at your executive sponsor Bobby Phil's new restaurant in downtown Manhattan. During design, a number of requirements were identified, such as sustaining oven temperature, color temperature ranges for all the guest areas, and required refrigerator and freezer temperatures. Now the project team is preparing to train employees for the new restaurant. On what document would the training requirements be identified?
a. Training and cutover requirements
b. Business requirements
c. Nonfunctional requirements
d. Transition requirements

28. Your sponsor, Alessandra, would like to use a tool that is contingent on peer pressure to resolve a conflict. You advise her:
a. This sounds a little Machiavellian and should not be a part of any sound method for resolving conflict.
b. We can use the Delphi tool.
c. The Nominal Group Technique would be an appropriate solution.
d. Take a step a back and think this through and evaluate other options, because peer pressure is not part of any sound method of resolving conflict.

29. Your sponsor, Callum, is familiar with the terms *quality assurance* and *quality control*. When he suggests that you identify the causes of metal failure in the company's bolts and recommend corrective action, he's referring to:
a. Quality analysis
b. Control charts

c. Quality control

d. Pareto Diagrams

30. Paladin is a business analyst for Anwen Foods Inc, a world-class manufacturer of kitchen equipment for pizza restaurants. They are planning to expand their products to the Korean barbecue market, and in the process, they would gain a better understanding of how competitors' products are used. What form of observation would best help them to experience how their product would be used and to ascertain the criteria for acceptance in the market place?

a. Active

b. Simulation

c. Passive

d. Participatory

31. Your sponsor, Zander, requested a meeting with you and the project manager, Carter, to discuss documenting the requirements change management process. As you prepare for the meeting with Zander, you and Carter agree:

a. As the business analyst, you'll take accountability for documenting the plan, and it will become part of the project management plan.

b. As the PM, Carter will document the plan as a part of the business analysis plan.

c. As the business analyst, you'll document the process and communicate it to the stakeholders.

d. You and Carter will jointly develop the approach and communicate it to all relevant stakeholders.

32. Miriam has just joined the project team in the lead role responsible for coordinating and monitoring testing as the third round of system integrated testing commences. The team is planning to conduct ten rounds of testing for this highly complex automatous vehicle. As the team progresses to later rounds of system integrated testing, they are questioning why new issues are emerging.

a. New issues are emerging as new functionality is being tested.

b. Prior to Miriam's arrival, the test plan was inadequate.

c. Although the team was conducting testing from the start, they were not charting their results.

d. Due to the complex nature of the testing, the team needed to purchase a tool to automate the testing, which has uncovered the new issues.

33. Alyssa has found that subject matter experts are unsupportive of the business analysis activities detailed within her plan, and there was a misunderstanding regarding their level of involvement for the requirements activities. How could Alyssa have avoided this situation?

a. She could have involved the subject matter experts when defining the timing and sequence of activities.

b. She could have involved the broader project team when defining all the activities.

c. She could have collaborated with the project manager on the approach to business analysis activities.

d. She could have sought approval of the business analysis plan.

34. After several weeks of effort, you have just completed the business analysis work plan for a highly complex project. Following a review session with the project team and key stakeholders, your project manager, Flynn, has expressed concern about full integration with the project management plan. To simplify matters, you suggest that:

a. The fine details can be tracked in a separate plan; summary information should be integrated with the overall project management plan.

b. Because a majority of the project is focused on requirements, all the details can be integrated, and as the business analyst, you will maintain the overall plan.

c. Considering the highly complex nature of the project, we can maintain two separate project plans.

d. To properly manage the project, both plans need to be fully integrated.

35. The business lead for the human capital thread of your program recommended a number of changes to the Change Control Board (CCB). With minimal supporting documentation, the board approved all changes. Now during system integrated testing, your team is uncovering issues. What could have been done to ensure the CCB knew the full picture?

a. The business lead could have conducted an assessment to understand the relationships and considerations to related development objects.

b. The program manager could have asked the business analysts to provide their recommendation to the Change Control Board in person rather than via email.

c. The Change Control Board should have deferred approving all change requests until after system integrated testing.

d. The Change Control Board should have adopted a policy of no changes unless they are foundational to the organization's achieving its business objectives.

36. Your subject matter expert for the order entry team, Preston, is describing how a button should work on a redesigned web page. How should these requirements be classified?

a. Business

b. Nonfunctional

c. On the requirements traceability matrix

d. Functional

37. You are the business analyst collaborating with your project manager, Graham, on a project that, when complete, will affect nearly all the residents in a community in Iowa. During your needs assessment, you recommend completing a RACI. Graham asks, what's a RACI? You explain:

a. It stands for Revitalization Assistance for Community Improvement, a document the sponsor requested be included in the charter.

b. It stands for Risk-Adjusted Complication Index, an analysis to determine exactly how many residents will be affected by the project.

c. It is a type of responsibility assignment matrix.

d. It stands for responsible, accountable, consulted, informed.

38. Your college intern, Maverick, inquires as to who can request a significant change to the approved requirements traceability matrix. You share:

a. Once documented, only the customer can request a significant change to the approved product.

b. The Change Control Board must review and vote on the change request.

c. Any stakeholder can request a significant change to the approved product, provided the request is documented.

d. Any team member can recommend a change, provided the request is documented.

39. Your sponsor, Selena, has offered to assist with your efforts to plan for solution evaluation. From the answer choices below, which should not be included as part of your efforts?
    a. Evaluation criteria, acceptance thresholds, how results will be analyzed and reported
    b. When and how often evaluation will be performed
    c. Special measurement tools not used as part of solution evaluation
    d. Focus groups, observations, surveys, qualitative and quantitative activities

40. Aryan is an experienced business analyst and highly skilled in elicitation. Tobias, a colleague leading a focus group that Aryan is attending, is having difficulty extracting key requirement baseline information from attendees. During a brief 15-minute break, Aryan suggests Tobias try:
    a. Using only closed-ended and context-free questions
    b. Primarily using contextual and context-free questions, with the occasional closed-ended question
    c. Concentrating to use only open-ended and context-free questions
    d. Using a combination of open-ended, contextual, context-free, and closed-ended questions

41. Collaborating with Rhett, your project manager, you have logically decomposed the total scope of work related to the product in a hierarchical manner. What was the output of your effort?
    a. Process model
    b. Work breakdown structure
    c. Capability table
    d. Salience Diagram

42. The VP of human resources, Septimus, convenes a meeting with Adalyn, Buster, Geneva, Madoc, Adrienne, and Cade, intending to discipline them for playing cards at their desk. What is the appropriate response?
    a. Septimus must refer to the corporate policy for playing poker during working hours.
    b. Adrienne and Cade suggest they were playing Iterative Manjaro, a card gamification technique to arrive at consensus.
    c. Geneva and Madoc offer that the cards weren't based on suits, but rather on the Fibonacci sequence, and were being used to estimate the project.
    d. Adalyn and Buster refer Septimus to the team's Scrum master for clarification, justification, and support of their activities.

43. You are the business analyst for Xgen, a global pharmaceutical company supporting the ERP upgrade project. You are four months from go-live and have spent $20 million USD to date. The estimated cost to complete the upgrade is $525,000 USD, all associated with training, as all the development is complete. Your company has just merged with a multinational conglomerate, which completed a similar project last year. You suggest to the project sponsor and CFO:
    a. The project should continue as planned; once live we can begin depreciating the capital.
    b. We should complete a needs assessment using the results to update the business case and project charter.
    c. Using virtual classrooms, training can be fast-tracked, thus delivering the project ahead of schedule and under budget, thereby allowing the team to devote time to the merger.
    d. The CFO should meet with the Technology Subcommittee of the Board of Directors to request approval to hire a consulting firm to provide their assistance with the merger.

44. While building the traceability matrix, Ada has found a number of development objects that seem to be related. In the process of conducting a dependency analysis, she documented:
    a. Implementation, subset, and benefit dependencies
    b. Value and implementation dependencies
    c. Benefit, subset, and value dependencies
    d. Implementation dependencies and related objects

45. You work for a clothing manufacturer that is looking to increase both market share and time to market, considering the recent delays in ocean freight. Your product manager, Ophelia, suggests exploiting 3D printers and selling the pattern code for the clothing, rather than the clothing itself. You think back to business school—is this a "Blue Sky," "Blue Ocean," or "Red Ocean" concept? Either way, to assess the viability, you decide to start by:
    a. Meeting with all internal stakeholders to conduct a Rationalized Group Technique (RGT) session and outline the opportunity.
    b. Conducting a study of the potential opportunity to determine the viability
    c. Developing a business case with Ophelia and presenting it your sponsor for consideration
    d. Conducting a proof of concept using the Monte Carlo Technique

46. Crispin is helping you build and maintain the stakeholder register for a highly regulated generic pharmaceutical. As the list is quickly growing, how can the register be organized to simplify the stakeholder communication and engagement activities?
    a. Crispin should first identify the target audience, then create an interest table.
    b. Crispin should create an interest table, followed by a power/interest grid.
    c. You should consult with the project office to determine how they would prefer the stakeholder register to be organized.
    d. Crispin could add designations and group stakeholders to simplify engagement.

47. In your role as a business analyst for Khaleesi Consulting LLC, you are talking with Jacob about the last iteration of software development. The team has just completed testing, and they are quite satisfied with the results. Who has accountability to document the results?
    a. Jacob should suggest that you document the results and post them to the team's collaboration site; this is a business analyst's responsibility.
    b. As the project manager, Jacob offers to take the lead and document that stakeholders are satisfied with the results.
    c. You suggest that Mary, the project coordinator and communication lead, update the information in the team's collaboration site.
    d. Jacob suggests that the subject matter experts who performed the testing have accountability for documenting and posting the results to the team's collaboration site.

48. Following a test cycle, Camille is working to understand the relationship that may exist between two variables. What type of analysis is Camille conducting?
    a. Quantitative analysis
    b. Data capture and logging analysis
    c. Regression analysis
    d. Organizational analysis

49. You are working with Yvaine and Fritz to create a presentation to leadership on the key elements of the service- and operational-level agreements. When building this presentation, you want to ensure that:
    a. Each slide has one topic, with six bullet points and a maximum of six words per bullet (1-6-6 Rule).
    b. Each slide is engaging, with text minimized.
    c. Each slide has one topic, with five bullet points and a maximum of five words per bullet (1-5-5 Rule).
    d. You consult with your communications department for guidelines on presentations.

50. Rhys believes his requirements documents are complete and is preparing to present them to the Project Steering Committee for review. Before doing so, what should he ensure that they all contain?
    a. All necessary business requirements
    b. Requirements identified that produce measurable outcomes, plus all necessary requirements
    c. All necessary requirements
    d. Labels and references to all figures; requirements identified that produce all necessary outcomes; responses identified for each input; and all necessary requirements

51. The project team has a very elaborate SharePoint site to post team documents and to collaborate, although it's referenced infrequently for requirements traceability information, because your spreadsheet has become the single source of truth. In addition, maintaining the SharePoint site is very time consuming, and it's become stale over the last few months. The Change Control Board (CCB) has just authorized the addition of new attributes that will further categorize business value. At a minimum, what should you do?
    a. Update your spreadsheet with the new attributes, plus all the associated projects as directed by the CCB.
    b. Update your spreadsheet with the new attributes, plus the associated projects as directed by the CCB, and communicate this update to your project team.
    c. Update your spreadsheet with the new attributes, plus the associated projects as directed by the CCB, and communicate this update to your stakeholders.
    d. Update your spreadsheet and the team SharePoint site with the new attributes, plus the associated projects as directed by the CCB, and communicate this update to your stakeholders.

52. Using actual data from a prior similar project, Kelsey is attempting to validate estimates provided to her by lead subject matter experts. What estimation technique is she using?
    a. Parametric
    b. Expert
    c. Relational
    d. Analogous

53. Your sponsor, Elian, and the VP of new product development, Fawn, would like to establish metrics to help evaluate whether the delivered solution is achieving its intended goals. What can be used to quantitatively evaluate the solution?
    a. Project management information system
    b. Key performance indicators

    c. Metrics and acceptance criteria

    d. Approved requirements documentation

54. Your organization has decided to undergo a business transformation to improve efficiency and eliminate waste. As part of this initiative, they are considering implementing optical character recognition (OCR) technology in their accounts payable department. Unfortunately, due to organizational turnover, no one truly understands the complete invoice/receipt-to-pay process. What steps can you take to learn more about the process?

    a. Create process models documenting the "as-is" state; conduct interviews with the AP Department; review any existing documentation.

    b. Observe the AP clerks; conduct interviews with stakeholders; develop documentation outlining the "to-be" state; create process models documenting the "as-is" state.

    c. Observe the AP clerks; conduct interviews with stakeholders; review any existing documentation; create process models documenting the "as-is" state.

    d. Observe the AP clerks; conduct interviews with the AP department; review any existing documentation; create process models documenting the "as-is" state.

55. Ace is an expert in business analysis and knows that feasibility is a key characteristic of a properly written requirements document. From the list provided below, what is one factor that can determine the feasibility of a proposed solution?

    a. Operational feasibility.

    b. Technology feasibility.

    c. Time and cost feasibility.

    d. Feasibility can best be determined by assessing a variety of factors.

56. Working with your project manager, Brooklyn, you have just delivered your fourth project in 36 months using a waterfall delivery method. You've now been asked to lead a highly visible project based on Scrum with nearly the same project team. In terms of change control, the team has agreed:

    a. That Brooklyn will document the process in the change management plan

    b. That they will adapt a flexible approach to change control, because the team anticipates that requirements will evolve as the project progresses

    c. To use the same process and templates as with all their other projects

    d. That they will defer to the project management office for guidance on the process

57. Sadie has a meeting later in the week to discuss solution requirements for a new line of organic thread at the textile mill. Prior to the meeting, where can Sadie look to gain insight into the organization and prior initiatives?

    a. The business analysis plan

    b. The project management information system

    c. The business analysis plan, requirements management plan, and plan for change control

    d. The requirements management plan

58. Your project manager, Ciaran, has suggested that the team consider using a tool based on two axes representing opposing viewpoints and interests, essentially creating a table with four cells to enable informed decision making. What tool is Ciaran suggesting the team use?

a. An options analysis

b. A decision table

c. A quadrant analysis

d. A strategic table

59. During a focus group at the law firm of Adelaide Nolan LLC, the team is having difficulty arriving at a decision between two options. To help the team evaluate a Yes/No decision based on the option pair, Jane, your subject matter expert in torts, suggests that they should consider using:

a. A mediator who can objectively assess each position and provide a recommendation

b. The Business Process Modeling Language (BPML) to optimize the process

c. A method that will enable the team to reach consensus, whereby each participant will document the advantages and disadvantages of their position

d. The requirements modeling language to visually show the solution requirements linked to the objectives and goals of the organization

60. Rebecca is preparing a project charter for a new line of dehydrated fruit. She has all the information from the needs assessment that you conducted, but she would also like to include information pertaining to costs and benefits, as well as to explore many of the aspects related to the recommendation. What should you do?

a. Work with a financial analyst to prepare NPV, IRR, and ROI assessments

b. Provide Rebecca with the business drivers, economic viability, and success criteria for the recommendation

c. Provide Rebecca with an impact analysis

d. Work with Rebecca to outline the business goals and objectives

61. Imogen, the VP of strategic marketing and customer engagement, would like to strengthen the relationship with customers and identify opportunities to improve the company's website. Wanting to focus externally, she commissions you to create a:

a. Ecosystem map

b. Customer ecosystem map

c. Value engineer map

d. User journey map

62. Your sponsor, Simba, is new to the concept of business analysis. Although you've explained that requirements are the sole justification for the existence of a project, he's asked for additional context. From the answer choices below, which would you not provide?

a. Business analysis reduces project risks.

b. Business analysis sets expectations with stakeholders as to activities that will be performed.

c. Business analysis activities will be planned upfront in sufficient detail for the duration of the project.

d. An overabundance of business analysis planning can negatively impact the project.

63. Aurelia would like to create several archetypes so that the project team can design an effective software solution. She would like to stay away from the typical outline commonly found in use cases. Which of the below could you recommend for a project in a hospital?

a. General practitioner doctor, general surgeon, specialist surgeon

b. Administrative support, clinical staff

    c. "Abigail Kian," "Eliana Bennett," "Savannah Knox"

    d. Overhead, revenue generating

64. Your organization is deploying a new, state-of-the-art website. Working with Penelope, an SME from marketing, you begin a process of outlining resources who will implement and support the system, along with those who will use and benefit from the website. You are:

    a. Identifying the transition and operations team, along with customers so that you can properly plan a go-live event

    b. Identifying system users who will support and use the system

    c. Identifying team members; as go-live approaches, all resources will be asked to provide additional support to ensure a smooth cutover

    d. Identifying individuals and groups who may be affected or perceive themselves to be effected by the launch of the website

65. Your consultant, George, has inquired if further review or approval is required as the payroll team submits their product documentation. To what artifact can you refer George?

    a. Product or solution management plan

    b. Scope management plan

    c. Change control plan

    d. Requirements management plan

66. Your subject matter experts, Zane and David, who are both certified public accountants, have a disagreement as to how the core financial management system should be structured. After lengthy discussions, it is determined that they would use the data warehouse for reporting. This is an example of what conflict management technique?

    a. Compromising

    b. Smoothing

    c. Agreeing

    d. Collaborating

67. Prior to meeting with stakeholders to estimate the work effort for a project, Bailey, an experienced business analyst, decides to review planning information from prior projects of similar size. What elicitation techniques could she rely upon?

    a. Document analysis

    b. Alternative analysis

    c. Interviews

    d. Focus group

68. Following a very long and sometimes challenging requirements confirmation process, you are now asking for sign-off. You anticipated that this would be a routine process, but subject matter experts Eleni, Fia, Behati, and Delta are unable to provide an opinion to Griffith, the senior VP of consumer plastics. Why are they unable to provide an opinion?

    a. While they signed off on the requirements, they were not involved in the confirmation sessions.

    b. They were not listed on the RACI matrix.

    c. They were not involved in any of the requirements sessions.

    d. They only participated in didactic interaction sessions.

69. In the development of a business case, you're working with Daisy, an experienced financial analyst, on the cost–benefit summation. Which financial valuation method addresses the time to recover a project investment?
    a. Net present value (NPV)
    b. Return on Investment (ROI)
    c. Internal rate of return (IRR)
    d. Payback period (PBP)

70. You're presenting a business case to your executive sponsor, along with your financial analyst, Olive, highlighting the financial valuation methods used to justify the investment. Your sponsor, Colton, asked that you expand the analysis to include the initial and ongoing costs, along with the projected annual yield of the investment. Upon returning with Olive, you decide to prepare:
    a. An internal rate of return (IRR) analysis
    b. A return on investment (ROI) analysis
    c. A present value versus future value (PV vs FV) assessment
    d. A net present value (NPV) analysis

71. Julius, a college intern, has inquired as to the differences between the requirements management plan and the business analysis plan. How would you best answer his question?
    a. The requirements management plan is a subsidiary plan of the overall project management plan, whereas the business analysis plan is a complementary artifact.
    b. The requirements management plan covers the project, whereas the business analysis plan is focused on the effort associated with business analysis.
    c. Both documents are complementary artifacts.
    d. The requirements management plan covers both the project and product, whereas the business analysis plan is focused on the effort associated with business analysis.

72. As part of the United States Antarctic Program, Callie is stationed at the Amundsen–Scott South Pole Station as a research scientist. Your team is in the process of building a software application that will improve the accuracy of weather modeling, with a significant contribution to geosciences. As you are based in the United States, and there is little opportunity to conduct live interviews to review evaluation metrics and acceptance criteria, what method of elicitation might you consider?
    a. Unstructured
    b. Asynchronous
    c. Structured
    d. Synchronous

73. Your team comprises very skilled individuals from 30 countries, all of whom work remotely, except for a semiannual company meeting. As the business analyst, while performing due diligence to ensure that the requirements remain aligned to the evolving needs of the organization, you want to focus on one key element:
    a. Addressing feelings of isolation
    b. Establishing tools for effective decision making
    c. Ensuring that the appropriate technology is in place
    d. Establishing parameters for effective and consistent communication

74. Natalia works as a business analyst for a highly regulated hedge fund, Kinsley & Associates LLC. She's profoundly aware of the need to link elements of the solution to the organization's policies and procedures. To do so, what model could Natalia use?
    a. Interrelationship diagram
    b. State table
    c. Decision table
    d. Entity relationship diagram

75. In an effort to leverage the significant investment in an enterprise software application, uncover pain points, and optimize the business processes, your CFO, Lydia Maxwell, requests that you perform what type of analysis?
    a. S.A.V.E. assessment
    b. Value engineering assessment
    c. Optimum value assessment
    d. Transformative value assessment

76. Azriel, your summer intern, is helping you plan the requirements approval and confirmation sessions. As you create the slide deck, you realize:
    a. These should be two distinct meetings, with approval occurring only after the solution is confirmed.
    b. In the interest of time, both topics can be discussed concurrently.
    c. The slides should be created following the 1-6-6 rule, each slide has one topic, with six bullet points and a maximum of six words per bullet.
    d. Your sponsor should start the session with a warm opening, setting the stage.

77. In your business case, you've included an analysis outlining the initial and ongoing costs, the projected annual yield of the investment, and an assessment addressing the time to recover the project investment. Prior to returning to meet with your executive sponsor, Hugo, Kate, your financial analyst, suggests that you include a section comparing the amount of the investment to the future value of the expected benefits. You are discussing a:
    a. Net present value (NPV) analysis
    b. Return on investment (ROI) analysis
    c. Payback period (PBP) analysis
    d. Internal rate of return (IRR) analysis

78. You are a business analyst for Daenerys Electronics Co, a sole-source supplier to Wolf Automotive Ltd, a hypercar company catering to the ultra-wealthy. Wolf has decided to move production from Frankfurt to Düsseldorf, Germany, some 232.2 km via the A3. Your factory is located in Hachenburg, Germany, geographically about halfway been the two factories, but there is an increase in travel time. How should the business analyst for Wolf address Daenerys Electronics' interests?
    a. As a project requirement, ensure that all the processes and conditions are met for a successful transition.
    b. As a component of nonfunctional requirements, ensure that there is no disruption to the just-in-time order process, by suggesting a buffer to the travel time.
    c. As part of stakeholder requirements, address the quantifiable interests of Daenerys.
    d. In the transition requirements plan, ensure that Daenerys is aware of the new ship-to address, and the relevant change in travel time.

79. Naomi is responsible for maintaining the requirements traceability matrix. At the end of each week, she would like to update the project team as to status and progress related to the development objects. How would this be most effectively accomplished?
    a. Documentation management
    b. Status reports
    c. Communications management
    d. Integrated change control

80. Ava is in the process of updating the configuration management systems with items that directly impact project work. From the list below, what should not be included?
    a. Building material from China will be delayed due to an inclement weather event.
    b. The HR team will not complete functional specifications in time due to competing priorities.
    c. The non-production application server sporadically reboots.
    d. A request to accelerate the timeline to secure a reduction in materials cost.

81. Your senior business analyst, Toulouse, is creating an entity relationship diagram to visually represent objects and their relationships. While building the diagram, you remind him to notate the ordinality. He agrees and proceeds to update the diagram by:
    a. Adding composite attributes to the strong and weak entities
    b. Adding derived attributes to the multivalued relationships
    c. Indicating the minimum number of times an instance in one entity can be associated with instances in a related entity
    d. Drawing the maximum number of times an instance in one entity can be associated with instances in a related entity

82. You're collaborating with Reese, an experienced project manager, on the business analysis plan. With regard to the section on change management, which of following plans addresses who will provide information to stakeholders following a meeting of the Change Control Board?
    a. Human resource management plan
    b. Stakeholder management plan
    c. Resource management plan
    d. Communications management plan

83. You work for a paper mill and have been asked to fill the role of PM/BA on a new self-sealing envelope project. Your lead subject matter expert, Milos, seems to be introducing requirements that are not in line with the paper mill's business needs. What should you do?
    a. Review the requirements with your sponsor
    b. Track the requirements on a matrix
    c. Ask Milos to provide the business justification
    d. As the project manager/business analyst, defer to Milos as the subject matter expert

84. In the process of reviewing requirements documents, Caroline uncovers several related objects with her subject matter experts. What is one way to visualize the order and reliance of objects?
    a. Caroline and her subject matter experts can build a process flow diagram.
    b. Caroline can create a dependency graph.
    c. Caroline and her subject matter experts can use a requirements traceability matrix.
    d. The team can use an Ishikawa/fishbone diagram.

85. Your executive sponsor, Enzo, asked that you collaborate with Cillian, a veteran project manager, on a key document that will outline the need for action, the root causes, and the main contributors of the problems, along with the rank order of the recommendations. Cillian suggests that Enzo is referring to a:
    a. Business case
    b. Situation statement
    c. Project charter
    d. Capability framework

86. Your data conversion lead, Wolfgang, has just reviewed the requirements traceability matrix and identified all aspects to consider, as per the organizational records and retention policy. Where should these requirements be identified?
    a. Business requirements document
    b. Stakeholder requirements document
    c. Transition requirements document
    d. Functional requirements document

87. Claire works for ZooZle, a complex internet search and news company. While documenting solution constraints, what might be something she needs to keep in mind?
    a. Access to the search engine and news may be limited based on the country of origin.
    b. The success of the solution is based on a future event.
    c. All product team members may not stay to witness the delivered solution.
    d. The iron triangle of quality as constrained by time, cost, and scope.

88. The Agile software development methodology is new to your organization, and you are very excited to be part of the team. You've been asked to work with Oscar, a user-interface expert, on the actions for each button that is to be part of the order-entry screen. When selecting the template to track your requirements, you choose:
    a. The W3C Standards template from the Open Web Platform, as this will ensure cross-platform compatibility
    b. A template from your project management office that tracks the behavior of the screen and buttons
    c. The agile burndown and burnup template
    d. A value stream map, as one of the guiding principles is to reduce customer wait time and eliminate non-value-added time

89. During an interview, your executive sponsor, Adelaide, is providing very detailed responses but does not seem to answer the questions in a refined, direct manner. In an effort to reframe the interview, you've requested that Adelaide respond based on validation. What result will this yield?
    a. Responses will be used as lead-ins to follow-up questions.
    b. It will produce responses based on forced-choice answers.
    c. Responses will be based only on the current topic of proposed solutions.
    d. Adelaide will quantitatively validate or substantiate her responses.

90. You've been invited to meet with the CEO of Blythe Consulting LLC, along with several other senior leaders of the product development team. The team has just completed the baseline of the

requirements, and Brian Blythe begins the meeting by reviewing the project justification. As the business analyst, where should you document this information?

a. In the meeting minutes

b. As part of the solution scope statement

c. In the business case

d. In the business requirements document

91. You are the subject matter expert representing the distribution logistics center of a textile company. Your business analyst, Nathaniel, is facilitating sessions with technicians Austin and Scarlett on the required training and associated operational changes for a proposed relocation. What type of requirements is Nathaniel eliciting?

a. Technical requirements, as the relocation will require moving several servers and phone systems

b. Operational requirements, as this is more than an IT initiative—the entire business unit is relocating

c. Transition requirements, as the logistics center considers the capabilities needed to migrate to the future state

d. Foundational requirements; essentially, the business process drives the training for the proposed operational changes

92. At the request of the operations business lead, the Change Control Board (CCB) approved a $250,000 USD workaround, based on information from the requirements documentation. Once the development team started to review the details of the workaround, their estimate nearly tripled. At an emergency meeting of the CCB, your sponsor, Charles, detailed the issues with the change approval process, asked for the team's suggestions on how to improve, and inquired whether there was any part that was working well. Why?

a. Charles wanted to prevent further cases of improper estimating.

b. The business units were exploiting a weakness in the change approval process.

c. Charles found the process to be too light and wanted to add additional rigor.

d. Charles was conducting a retrospective.

93. You work for an IT service organization that was recently certified as ISO 9004 compliant. As a result, your director, Tuesday, would like to update service quality, performance, and response guidelines for both new and existing customers. What artifact describes these metrics?

a. Service-level agreements

b. ISO 9004 requirements

c. Quality management plan

d. Requirements management plan

94. Elise, a senior project manager who is also filling the role of business analyst, is collaborating with Atticus, the subject matter expert from the mayor's office, on a document that will outline the compliance requirements for a new procurement application. What document are they working on?

a. Solution requirements

b. Quality requirements

c. Business requirements

d. Stakeholder requirements

95. Following a meeting with senior leadership, your executive sponsor, Brennan, shared privately that the company was being acquired by another firm. He expressed concern that this action could have significant implications on your product and didn't want to spend money unnecessarily on the project. As a result of the meeting, you decide to:
    a. Collaborate with the project manager on ways to inform the team and place the project on hold pending further guidance from your sponsor
    b. Review the project management plan
    c. Continue the project until the information is made public
    d. Convene a meeting of the Change Control Board to discuss the implications

96. You're facilitating a workshop for Alexandra and several other key stakeholders at QQQ Manufacturing Incorporated. You've just completed outlining all the key stakeholder needs and are now trying to determine the critical characteristics for a new product. What technique are you most likely using?
    a. Quality function deployment
    b. Voice of the customer
    c. Quality assurance
    d. Joint design/development

97. Ziggy is preparing to present the business analysis plan to key stakeholders for final approval following several rounds of meetings, during which the attendees continually questioned the rationale and were having difficulty justifying the resource commitments. What could Ziggy have done to prepare stakeholders prior to the initial approval meeting?
    a. Sent out an advance copy to all invited attendees requesting their support
    b. Sent out an advance copy to all invited attendees requesting they send you any questions in advance
    c. Met with the project manager to review a draft copy of the plan
    d. Met with his sponsor to review a draft copy of the plan, along with a few bullet points outlining the justification

98. Caius is working to categorize the models that will be used while the team is drafting specification documents. Entity relationship diagrams, data dictionaries, and state tables are all considered:
    a. Models
    b. State models
    c. Data models
    d. Process data models

99. Working with Ezekiel, your executive sponsor and VP of distribution and logistics for a pet food company, you've both agreed there are two options for addressing the opportunity. Ezekiel suggests using a method whereby team members can vote to determine the most suitable or preferred option. You suggest:
    a. A matrix in which options are weighted and ranked, with criteria that align with objectives set forth in the needs assessment
    b. Planning poker, allowing stakeholders to vote and score options
    c. A needs assessment
    d. A scored capability table in which each team member's vote is weighted in proportion to their role in the organization

100. At what appears to be an interval of every 15 days, your product owner is reprioritizing the work to be addressed in the next iteration. What do they hope to achieve?
    a. They are performing a reprioritizing analysis.
    b. They are performing a backlog reprioritizing analysis.
    c. They are grooming the backlog to manage scope.
    d. They are preparing a sprint analysis to manage scope.

101. Aria, the subject matter expert for a new line of autonomous vehicles, is impressed with all the features and highlighted functionality. However, after the vehicle started, the windows began to roll up and down by themselves and the radio's volume was acting erratically. When she attempted to turn the car off, the sunroof opened. As the SME, is Aria more concerned about grade or quality?
    a. Quality: all the defects render the product useless.
    b. Grade: all the defects render the product useless.
    c. Quality and grade: all the defects render the product useless.
    d. Neither: Aria is concerned about precision and accuracy.

102. Pippin is the business analyst on a Kanban project. What tool can she use to report the remaining items to be addressed?
    a. Burndown chart
    b. Estimate at completion
    c. Estimate to complete
    d. Variance report

103. Amelie, a recent college graduate, is assisting you in identifying the cardinality and multiplicity of relationships. What is a common manner of displaying these relationships?
    a. Entity relationship diagram
    b. Crow's foot notation
    c. Context diagram
    d. Wireframe

104. Connor, the chief engineer for a microprocessor that will be used in a new line of autonomous bicycles in New York City, documents a problem during the last round of testing prior to go-live. Connor is very outspoken, and many team members defer to his opinion. As the business analyst, you:
    a. Interview each of the stakeholders involved with testing and document their results.
    b. Repeat the system and integrated testing cycles.
    c. Facilitate a MultiVoting session.
    d. Consult with the project manager and escalate the concern to the project sponsor.

105. You're collaborating with your project manager, Caspian, on creating an illustration that supports the prioritization of solution elements based on their value to the business. Impressed by your well-thought-out model, your sponsor, Cecilia, inquired as to the name of the model. You advise:
    a. Business Objective Model
    b. Business Rule Model
    c. Business Value Model
    d. Business SWOT Model

106. It would appear that the team has reached an impasse and is struggling to reach a consensus on the best approach to address a defect in the manufacturing process. Although each proposed solution has merit, the team can only select one. Ciaran, a trained business analyst, proposes the nominal group technique. To begin, he explains that the first step is to:
   a. Rank and score the concepts, then quantitatively agree upon a solution.
   b. List only designs that are relevant, so the team can rank and score the ideas.
   c. Quantitatively agree upon a solution, based only on ideas that are relevant.
   d. Identify all the possible theories, even listing ones that may appear to be unrealistic.

107. While giving a presentation, Brody states, "Although risks can become issues, not all issues started as risks. And assumptions are . . ."
   a. Practices that should be avoided, because they are not valid or warranted.
   b. True, real, or certain without required proof.
   c. For the most part, may or may not hold true over the life of the project.
   d. Should not be documented, because they are baseless.

108. Your project is part of a larger program that is significantly ahead of schedule. In the interest of the customer, Navy Kate LLC, your lead developer, Mabel, decided to build a new module that the customer had mentioned during the needs assessment, even though the module never made it to the approved requirements traceability matrix. She worked on this module for 90 days, with the project coordinator's knowledge, before the other components caught up for system testing. Because the project coordinator was aware of the work effort and reported the status each week to the PMO, was this acceptable?
   a. Yes, the developer built the module in the interest of the customer, and status was reported to the PMO.
   b. No, this is considered scope creep.
   c. Yes, because the developer received permission from the customer.
   d. No, this is considered gold plating.

109. Your organization is considering a major investment in redesigning its manufacturing line. Your sponsors, Emilia and Caroline, have requested that you conduct an assessment to determine the organization's capability, capacity, willingness, and commitment to the project. What type of analysis will you perform?
   a. Organizational readiness assessment
   b. Capability and capacity assessment
   c. Quadrant analysis assessment
   d. Organizational factors analysis

110. Alistair, the lead for your time-and-attendance system, hung a whiteboard outside his cubical, with columns representing the project phases. What was the intent?
   a. To establish a platform on which to base discussions
   b. To illustrate the requirements lifecycle
   c. To provide the basis for elicitation
   d. To provide alternative analysis for satisfying the business need

111. Aurelia is working with stakeholders to determine which project requirements will be deferred to a future phase of her modernization initiative at the bakery plant. The team cannot agree on the decision-making process, their overall authority, or how project requirements are to be prioritized. To what document can they refer?
    a. Requirements management plan
    b. Requirements matrix
    c. Project plan
    d. Business analysis plan

112. Wesley works for a consumer electronics company that is focused on building innovative games. Because of the constantly changing dynamic of the industry, Wesley is keenly aware that requirements can lose relevance overnight. To estimate the effort required to perform due diligence as to the relevance of the solution, where would Wesley not look?
    a. A competitor's press release
    b. The business analysis plan
    c. Lower-level components of the work breakdown system
    d. The project management information system (PMIS)

113. Sam is the chief project officer of a consulting company that is aspiring to achieve greatness and knows that to deliver solutions on time, within budget, and aligned to stakeholder expectations, they need to be really good at business analysis. To be good at business analysis, what else should they have expertise in?
    a. Requirements management
    b. Project delivery methodologies
    c. Organizational change control
    d. Resource management (human, financial, equipment)

114. Your executive sponsor, Malachi, asked for your help preparing a document that will formally authorize the project to colonize Mars. During your conversation, he referenced statements of work (SOWs), government regulations, lessons learned, and the business case. What project document was Malachi referring to?
    a. Project scope statement
    b. Project charter
    c. Project situation statement
    d. Project recommendation

115. Melissa, an expert in interface design, is creating a model to demonstrate the precise user interactions within a global financial consolidation system. The development team prefers this model, because it places the requirement statements associated with each element on the screen. What model does Melissa's development team prefer?
    a. On-screen contextual model
    b. Display-action-response model
    c. Wireframe model
    d. High-fidelity prototype model

116. Your sponsor, Augustus, has just signed off on the Harriet–Keira project, a transformational initiative that has modernized a previously mothballed manufacturing facility. Prior to shutting down the existing facility, in your role as business analyst, you need to:
    a. Notify the city that work is shifting as part an existing labor agreement.
    b. Conduct a cost–benefit realization analysis.
    c. Collaborate with the project manager to document the steps required to move from the old to the new facility.
    d. Ensure that there are adequate resources to conduct a risk assessment the day the facility opens.

117. You are the business analyst for a hotel loyalty program; Violet, your subject matter expert for guest reservations, is describing requirements for family travelers and requirements for those that travel on business. What is she describing?
    a. A subset
    b. An implementation relationship
    c. A value dependency
    d. A rationale dependency

118. You're the office manager for a construction company that is six months into a five-planned-year development of a large retirement community. Levi, the district manager for your primary supply company, stops by the site office and offers a substantial discount on material if you can commit to a regular cadence of orders. You determine:
    a. With four and a half years to go, this is amazing—the money saved can be shared with the team as a bonus.
    b. A needs assessment should be conducted to analyze the opportunity.
    c. The discount will amount to over $3 million USD when analyzing the proposal in your job-costing software.
    d. The results from a Monte Carlo analysis suggest that construction will slow at certain points in time, and you will not be able to maintain the order thresholds and cadence.

119. As the project manager for a newly implemented software solution, you are about to commence the close activities. Before doing so, your sponsor, Emmett, has inquired whether the delivered solution is providing business value. In turn, you ask Faye, the project business analyst, to:
    a. Review the results from the nominal group technique session along with the signoff documents.
    b. Review the project's acceptance criteria.
    c. Measure and validate the solution referring to SLAs and OLAs.
    d. Evaluate the delivered solution and conduct a net promoter survey.

120. While assisting you with writing project documentation, your intern from a local community college, Dante, inquired as to the principle behind solution requirements. How would you best respond to his question?
    a. They encompass all the requirements as identified by the stakeholders.
    b. They describe the features and functions to fulfill the stakeholder requirements.
    c. Solution requirements address the foundational needs of the organization.
    d. Solution requirements address the KPIs to determine a successful outcome.

121. You work for Aoibhinn Medical Supply, a manufacturer of equipment for first responders and paramedics. Because your customers have expressed concern with your stretchers, your company is undertaking a complete redesign. In an effort to actively engage with your stakeholders and understand how they are using your products, you offer to participate in a disaster simulation. During this simulation exercise, you seek clarification of their activities, ask what they like about your current product line, and ask how the product can be improved. You are participating in what form of observation?
    a. Simulation
    b. Active
    c. Passive
    d. Participatory

122. Collaborating with your project manager, Greyson, you've defined the business need and solution scope and ensured that the product is aligned with the goals and objectives of the business. What should you do next?
    a. Understand stakeholders' interests so that the requirements can be baselined.
    b. Use valuation techniques such as payback period (PBP), return on investment (ROI), internal rate of return (IRR), or net present value (NPV) as inputs to the business case.
    c. Develop the solution scope statement based on either the RCA or the opportunity analysis.
    d. Develop an interrelationship analysis so that stakeholders can visualize the proposed solution scope.

123. In planning for the implementation of a new enterprise software application, Will, the director of IT, would like to outline the skills, competencies, and requirements for each of his staff, because their responsibilities will be changing both during the initiative and at go-live. As the business analyst assigned to support the IT team, what can you suggest?
    a. The output should detail the expectations of each staff member, all required training, and any suggested certifications; these can also be used for any future job postings.
    b. He should refer to the Skills Framework for the Information Age.
    c. The output should detail how staff will be managed, appraised, and trained.
    d. We should contact our human resource business partner to assist with assessment.

124. From your experience on iterative projects, you recognize the value of maintaining a spreadsheet to manage requirements. You've been tasked with supporting a project that is using an adaptive lifecycle. When team members submit user stories, how might you track them?
    a. Using a requirements tractability matrix
    b. Adding them to the epic log
    c. Tracking them in the user story matrix
    d. Adding them to a backlog

125. Following your second sprint, your product manager, Innes, inquired as to the team's performance. In terms of metrics, what would you provide him?
    a. Variance
    b. Vertical velocity variance
    c. Mean variance velocity
    d. Velocity

126. Your Change Control Board recently voted on several requests; three were approved, two were denied, and one was deferred pending further information. Upon leaving the room with Finn, your project manager, you both agreed:
    a. In your role as business analyst, you would only communicate the approved and denied requests. Pending the outcome of the next meeting, you would share the decision on the deferred request.
    b. Finn, as the project manager, would communicate the decisions on all requests to all stakeholders.
    c. In your role as business analyst, you would communicate the decisions regarding all requests to all stakeholders.
    d. Finn, as the project manager, would only communicate the decision regarding the deferred request, as representatives for the five requests were present.

127. Your primary stakeholders are having a difficult time understanding the problems and the complex relationships among all the components. To simplify the relationships, your sponsor, Aveline, suggests that you:
    a. Conduct a 5 Whys analysis to help all stakeholders gain a better understanding of the problem and complex relationships.
    b. Develop a cause-and-effect diagram to trace the problem to complex relationships among all the components.
    c. Present the variables in the form of an interrelationship diagram.
    d. Create an opportunity analysis diagram to simplify the relationships.

128. A key team member, Mira, has recommended that the team conduct a brand-satisfaction and loyalty survey as the team celebrates the one-year anniversary of the product launch. How will participants be grouped?
    a. Into three categories: promoters, satisfied, and detractors
    b. Into two categories: those who responded and those who did not
    c. Into three categories: brand loyal, promoters, and neutral
    d. Into three categories: neutral, satisfied, and passive

129. Your sponsor, Esmeralda, has requested assistance in preparing an artifact that will contain a list of stakeholders, milestones, assumptions, constraints, and success criteria. You will be assisting with preparation of the:
    a. Solution scope statement
    b. Value proposition statement
    c. Charter
    d. Product solution and value justification

130. You work in a fairly stable industry, and projects are generally completed on time and within budget. In working with your sponsor, Nala, she'd like to gauge project capriciousness over the planned lifecycle. What is one measurement that can be considered?
    a. Assessing the items presented at change control
    b. Monitoring the status of requirements
    c. Evaluating the defects and resolution time
    d. Measuring the efficiency of the project

131. Maeve is working on a questionnaire that she intends to distribute to a panel of experts in hopes of soliciting their feedback based on the last round of testing. What tool is Maeve using to protect the anonymity of respondents?
    a. The net promotor tool
    b. Planguage
    c. Entry and exit criteria for expert opinion on leading practice
    d. Delphi

132. At lunch, you're conversing with Aspen, the project manager assigned to your initiative, about the dynamics of the enterprise project. It seems that from month to month, your sponsor, Eero, is slightly altering scope to appease various business units. What is one benefit of this approach?
    a. Reduction in risk
    b. Stakeholder satisfaction
    c. Faster depreciation
    d. Dynamic scheduling

133. Your CIO, Zenobia, is negotiating a contract with a solution provider to implement a large-scale software application. Including contingency and management reserves, the project budget is $4,000,000 USD. However, the solution provider is estimating about twice the budget to deliver the work. Without sacrificing scope or quality and still meeting the deadline, Zenobia requests the vendor to present:
    a. A work plan elaborated through the first phase of the project
    b. A quote using off-shore resources
    c. A comparative quote based on adaptive and iterative lifecycles.
    d. An incentive-based contract.

134. Your sponsor, Ansley, has scheduled a meeting for you to present the outline of scope baseline to the Executive Steering Committee for the next generation of high-speed passenger rail cars that will be used in the change control process. What will you present?
    a. An explanation and tiered decomposition of the scope, including limitations, and a supporting dictionary
    b. The project management information system and the configuration management system
    c. The requirements traceability matrix, the business analysis plan, the BA work plan, and the change management plan
    d. The change management plan, the requirements management plan, and the requirements traceability matrix

135. Declan, an intern working on your project, has inquired as to the key processes associated with solution evaluation. You candidly tell Declan, "This isn't my first time at the rodeo." They are:
    a. Tracing and tracking all requirements in the team wiki
    b. Posting the results to the team wiki, selecting techniques that will be used during the evaluation process, and hosting a demonstration for the key stakeholders
    c. Hosting a demonstration following each phase or sprint
    d. Validating frequently and thoroughly

136. Your executive sponsor, Constance, has asked for your help with an assessment; to date only the approach to the initiative has been approved. What should you begin working on?
    a. The solution scope statement
    b. The project scope statement
    c. The project charter
    d. A cost–benefit analysis

137. After creating the work breakdown structure (WBS), Oswin is estimating the overall project duration by totaling the estimates of all the lower-level components. What estimating technique is Oswin using?
    a. Three-point
    b. Parametric
    c. Analogous
    d. Bottom-up

138. Collaborating with your project manager, Marcus, you've completed a thorough and complex stakeholder identification process. Your sponsor, Garrett, is quite impressed and suggests:
    a. A regular review of the quadrant analysis.
    b. His name can be removed, because Marcus will be leading the project once approved.
    c. The stakeholder register not be shared with others, because it contains confidential information and should not be updated beyond this point.
    d. The stakeholder register be published to the company intranet to foster collaboration among team members.

139. Your project director, Beatrice, completed a project RACI and noted that Soren, the project manager, was responsible for managing quality requirements and scope. Is this correct?
    a. No, it is a collaborative effort between the business analyst and the project manager to manage project scope.
    b. No, the management of project scope is the responsibility of the business analyst.
    c. Yes, Soren is responsible for managing project scope.
    d. No, the project stakeholders have accountability to manage and control scope.

140. Lucy, the product manager for Lucky Rabbit Food Inc, expressed concern that the delivered product might not meet stakeholder expectations. She requests that you hire a consulting firm that has not been associated with the project to:
    a. Conduct an independent verification and validation of the rabbit toys
    b. Conduct day-in-the-life testing and observe rabbits playing with the toys
    c. Use a fishbone diagram to determine why rabbits of certain breeds are not playing with the toys
    d. Conduct a Delphi method vote

141. In a meeting with your sponsor, Jeff, he asks that you lead a brainstorming session to identify the reasons product management submitted a proposal to redesign the ketchup bottles that have been used for over 100 years. Your session will focus on:
    a. Transition requirements
    b. Solution requirements
    c. The requirements of product management
    d. Business requirements

142. Braelyn, your subject matter expert, is having difficulty estimating the work effort for a number of activities. In addition, depending on the software developer, there could be additional activities that are simply not known at this time. As the business analyst, how should you note this on the work plan?
    a. Add a reserve for both items at the activity level, and update the work plan as more information becomes available.
    b. Collaborate with your project manager and use a consistent approach across the project.
    c. Add a contingency reserve for the known activities and a management reserve for those activities that are unknown at this time.
    d. Use a management reserve at the activity level for estimating both the work effort and the unknown activities that are not known at this time, then update the work plan as more information becomes available.

143. You are conducting a focus group to build out the requirements traceability matrix. While doing so, you are having difficulty tracing a few items to business goals and objectives. Your sponsor, Jacob, is not concerned and suggests that they are all relevant. What should you do?
    a. Add a short textual description, stating "requirement provided by sponsor."
    b. Allow the project manager to manage these items, as they pertain to project management.
    c. In some circumstances, it is acceptable not to track requirements to business goals and objectives.
    d. Ask Jackson, your associate, to conduct an analysis to determine the relevance of the requirement.

144. Agnes, the VP of the project management office (PMO), requests that you complete a feasibility study for the proposed recommendation. Why?
    a. A feasibility study will help determine the extent to which the proposed solution addresses the identified opportunity or problem in an effort to ensure that the correct project is implemented.
    b. A feasibility study will help determine the extent to which the proposed solution addresses the identified opportunity.
    c. A feasibility study will outline potential alternatives.
    d. A feasibility study will help determine the extent to which the proposed solution addresses the identified opportunity or problem in an effort to ensure that the correct solution is implemented.

145. Carson is creating screen mockups to classify page elements and their associated functions. As the mockup is decomposed to user interface elements, how are they described?
    a. As unique development objects
    b. As individual requirement objects
    c. From the standpoint of displays and behaviors
    d. From the perspective of a wireframe and relationship to requirement

146. During user acceptance testing, your subject matter expert, Phoebe, signs off that there are no defects with the program and the documentation is acceptable. However, she indicates that she is not satisfied with the software application as delivered, commenting that it is missing some features. As a result, you:
    a. Agree, the software does not fully meet the business requirements.
    b. Suggest Phoebe review the business requirements documents.
    c. Advise that the software meets the business requirements; she is concerned with the low quality.
    d. Advise that the software meets the business requirements; she is concerned with the low grade.

147. Trinity, VP of human resources and head of your Executive Steering Committee, asked that, in collaboration with Ivor, your project manager, you present artifacts that can influence scope management for the new human capital management system. What will you present?
    a. Scope management plan, requirements management plan, quality management plan
    b. Work breakdown structure and work breakdown structure dictionary
    c. Scope management plan, scope baseline, and requirements traceability matrix
    d. On-boarding, transfer, promotion, and off-boarding procedures; travel reimbursement policies and detailed information on the applicant portal

148. You are facilitating a session with product management, engineering, and manufacturing to understand the high-level causes for problems with the floating-point calculation in a newly designed microprocessor. You begin the session by drawing a fishbone diagram and place the problem to be addressed:
    a. At the head of the fish, which is facing right
    b. At the head of the fish, which is facing left
    c. Along the spine of the fish
    d. At the head of the fish, facing either left or right

149. Team members Beckett and Eliza have outlined the steps to (a) ensure that quality standards are effectively used during the project, (b) document how the project will demonstrate adherence to standards, and (c) recommend any necessary changes. At the same time, team members Sophia and Christian are focused on communications management. What elements are Beckett and Eliza not concerned with:
    a. Developing the communications management plan, ensuring that all stakeholders receive timely and relevant information about the project
    b. Performing quality assurance, quality management planning, and controlling quality
    c. Performing quality management
    d. Controlling quality

150. Your organization is making a significant investment in a new enterprise resource planning (ERP) system. This system is purpose-built for your industry and is highly configurable. Your sponsor, Seraphina, has requested that you use the requirements modeling language to assist with the refinement of requirements. To what benefit?
    a. These models will focus on the system design to maximize value.
    b. These models will visually model the requirements for easy consumption by stakeholders.
    c. These models will provide the basis for business transformation.
    d. These models will focus on the project's goals and objectives for the purpose of transformation.

151. Henley is the business analyst for construction of a high-rise tower in lower Manhattan; she is accountable for providing the estimates for installing carpet and painting the walls and trim. Based on a recent project, she knows the carpet team can install 5,000 square feet per day. The current project calls for 35,000 square feet of carpet, so Henley estimates the work effort at seven days. What technique did she use to arrive at this estimate?
    a. Analogous
    b. Parametric
    c. Bottom-up
    d. Three-point

152. Isla, a skilled business analyst, would like to use a tool that can objectively compare solution options for her subject matter experts. What can you recommend?
    a. Multi-criteria weighted ranking
    b. Nominal group technique
    c. Business Process Model and Notation (BPMN)
    d. Requirements modeling technique

153. The Executive Steering Committee for your project has just approved your business plan. Unfortunately, they've only provided you with four months to complete the requirements evaluation process and solution development. As the business analyst, how do you approach this constraint?
    a. Perform a MoSCoW analysis to determine the features for the release window.
    b. Work with the time window provided to complete the requirements and development efforts; as needed, request more time in the form of a change request.
    c. Analyze the team's capability and capacity within the defined time period, conduct a MoSCoW analysis, and prioritize the work accordingly.
    d. Conduct a feasibility assessment based on technology, system, and cost effectiveness.

154. You are establishing the process for securing approval of solution requirements for a project using extreme programming (XP). Collaborating with your project manager, Sasha, what format would your offshore developers prefer?
    a. Business requirements document
    b. Functional specification
    c. User story
    d. Use case

155. Lauren, an experienced project manager and business analyst, is building her work plan for the construction of the new kindergarten wing at the local elementary school. She has identified a few instances in which successor activities cannot begin immediately following the completion of predecessor activities. For example, the desks cannot be moved into the classrooms until all the tile floors have cured. How can Lauren best represent a two-day delay on her work plan for this activity?
    a. FF +2
    b. FS −2
    c. FS +2
    d. SF −2

156. While building his work plan, Jonathan noted several dependencies in the real estate module of the ERP. Based on information learned at a recent seminar by experts in the field, how can these best be catalogued?
    a. Discretionary
    b. External preferred
    c. Internal preferred
    d. Mandatory

157. Frederick, a professor at the local university, is explaining the principle of requirements containment. For a requirement document to be "self-contained," what must be true?

a. It must contain all the requirement information for the development team to design, build, and test the development object.

b. It must not be dependent on any other requirement document.

c. It must be measurable, feasible, complete, consistent, and correct.

d. It must contain all the information based on the stakeholders' requirements.

158. Jinx, the VP of supply chain for a regional hospital, has established the need for several highly customized workflows. The project team has established the process for these requirements to be documented; however, before approving the onshore specialty development, what can the Project Steering Committee suggest?

a. Request a detailed review of the solution requirements with the supply chain team and developer.

b. Require that the supply chain lead present a cost–benefit analysis to the Change Control Board.

c. Review the customized workflows with the VP of supply chain from a competitor, conduct a detailed review with the supply chain team, and require a cost–benefit analysis be presented to the Change Control Board.

d. Ask the VP of supply chain from a competitor to review the business requirements documents and provide their opinion.

159. You are collaborating with Colleen, your project manager, on a project artifact that outlines how requirements will be analyzed, documented, and managed. What are the required inputs to aid in creation of this document?

a. Scope management plan, requirements management plan, stakeholder register

b. Project charter, project management plan, scope management plan

c. Enterprise environmental factors, organizational process assets, charter, project management plan

d. Enterprise environmental factors, organizational process assets, scope management plan, requirements management plan, stakeholder register

160. Your sponsor, Tabitha, and project manager, Braxton, are quite impressed with the significant number of ideas that have been generated during the facilitated elicitation session. To organize these ideas, Tabitha suggests:

a. You organize them using an affinity diagram.

b. You organize them using an interrelationship diagram.

c. They all be included in the work breakdown schedule (WBS), which is part of the project management plan.

d. They be outlined in the solution statement.

161. Your project manager, Raelyn, is establishing a platform that will serve as the central repository for project plans, requirements documentation, and stakeholder lists that will be used over the life of the project. What is this platform considered?

a. A project management information repository

b. A project management collaboration site

c. SharePoint or team wiki

d. A project management information system

162. You work for an advertising agency in a highly competitive market. In an effort to increase market share across several product lines, you've been asked to support Emily, a strategic marketing expert. During your initial meeting with Emily, you provided her with statistics from the web

development team that included number of visitors, time on each page, transaction volume, and source of traffic. Although impressed with the data, Emily also wants to understand market share, competition, industry cycle times, and an overall competitive analysis. What can you suggest?

    a. Advise Emily that the Sherman Act of 1890 has very strict antitrust guidelines, and further that what she is asking for is banned by the Fair Trade Commission Act.

    b. Recommend hiring a third party to conduct the competitive analysis, looking for both public and nonpublic data.

    c. Advise Emily that the data cannot be reasonably gathered, much less analyzed.

    d. Suggest to Emily that you conduct the analysis, looking only for data that has been made publicly available.

163. Jacob is working on the development of the business case for your project. You just completed the initial needs assessment, when you were asked to provide the situation statement. To clearly communicate the nature of the problem, you construct the statement in the following format:

    a. Problem–cost–benefit

    b. Opportunity–effect–cost

    c. Opportunity–outcome–impact

    d. Problem–effect–opportunity

164. Who has accountability for determining the extent to which the project will trace and track requirements?

    a. It is the project manager's responsibility to assess and determine the extent to which the project will trace and track requirements.

    b. It is the business analyst's responsibility to assess and determine the extent to which the project will trace and track requirements.

    c. The business analyst should discuss this topic with the project manager over lunch.

    d. It is the stakeholders' responsibility to assess and determine the extent to which the project will trace and track requirements.

165. Rayden is new to business analysis, and while helping you draft several key documents, she asks you to provide examples of organizational process assets. You suggest:

    a. Work authorization systems, project management information system

    b. Guidance from consultants, industry professionals, and technical associations who are available as resources to the performing organization

    c. Communication channels influenced by organizational culture and structure

    d. Artifacts from previous projects, organizational polices, procedural documents, checklists, or templates

166. Your organization is considering moving from an on-premise solution to one that is cloud based with Icarus Software Ltd. As the senior business analyst, you've been asked to assist with documenting the requirements for the transition. Your team is accustomed to 10-minute response time from an on-call resource, 24/365. In working with your counterpart from Icarus, how would you classify performance, continuity, and availability requirements?

    a. Nonfunctional requirements

    b. Service-level agreement

    c. Business requirements

    d. Stakeholder requirements

167. Tamsin, a business analyst for an international cargo carrier, has been analyzing data in an attempt to determine the leading causes of variation at the shipping port. Her hypothesis is that there are most likely only a few factors driving the majority of the defects. What tool can she use to validate her assumption?
    a. Control chart
    b. Pareto chart
    c. Process model
    d. Trend analysis

168. You are working with your IT security manager, Sienna, on a project to implement a new firewall for a pharmaceutical company whose two primary customers are the Centers for Disease Control and Prevention (CDC) and the Food and Drug Administration (FDA). While planning the kick-off meeting, you suggest collaborating on a responsibility assignment matrix. Why?
    a. A responsibility assignment matrix is necessary because the company operates in a highly regulated industry.
    b. Because of the sensitive nature of the data and the risk associated with the project, the CDC and the FDA require a responsibility assignment matrix as part of the statement of work.
    c. The RACI will identify those who are either affected or perceived to be affected by the project.
    d. The completion of a reliability and maintainability matrix is part of the Statement on Standards for Attestation Engagements (SSAE16) for auditing purposes.

169. What tool can Amias use to ensure that his subject matter experts are creating requirements documents of high quality?
    a. A checklist
    b. The requirements traceability matrix
    c. Planguage
    d. Requirements acceptance criteria

170. Elyse is designing an interactive website for customers at a dine-in restaurant to place food and beverage orders. Her primary stakeholder, Leonidas, requested that she create a model that plots how customers will navigate the screens and the system's responses based on their selections. What model should she use?
    a. Process-flow diagram
    b. System interface flow
    c. Interoperability diagram
    d. User-interface flow

171. It's been nearly ten years since your organization optimized its processes. Gabriel, the VP of strategic initiatives, has requested that you model the existing processes to help prepare the organization for a transformative initiative. To accomplish this task, what tool would you use?
    a. Unified Modeling Language (UML)
    b. System Modeling Language (SysML)
    c. Business Process Modeling Notation (BPMN)
    d. Process-flow diagram

172. As an experienced business analyst, Thea suspects that there may be gaps in the business objective model that she developed for a local pet store. She would like to create a model detailing the lifecycle of an object through various conditions. What tool can Thea use to uncover any potential gaps?
    a. An entity relationship diagram further detailing the workflows from the business objective model
    b. A pair-wise table, which evaluates the initial and target states for object
    c. A process-flow diagram aligned to industry leading practice
    d. A system diagram aligned to industry leading practice

173. Stefan is having difficulty explaining the motivations, behaviors, and ultimate goals for the targeted audience of the newly designed website. To aid in understanding their characteristics, as the business analyst, you:
    a. Recommend creating a short narrative about each user group
    b. Suggest creating a user analysis based on the targeted user groups
    c. Determine models that can be used to aid in the analysis of user groups
    d. State each of the elements in the format of: As a <role or type of user>, I want <goal/desire>, so that <benefit>, to help articulate the needs of the targeted audience

174. You are serving as both project manager and business analyst on a small project in a heavily regulated industry. Your sponsor, Thaïs, would like to implement a formal approach to tracking changes to project artifacts. Where should this requirement be identified?
    a. In the configuration management system and version control system
    b. In the business analysis plan
    c. In the requirements management plan
    d. In the scope management plan

175. Your project manager, Mila, is concerned about scope creep at both the project and the product level. To address this concern, you both agreed to institute a requirements traceable matrix. After the matrix is created and approved, what can you do to set boundaries?
    a. Present the matrix to your stakeholders for approval on the approach.
    b. Perform a dependency analysis.
    c. Establish a process for monitoring and controlling scope.
    d. Establish a baseline.

176. Your subject matter expert, Jennifer from the radio repair shop, recently submitted her requirements documents for the new wire reel. Although they are all very well written, they seem to be missing a key component:
    a. There were no preconditions to enable testing.
    b. They are missing the actors and personas.
    c. Some requirements were labeled TBD because the team was still addressing some elements.
    d. They should have been in the format of a user story or epic.

177. You are an experienced project manager, and you have been asked to fill the key business analyst role on a high-profile project. Shortly after starting the requirements sessions, you've noticed considerable changes to the requirements documents. To assist with your efforts, you:

    a. Ask the subject matter experts to add versioning control to each document.

    b. Add the requirements documents to a control system.

    c. Conduct an analysis to determine the material nature of the differences.

    d. Provide your technical resources with only the most current version, based on the control date.

178. You work for Porcine Lifesciences LLC, the leading producer of aortic pig valves and instruments for use in human heart valve replacement surgeries. To gain a better understanding of the operation and the related steps involved with surgery, a local hospital has invited you to view a surgery from the observation suite. What type of observational activity is this considered?

    a. Passive

    b. Participatory

    c. Active

    d. Simulation

179. What is the benefit of using the iterative planning technique of rolling wave planning?

    a. Team members skilled at planning can provide their input to the plan creation.

    b. The work plan can be decomposed to varying levels of detail.

    c. Rolling wave planning is based on the WBS, which contains a hierarchical decomposition of all the work to be completed as part of the project.

    d. Rolling wave planning highlights all the significant mandatory events in the project.

180. Skyler holds certifications from the Project Management Institute in both project management and business analysis. He is presently working to estimate construction times for a multi-seasonal resort in New Hampshire. Due to the weather uncertainty, he is using a technique that takes into account pessimistic, most likely, and optimistic estimates from his subject matter experts. What estimating technique is Skyler using?

    a. Analogous

    b. Three-point

    c. Parametric

    d. Bottom-up

181. Cassian is working for a digital advertising agency and is attempting to articulate how customers would interact with a timed advertisement that would load prior to free content on a client's website. How could Cassian best explain the concepts and interactions in an attempt to solicit feedback?

    a. Cassian could conduct a series of elicitation sessions focused on soliciting feedback.

    b. Cassian's team could create a high-fidelity prototype of the finished product.

    c. Cassian's team could create an evolutionary prototype of the finished product.

    d. Cassian could create graphical illustrations, which would eventually be discarded.

182. To ensure project consistency and completeness, your project manager, Tom, suggests that he will manage the requirements traceability matrix, because he has greater insight into the tasks and deliverables associated with each project phase. Is this approach acceptable?

    a. Yes. For fundamental accuracy, Tom as the PM should maintain the matrix.

    b. No. Muriel, as the organizational change management lead, should maintain the matrix to ensure consistency and completeness. Her utilization is at 50%.

    c. No. The business analyst, Woodrow, who is a part-time resource and new to the organization, should manage the matrix to ensure consistency and completeness.

    d. Yes, the approach is acceptable.

183. Brock is a well-respected business analyst for a Fortune 200 chemical company. He receives praise from senior leadership for providing support, and he's very successful in terms of mediation and conflict resolution. While conducting his 360-degree peer review, you were mostly in agreement with leadership, just adding the following comments:

    a. It is a pleasure to work with Brock; he is both a friend and mentor.

    b. Although it's a pleasure to work with Brock, if he were to focus more on understanding the motivations, goals, and objectives of the team, he would be more successful in several aspects of business analysis.

    c. Although Brock is skilled in many areas, he needs to exhibit less authority and more influence; in addition, focusing on the business processes could help identify opportunities for further improvement in the organization.

    d. It's a pleasure to work with Brock. In terms of improvement, he would be more valuable to the organization if he were to focus more on the business processes, which in turn could help identify opportunities for further improvement. As a leader, he needs to demonstrate slightly more consigned authority.

184. Your project manager, Prescott, has recently successfully completed a $150 million USD software project for a major healthcare customer, for which it won the PMI Project of the Year Award. When talking about your new $50,000 USD project, he offers to share his requirements traceability matrix, stressing that you must use this tool on your project. You decide:

    a. To use his requirements traceability matrix—if it helped to win Project of the Year, it's a tool and technique to use.

    b. To use his requirements traceability matrix—it's a PMO process asset.

    c. To review his matrix, but adapt it for your needs to trace requirements to the business mandate.

    d. To review his matrix, but adapt it for your needs to trace a list of actions.

185. Koda, your project manager, is working to decompose the elements of the project into logical and manageable portions so that you can begin creating the requirements traceability matrix. Once complete, what will she have created?

    a. The project management plan

    b. The details of all project and product work

    c. An outline of the requirements traceability matrix

    d. The scope baseline

186. Achilles is an experienced business analyst and an expert facilitator. While working with Hayley, a subject matter expert for the seed division of the Ministry of Agriculture & Farmers, to uncover objects that are related to or may have dependencies with other objects, he inquires as to the cardinality relationships. To what is Achilles referring?

    a. The minimum number of times an instance in one entity can be associated with instances in a related entity

    b. A line notation

c. The maximum number of times an object in one entity can be associated with objects in a related entity

d. Business data objects

187. Cecelia is in the process of developing her elicitation plan for a meeting with Gitchi Goo, the owner of a local gallery and a famous artist. What are some of the elements that should be included in the elicitation plan?

a. What does Cecelia need to know to answer the question, what methods will be used to elicit information from Gitchi, in what order should activities be conducted?

b. How and when elicitation will be conducted with Gitchi.

c. In what order should communication activities be conducted, and what methods will be used to elicit information from Gitchi regarding her paintings?

d. The source of information, activities to be conducted, and methods of communication.

188. Aries was contracted to develop several queries and reports for the legal department of a utility company. To ensure that requirements were appropriately addressed, your subject matter expert, Kiera, created a prototype along with a model that establishes the details for each of the reports. What type of model did Kiera provide Aries?

a. Interface model

b. Business intelligence model

c. Laszlo report table

d. Report table

189. Although Tristan received approval from his sponsor and primary stakeholders for his business analysis plan, when the time arrives to conduct a test cycle, the lead resources were unable to perform their assigned tasks. What could be the root cause?

a. The system wasn't sufficiently complete to conduct testing.

b. Those performing the testing didn't participate in the review of the plan.

c. The plan didn't outline the testing components.

d. Test scripts were not prepared for the cycle.

190. Upon returning from a PMI-PBA® Boot Camp, your sponsor, Alina, has requested that the team begin using the Planguage tool. What is the benefit of using this tool, which was developed by Tom Gilb?

a. The tool is used to address concerns with ambiguous and incomplete functional requirements.

b. The tool is used to address concerns with ambiguous and incomplete nonfunctional requirements.

c. The tool provides a standard set of templates for solution requirements.

d. The tool ensures consistency across all requirement types.

191. Aurelia is in the process of auditing the project's quality requirements to ensure that the output meets the stakeholder requirements and expectations. What process is she performing?

a. Perform quality assurance

b. Monitor and control quality

c. Update quality checklists

d. Conduct quality control review

192. You are the program manager for a new line of nanorobotic vehicles. You ask Lou, a senior business analyst and board-certified occupational therapist, to conduct a series of sessions with potential customers, suppliers, and strategic business partners to better understand their needs. Lou is going to conduct a:
    a. Needs assessment workshop
    b. Stakeholder requirements session
    c. Business requirements session
    d. Nominal group technique workshop

193. Your sponsor has just recommended a functional requirement that was not previously identified during the needs assessment or included in the approved requirements snapshot. Your project manager suggests you consult:
    a. The scope management system (SMS)
    b. The configuration management system (CMS)
    c. The requirements management system (RMS)
    d. The service knowledge management system (SKMS)

194. You're meeting with Jackie, the project manager for XYZ Soap Company. In your role as business analyst, she's asking that you document the product quality conditions required for the new line of fragrance-free soap to be effective. You agree, stating that these qualities will be included as part of the:
    a. Business quality requirements
    b. Stakeholder requirements
    c. Solution and nonfunctional requirements
    d. Functional and solution requirements

195. You work for the Blue Ball Point Pen Company. Grayson is in the process of prioritizing the features for a new line of translucent pens, based on user stories submitted by the team members. What role does Grayson have?
    a. The business analyst
    b. Your product owner
    c. The scrum master
    d. The project manager

196. Shortly after go-live, your internal auditor, Maddox, identified a list of concerns and observations. Management agreed with all the audit findings, which have been classified as high-priority defects. Who has accountability to see these concerns to resolution?
    a. The project team, because they will be correcting the defective code
    b. The business analyst, despite rolling off to another project
    c. The sponsor, who has yet to sign off on the project and release final payment
    d. The stakeholders, as they are the ones most impacted by the defective code

197. Malakai, an experienced application developer, is working on a complex interface between two electronic health record (EHR) systems. Because he's a contractor working for your system integrator and will roll off once you're live, you'd like him to create an artifact that notes all the attributes and subsequent details of each interface. What template can you suggest?

    a. Interface flow table

    b. System interface table

    c. Interface schematic

    d. Interface element table

198. Your project intern, Miles, hears you discussing the traceability matrix with Karen, the lead assigned to manage requirements for the ERP initiative. After the discussion, he inquires as to the benefit of the tool. How would you respond?

    a. The requirements traceability matrix helps to control and monitor project scope.

    b. The requirements traceability matrix helps to manage sponsor expectations.

    c. The requirements traceability matrix helps to control and monitor product scope.

    d. The requirements traceability matrix helps to manage customer expectations.

199. As a business analyst, what project artifact can Kaison refer to in order to plan and assess project performance?

    a. The requirements management plan

    b. The work breakdown structure

    c. The requirements traceability matrix

    d. The time management plan

200. Chanel is working with stakeholders to determine the metrics that will be used during solution evaluation, and also to establish the definition of success. Which of the elements below can influence project costs?

    a. Service-level agreements

    b. Acceptance criteria

    c. Scope baseline

    d. Operational-level agreements

Answer Sheet

	Answer Choice				Correct	Incorrect	Predicted Answer			
							90%	75%	50%	25%
1	a	b	c	d						
2	a	b	c	d						
3	a	b	c	d						
4	a	b	c	d						
5	a	b	c	d						
6	a	b	c	d						
7	a	b	c	d						
8	a	b	c	d						
9	a	b	c	d						
10	a	b	c	d						
11	a	b	c	d						
12	a	b	c	d						
13	a	b	c	d						
14	a	b	c	d						
15	a	b	c	d						
16	a	b	c	d						
17	a	b	c	d						
18	a	b	c	d						
19	a	b	c	d						
20	a	b	c	d						

Answer Sheet

	Answer Choice				Correct	Incorrect	Predicted Answer			
							90%	75%	50%	25%
21	a	b	c	d						
22	a	b	c	d						
23	a	b	c	d						
24	a	b	c	d						
25	a	b	c	d						
26	a	b	c	d						
27	a	b	c	d						
28	a	b	c	d						
29	a	b	c	d						
30	a	b	c	d						
31	a	b	c	d						
32	a	b	c	d						
33	a	b	c	d						
34	a	b	c	d						
35	a	b	c	d						
36	a	b	c	d						
37	a	b	c	d						
38	a	b	c	d						
39	a	b	c	d						
40	a	b	c	d						

Answer Sheet

	Answer Choice				Correct	Incorrect	Predicted Answer			
							90%	75%	50%	25%
41	a	b	c	d						
42	a	b	c	d						
43	a	b	c	d						
44	a	b	c	d						
45	a	b	c	d						
46	a	b	c	d						
47	a	b	c	d						
48	a	b	c	d						
49	a	b	c	d						
50	a	b	c	d						
51	a	b	c	d						
52	a	b	c	d						
53	a	b	c	d						
54	a	b	c	d						
55	a	b	c	d						
56	a	b	c	d						
57	a	b	c	d						
58	a	b	c	d						
59	a	b	c	d						
60	a	b	c	d						

Answer Sheet

	Answer Choice				Correct	Incorrect	Predicted Answer			
							90%	75%	50%	25%
61	a	b	c	d						
62	a	b	c	d						
63	a	b	c	d						
64	a	b	c	d						
65	a	b	c	d						
66	a	b	c	d						
66	a	b	c	d						
68	a	b	c	d						
69	a	b	c	d						
70	a	b	c	d						
71	a	b	c	d						
72	a	b	c	d						
73	a	b	c	d						
74	a	b	c	d						
75	a	b	c	d						
76	a	b	c	d						
77	a	b	c	d						
78	a	b	c	d						
79	a	b	c	d						
80	a	b	c	d						

Answer Sheet

	Answer Choice				Correct	Incorrect	Predicted Answer			
							90%	75%	50%	25%
81	a	b	c	d						
82	a	b	c	d						
83	a	b	c	d						
84	a	b	c	d						
85	a	b	c	d						
86	a	b	c	d						
87	a	b	c	d						
88	a	b	c	d						
89	a	b	c	d						
90	a	b	c	d						
91	a	b	c	d						
92	a	b	c	d						
93	a	b	c	d						
94	a	b	c	d						
95	a	b	c	d						
96	a	b	c	d						
97	a	b	c	d						
98	a	b	c	d						
99	a	b	c	d						
100	a	b	c	d						

Answer Sheet

	Answer Choice				Correct	Incorrect	Predicted Answer			
							90%	75%	50%	25%
101	a	b	c	d						
102	a	b	c	d						
103	a	b	c	d						
104	a	b	c	d						
105	a	b	c	d						
106	a	b	c	d						
107	a	b	c	d						
108	a	b	c	d						
109	a	b	c	d						
110	a	b	c	d						
111	a	b	c	d						
112	a	b	c	d						
113	a	b	c	d						
114	a	b	c	d						
115	a	b	c	d						
116	a	b	c	d						
117	a	b	c	d						
118	a	b	c	d						
119	a	b	c	d						
120	a	b	c	d						

Answer Sheet

	Answer Choice				Correct	Incorrect	Predicted Answer			
							90%	75%	50%	25%
121	a	b	c	d						
122	a	b	c	d						
123	a	b	c	d						
124	a	b	c	d						
125	a	b	c	d						
126	a	b	c	d						
127	a	b	c	d						
128	a	b	c	d						
129	a	b	c	d						
130	a	b	c	d						
131	a	b	c	d						
132	a	b	c	d						
133	a	b	c	d						
134	a	b	c	d						
135	a	b	c	d						
136	a	b	c	d						
137	a	b	c	d						
138	a	b	c	d						
139	a	b	c	d						
140	a	b	c	d						

Answer Sheet

	Answer Choice				Correct	Incorrect	Predicted Answer			
							90%	75%	50%	25%
141	a	b	c	d						
142	a	b	c	d						
143	a	b	c	d						
144	a	b	c	d						
145	a	b	c	d						
146	a	b	c	d						
147	a	b	c	d						
148	a	b	c	d						
149	a	b	c	d						
150	a	b	c	d						
151	a	b	c	d						
152	a	b	c	d						
153	a	b	c	d						
154	a	b	c	d						
155	a	b	c	d						
156	a	b	c	d						
157	a	b	c	d						
158	a	b	c	d						
159	a	b	c	d						
160	a	b	c	d						

Answer Sheet

	Answer Choice				Correct	Incorrect	Predicted Answer			
							90%	75%	50%	25%
161	a	b	c	d						
162	a	b	c	d						
163	a	b	c	d						
164	a	b	c	d						
165	a	b	c	d						
166	a	b	c	d						
167	a	b	c	d						
168	a	b	c	d						
169	a	b	c	d						
170	a	b	c	d						
171	a	b	c	d						
172	a	b	c	d						
173	a	b	c	d						
174	a	b	c	d						
175	a	b	c	d						
176	a	b	c	d						
177	a	b	c	d						
178	a	b	c	d						
179	a	b	c	d						
180	a	b	c	d						

Answer Sheet

	Answer Choice				Correct	Incorrect	Predicted Answer			
							90%	75%	50%	25%
181	a	b	c	d						
182	a	b	c	d						
183	a	b	c	d						
184	a	b	c	d						
185	a	b	c	d						
186	a	b	c	d						
187	a	b	c	d						
188	a	b	c	d						
189	a	b	c	d						
190	a	b	c	d						
191	a	b	c	d						
192	a	b	c	d						
193	a	b	c	d						
194	a	b	c	d						
195	a	b	c	d						
196	a	b	c	d						
197	a	b	c	d						
198	a	b	c	d						
199	a	b	c	d						
200	a	b	c	d						

Answer Key

1. You are collaborating with Blake on the sequencing of activities for the next generation of air transport that your company is pioneering. As the project manager, she is inquiring about any dependencies that may exist in your product. You describe dependencies that are:
    a. Internal and external to the proposed solution
    b. Mandatory and discretionary to the proposed solution
    c. Mandatory external and discretionary internal to the proposed solution
    d. Represented by nodes and graphically linked to logical relationships

    The correct answer is: **C**

    As you build the requirements traceability matrix, it's important to investigate and record dependencies during the sequencing of business analysis work. Having this information will improve development efforts and walkthroughs, concentrating efforts on those requirements that are mandatory to the project. There are four possible combinations of dependencies for each requirement: (a) mandatory internal, (b) mandatory external, (c) discretionary internal, and (d) discretionary external.

    Answer Choice A: This answer choice is partially correct. Internal dependencies are most often within the control of the project team and involve predecessor relationships, whereas external dependencies are most often outside the control of the project team and involve interactions between project and non-project activities. However, it is missing mandatory and discretionary.

    Answer Choice B: This answer choice is partially correct. Mandatory dependencies either are determined to be bound by contractual, legal, or statutory obligations or are essential to delivering the solution, whereas discretionary dependencies are based on leading practice in a particular subject area. They are often referred to as preferred or soft logic.

    Answer Choice D: This is a distractor and made-up answer choice.

    A Guide to the Project Management Body of Knowledge (PMBOK®), Section 6.3.2.2, "Dependency Determination."
    PMI Professional in Business Analysis (PMI-PBA)® Examination Content Outline, 2013, "Planning," Task 2.

2. As the business analyst for a company that manufactures fire suppression systems, you're collaborating with a project manager on the development of the project management plan for a new state-of-the-art fire suppressor. It's the project manager's responsibility to ensure that requirements-related activities are performed on time and:
    a. Are completed within budget
    b. Are within the boundaries of the charter
    c. Are within budget and deliver value
    d. Are approved by the project management office (PMO)

    The correct answer is: **C**

    When working on projects, the business analyst and project managers collaborate to deliver a solution that meets the needs of the stakeholders, as established in the business case. From the perspective of the responsibility assignment matrix, the assigned project manager is responsible for ensuring that requirements-related work is accounted for in the project management plan and that these activities are performed on time, are within budget, and deliver value.

Answer Choice A: This answer choice is partially correct; however, it is missing "deliver value," which is the link to the business case and the value proposition.

Answer Choice B: This answer choice is the distractor; the charter is the artifact that authorized the project.

Answer Choice D: This answer choice is too generic.

Business Analysis for Practitioners: A Practice Guide, Section 1.7.1, to review "Responsibility for the Requirements."

PMI Professional in Business Analysis (PMI-PBA)® Examination Content Outline, 2013, "Needs Assessment," Task 5.

3. In your role as business analyst for a company that manufactures roller coasters, you are preparing for an elicitation session with your primary stakeholders to discuss changes to an existing model. During these sessions, you will want to cover the following types of requirements:
    a. Functional and business
    b. Functional, service quality, solution, and business
    c. Nonfunctional, transition, business, quality, stakeholder, solution, and functional
    d. Business, functional, stakeholder, solution, transition, and other requirements that describe conditions or qualities required for the product to be effective.

The correct answer is: D

Business processes drive product, quality, and stakeholder requirements, which spell out and articulate a business need or a required capability of a solution. There are four categories of requirements: (a) product, (b) stakeholder, (c) quality, and (d) project. The business analyst will collaborate with the project manager on stakeholder and quality requirements and will have the lead on product requirements. The project manager will solicit input from the business analyst on project requirements but will have ultimate accountability. Table 1.5: Requirement Types (page 12) and Table 1.6: Requirement Examples (page 13) further explain the categories that are likely to appear on the exam.

Answer Choice A: This answer choice is partially correct; however, it omits stakeholder, solution, transition, and requirements.

Answer Choice B: This answer choice is partially correct; the distracter is service quality. ***This is a confusing topic on the exam***—nonfunctional requirements are quality requirements that describe the quality characteristics of the solution, whereas quality of service describes the quality of the delivered product or solution.

Answer Choice C: This is a distractor—nonfunctional and quality are synonymous.

Business Analysis for Practitioners: A Practice Guide, Section 1.7.2, "Requirements Types."

PMI Professional in Business Analysis (PMI-PBA)® Examination Content Outline, 2013, "Needs Assessment," Task 5.

4. Your colleague, Alice, has asked for help defining objectives and goals that define context, provide direction, and are easily understood. From the samples provided below, which would you recommend be included in the business case?
    a. To meet the organizational goal of reducing its carbon footprint by 10% next year, solar panels will be installed over all the parking lots and all the office buildings.
    b. To reduce organizational carbon footprint by 10%.

c. To meet the organizational goal of reducing its carbon footprint by 10% next year, solar panels will be installed over all the parking lots by the end of Q2 and all the office buildings by the end of Q4.

d. To install solar panels over all the parking lots by the end of Q2 and all the office buildings by the end of Q4.

The correct answer is: **C**

This question tests your understanding and application of the SMART principle, in which established goals and objectives are **s**pecific, **m**easurable, **a**greed to by all parties, **r**ealistic to achieve in the allotted schedule, and **t**ime bound. Answer choice C most closely addresses all the principles of SMART goals and objectives.

Answer Choices A, B, & D: While each answer choice offers components of SMART goals, choice C more closely addresses all the aspects.

Business Analysis for Practitioners: A Practice Guide, Section 2.4.1.2, "SMART Goals and Objectives."

PMI Professional in Business Analysis (PMI-PBA)® Examination Content Outline, 2013, "Needs Assessment," Task 3.

5. Cassius has been designated as your subject matter expert from the poultry processing plant. It's been incredibly difficult to schedule time with him to establish the evaluation metrics and measurement tools for the newly designed solution to euthanize turkeys. Rather than delay this meeting any longer, as the business analyst, you decide to:

a. Review the existing service-level agreement.

b. Visit the plant unannounced and request a meeting.

c. Conduct a review of the consensus criteria with the designated representatives from the plant.

d. Perform an event analysis to first understand any potential concerns with the existing evaluation metrics and acceptance criteria, then schedule a meeting with Cassius.

The correct answer is: **A**

Business analysts may often encounter unforeseen challenges and issues when conducting elicitation sessions. To be effective in their roles, they need to think outside the box to overcome these challenges. This practice guide lists four common challenges, and this practice question focuses on stakeholder accessibility when attempting to establish evaluation metrics and success criteria. Applicable measurement tools include service and operational-level agreements and Planguage. An acceptable interim solution is to conduct a documentation review, focusing on information and not on the individual resource.

Answer Choice B: This is a distractor; unannounced meetings are often unwelcome.

Answer Choice C: This is a made-up answer choice.

Answer Choice B: This is distractor; an event analysis is a structured review and would not be applicable in this scenario.

Business Analysis for Practitioners: A Practice Guide, Section 4.8, "Elicitation Issues and Challenges."

PMI Professional in Business Analysis (PMI-PBA)® Examination Content Outline, 2013, "Analysis," Task 8.

6. The executive sponsor of your wind turbine project asked that you present an analysis comparing potential solution options for the next generation of rotor blades made of carbon fiber that would be commercially available in five years. Your analysis is based on which of the following?

a. Are the options presented aligned with the organization's requirements for sustainability and reliability; can the organization acquire the required technology; is it financially feasible; and can the solution be delivered in the timeframe outlined?
b. Would the project be based on the project management iron triangle considerations of quality as constrained by time, cost, and scope?
c. Would the analysis be based solely on business requirements, stakeholder requirements, and solution requirements?
d. Can the organization acquire the technology; is it financially feasible; and can the solution be delivered in the timeframe outlined?

The correct answer is: **A**

When considering solution options, business analysts may often need to assess their feasibility so the organization can determine the preferred option. Feasibility studies often consider (a) operational, (b) technology/system, (c) cost-effectiveness, and (d) time constraints associated with each option. In this scenario, the business analyst is tasked with conducting a complete feasibility analysis. Organizations will often have "blue sky" ideas; given the current environment, we need to determine the "art of the possible."

Answer Choice B: This is a distractor; an analysis of the iron triangle would not help in the assessment of potential solution options.

Answer Choice C: This is a distractor; requirements are based on processes. An analysis of business, stakeholder, and solution requirements would not help in the assessment of potential solution options.

Answer Choice D: While partially complete, it is missing nonfunctional operational feasibility considerations, which can include maintainability, reliability, sustainability, and supportability.

Business Analysis for Practitioners: A Practice Guide, Section 2.5.4, "Assess Feasibility and Organizational Impacts (Operations, Technology/System, Cost-Effectiveness, & Time)."

PMI Professional in Business Analysis (PMI-PBA)® Examination Content Outline, 2013, "Needs Assessment," Task 2.

7. Your company is working on a redesign of a motorized scooter, marketed for active individuals in their golden years. To better understand the concerns and limited adoption of the current model, you decide to take the scooter to the local supermarket and then to the mall for the afternoon. How would this activity be best characterized?
   a. Passive
   b. Participatory
   c. Simulation
   d. Active

The correct answer is: **C**

This question tests your knowledge of the observation concept, which allows individuals to gain a greater understanding of the situation under real-world conditions. There are four categories of observation: (a) passive, (b) active, (c) participatory, and (d) simulation. When using the simulation technique, the observer mimics or recreates the activities from the perspective of the end user. In this scenario, by test driving the scooter, the business analyst is simulating its use by potential customers. This can also be performed in test or training facilities.

Answer Choice A: When using passive observation, there is absolutely no interruption with the individual performing the process.

Answer Choice B: When conducting a participatory observation, the observer would jointly perform the activities with the individual performing the process.

Answer Choice D: When using active observation, the observer would ask questions of the individual performing the process.

Business Analysis for Practitioners: A Practice Guide, Section 4.5.5.6, "Observation."

PMI Professional in Business Analysis (PMI-PBA)® Examination Content Outline, 2013, "Analysis," Task 3.

8. You are mediating a session with Delaney, a well-respected customer, and Meilani, the head of the solution team, trying to come to terms on the final requirements. As a well-trained business analyst, you know there are several techniques for conflict resolution, some of which include compromising, withdrawing, and accommodating, but there can also be others; which one is crucial and most important?
   a. Accommodating
   b. Compromising
   c. Withdrawing
   d. Collaborating

The correct answer is: **D**

To build consensus and resolve conflict, there are several techniques a well-trained business analyst can use. Often, the most important technique is collaborating or arriving at a win–win solution that satisfies the interests of both parties; this is sometimes referred to as *confronting* or *problem solving.* Other conflict management and resolution techniques can include (a) forcing, which is also known as *competing;* (b) compromising; (c) withdrawing or avoiding; and (d) smoothing, also known as *accommodating.*

Answer Choices A–C: While these are all valid conflict management and resolution techniques, collaborating is crucial to ensuring the success of the project, as all parties have a vested interest in the solution.

Business Analysis for Practitioners: A Practice Guide, Section 4.15, "Resolve Requirements-Related Conflicts."

PMI Professional in Business Analysis (PMI-PBA)® Examination Content Outline, "Knowledge and Skills," #7, Conflict Management and Resolution Tolls and Techniques.

PMI Professional in Business Analysis (PMI-PBA)® Examination Content Outline, 2013, "Analysis," Task 5.

9. You've requested that Charity draft a document outlining how changes to product elements will be completed and communicated over the life of the project. The document is a:
   a. Requirements management plan
   b. Requirements change management plan
   c. Requirements traceability matrix
   d. Requirements work plan

The correct answer is: **B**

Over the life of the project, because change is inevitable, the requirements change management plan will outline and establish how changes to product elements will be evaluated, communicated, and addressed. The delivery methodology will influence the level of rigor and formality around the change process. For example, adaptive lifecycle projects anticipate and welcome change, whereas predictive and iterative projects may look for a more formal process. In either case, the change management plan establishes the framework and guidelines for the change process.

Answer Choice A: This is a distractor. Although the change management plan is a component of the requirements management plan, answer choice B addresses the question in a more complete manner.

Answer Choice C: The traceability matrix is used for tracing and tracking requirements over the duration of the project.

Answer Choice D: The work plan establishes the required tasks to deliver the solution; it will not establish how changes to product elements will be completed and communicated over the life of the project.

Business Analysis for Practitioners: A Practice Guide, Section 3.4.14, "Define the Requirements Change Process."

PMI Professional in Business Analysis (PMI-PBA)® Examination Content Outline, 2013, "Planning," Task 4.

10. You are leading an approval session and are asking for formal signoff on the requirements documents. At a minimum, from whom should you seek formal signoff?
    a. Sponsor, product owner
    b. Product owner, solution team lead
    c. Business analyst, product owner, head of product development
    d. Sponsor, project owner, solution team lead

The correct answer is: **C**

The culture of the organization and the formality of approval sessions will influence the required number of signatures. During project planning, the responsibility assignment matrix (RACI) was created, which established accountability for requirements signoff. This matrix was included as part of the requirements management plan and business analysis plan. Some organizations may require physical (aka *wet*) signatures on the requirements documents, while others will allow for electronic signatures or voting. Remember, formal signoff was outlined during the planning phase of the project.

Answer Choices A, B, and D: At a minimum, the business analyst, business owner, and the person responsible for building the solution requirement should sign off on the requirement documents. Answer choice C completely addresses this by including the business analyst, product owner, and head of product development.

Business Analysis for Practitioners: A Practice Guide, Section 4.14, "Approval Sessions."

PMI Professional in Business Analysis (PMI-PBA)® Examination Content Outline, 2013, "Analysis," Task 5.

11. In a meeting with your sponsor, Ellie, she had recommended that you complete a RACI for the two highly visible projects that you're supporting for the multinational agricultural company. As you complete the RACI, who has accountability for providing input to the current state of the company and identifying any opportunities associated with the project?
    a. Your sponsor, Ellie
    b. Your sponsor, Ellie, and you in your role as business analyst
    c. Your sponsor, Ellie, product management, and you in your role as the business analyst
    d. Your sponsor, Ellie, product management, project management, and you in your role as business analyst

The correct answer is: **A**

RACI charts, also known as *responsibility assignment matrices* (RAMs) and *linear responsibility charts* (LRCs), are incredibly useful tools, with a broad range of applicability. ***On the exam, you can expect a few questions pertaining to this topic and tool.*** In this scenario, the sponsor, as the senior executive, would be accountable for providing input to the current state of the company, identifying any opportunities, and overall approving the needs assessment and business case. For each task or deliverable, there can be only ONE person designated as "accountable." Remember the saying, "Who's Got the 'A'?"

Answer Choices B–D: These answer choices offer invalid combinations of roles. The business analyst is responsible for conducting the needs assessment and creating the business case; other team members would be consulted about the current state of the company and identifying any opportunities associated with the project.

Business Analysis for Practitioners: A Practice Guide, Section 2.3.1, "Identify Stakeholders."

PMI Professional in Business Analysis (PMI-PBA)® Examination Content Outline, 2013, "Needs Assessment," Task 4.

12. What tool or technique can be used to determine if a requirement cannot be satisfied without another requirement being present, and what is a potential outcome?
    a. Dependency impact assessment, product documentation
    b. Requirements traceability matrix, issue management
    c. Dependency analysis, traceability tree
    d. Requirements traceability matrix, product documentation

The correct answer is: **C**

A dependency analysis is a technique used to determine whether a requirement cannot exist without the presence of another. Some tools show these relationships visually in the form of ws. There are four possible combinations of dependencies for each requirement: (a) mandatory internal, (b) mandatory external, (c) discretionary internal, and (d) discretionary external.

Answer Choice A: This answer choice tests your knowledge of the tools and techniques. Impact assessments are used to evaluate changes. Dependency impact assessment is a made-up tool.

Answer Choices B, D: While the requirements traceability matrix will record the dependency, a dependency analysis would first determine whether the requirement could exist without the presence of another. Product documentation and issue management could be used as an input during this assessment.

Business Analysis for Practitioners: A Practice Guide, Section 5.3, "Relationships and Dependencies."

PMI Professional in Business Analysis (PMI-PBA)® Examination Content Outline, 2013, "Traceability and Monitoring," Task 3.

13. It's been requested that you work with Arabella, the head of zipper manufacturing, to address nonconformance concerns with the product. Upon meeting with Arabella and her team, you recommend:
    a. Preventative measures, focusing primarily on training
    b. Preventative measures, addressing the time to manufacture each zipper
    c. Reviewing the 7QC tools to determine which ones are most appropriate within the context of the Deming Cycle (Plan, Do, Check, Act)
    d. Creating a SIPOC diagram to identify the suppliers, the process inputs and outputs, and the customer's requirements

The correct answer is: **C**

To analyze and address quality issues, business analysts can use one of the seven basic quality tools, also known as *7QC tools*, within the framework of the repetitive four-stage cycle of Plan, Do, Check, Act. Developed by Edwards Deming, it is sometimes referred to as the *Shewhart Cycle*. The tool later evolved to PDSA, replacing "Check" with "Study" to focus on analysis. The tools include cause-and-effect diagrams, flowcharts, checklists, Pareto diagrams, histograms, control charts, and scatter diagrams.

Answer Choices A and B: This answer choice is a distractor, as preventive measures could potentially be an output of the 7QC tools.

Answer Choice D: SIPOC diagrams list the suppliers relative to the process, the inputs to the process, the associated process tasks, the process outputs, and the customers of the process. This mapping technique would not be helpful to address nonconformance concerns with a product.

A Guide to the Project Management Body of Knowledge (PMBOK®), Section 8.1.2.3 to review the Seven Basic Quality Tools.

PMI Professional in Business Analysis (PMI-PBA)® Examination Content Outline, 2013, "Evaluation," Task 2.

14. Your sponsor, Wyatt, suggested spending additional money during the project to avoid production failures. What are the elements associated with conformance costs?
    a. Building a quality product and destructive testing loss
    b. Prevention costs and liabilities
    c. Testing and liabilities
    d. Testing, liabilities, and rework

The correct answer is: **A**

The cost of quality (COQ) includes all the costs associated with conformance (prevention and appraisal) and nonconformance (internal/external failure). Prevention (relates to creating a quality product: training, documentation, equipment, and allocating the time necessary to correctly perform the process) and appraisal (evaluating the quality: testing, destructive testing loss, and inspections). In this scenario, the elements associated with conformance costs are (a) building a quality product (prevention) and (b) destructive testing loss (appraisal).

Answer Choice B: This answer choice is partially correct. Prevention costs are accurate; however, liabilities are associated with external failure costs, which relate to nonconformance.

Answer Choice C: This answer choice is partially correct. Testing costs are accurate; however, liabilities are associated with external failure costs, which relate to nonconformance.

Answer Choice D: This answer choice is partially correct. Testing costs are accurate; however, liabilities are associated with external failure costs, and rework is associated with internal failure costs, which are both related to nonconformance.

A Guide to the Project Management Body of Knowledge (PMBOK®), Section 8.1.2.2, "Cost of Quality," can be categorized as Conformance and Nonconformance.

PMI Professional in Business Analysis (PMI-PBA)® Examination Content Outline, 2013, "Evaluation," Task 2.

15. Amara has started to build the requirements traceability matrix, beginning with high-level items and filling in the details as they become known. What is this process referred to as?
    a. Work breakdown structure (WBS)

b. Traceability matrix requirements

c. Detailed traceability

d. Progressive elaboration

The correct answer is: **D**

As details become known, the business analyst can update the matrix to a greater level of detail. The concept is that the matrix is built hierarchically, and the logical order is not disrupted as new requirements are added; furthermore, traceability can start immediately, rather than waiting for the lowest level of detail. This technique is known as *progress elaboration*.

Answer Choice A: The work breakdown structure is the logical decomposition of work, not a process.

Answer Choices B & C: This are made-up answer choices.

Business Analysis for Practitioners: A Practice Guide, Section 5.2.3.2.

PMI Professional in Business Analysis (PMI-PBA)® Examination Content Outline, 2013, "Traceability and Monitoring," Task 3.

16. Navi, an experienced programmer, has requested that Mahala, the software quality assurance lead, check her conceptual logic in a very complicated algorithm, asking that she manually record the results in hopes of uncovering any problems before she begins coding, which is expected to take several months. What technique will be Mahala be performing?

a. Desk checking

b. Algorithm review

c. Conceptual code review

d. Planguage review

The correct answer is: **A**

Desk checking is an informal, manual technique, often used by programmers to verify code and logic. Often performed prior to coding, teams can perform desk checks, whereby a peer will act like a computer and carefully review and manually notate the logic and results.

Answer Choices B–D: These are all made-up answer choices.

Examination Content Outline, Knowledge, and Skills, #39, "Verification Methods and Techniques."

PMI Professional in Business Analysis (PMI-PBA)® Examination Content Outline, 2013, "Analysis," Task 8.

17. Scarlett is writing an acceptance test for a user story related to a debit card transaction at a gas station. What can you suggest as the ideal format?

a. Step-by-step verification, based on the Agile Manifesto.

b. Verification based on the actors and personas.

c. As a checklist for actions related to actors, personas, events, and logic.

d. Given that a bank account has a positive balance, when a user processes a transaction less than the current balance, the transaction should complete without any error conditions.

The correct answer is: **D**

User stories are commonly associated with the agile software development methodology. They are no more than a few sentences long and are essentially a conversation about the desired functionality.

Large user stories are known as *epics*, which are decomposed into manageable pieces. The preferred logic to validate a user story is based on the format given-when-then. Answer choice D is a properly formatted user story.

Answer Choices A–C: These are all made-up answer choices.

Examination Content Outline Knowledge and Skills, #37, "Validation Tools and Techniques."

PMI Professional in Business Analysis (PMI-PBA)® Examination Content Outline, 2013, "Analysis," Task 8.

18. Greta is hosting a video conference to elicit information so that she may thoroughly document the solution requirements. Interviews performed in this manner are:
    a. Asynchronous
    b. Unstructured
    c. Synchronous
    d. Structured

The correct answer is: **C**

This question tests your knowledge pertaining to interviewing techniques, which can be either formal or informal, to elicit responses from stakeholders. Interviews can be either structured or unstructured, and can be conducted synchronously or asynchronously. This is an example of a synchronous interview, which is conducted in real time, either in person or via audio or video conference. With the proliferation of collaboration tools and globalization of our workforce, it's quite common for synchronous interviews to be held in formats other than in person. However, if provided the opportunity, in-person interviews would be the preferred method.

Answer Choice A: Asynchronous interviews are not conducted in real time; they are often recorded or conducted via email.

Answer Choice B: *Unstructured* refers to an interview type in which the interviewer has a list of questions, but it is conducted in a more free-flowing format.

Answer Choice D: *Structured* refers to an interview type in which the interviewer has a defined set of questions, and the elicitation is conducted with the intent of asking all the questions on the list.

Business Analysis for Practitioners: A Practice Guide, Section 4.5.5.5, "Interviews."

PMI Professional in Business Analysis (PMI-PBA)® Examination Content Outline, 2013, "Analysis," Task 1.

19. Nyla, the business analyst for a hedge fund, has team members located in 12 countries on four continents. Which of the following should not be a concern for her global team?
    a. Will team members work at locations other than their primary location, and what is a typical work day?
    b. How will team members contribute to the decision-making process and address conflict?
    c. What is the approach to documentation analysis?
    d. How will requirements be elicited, managed, maintained, and approved?

The correct answer is: **B**

This scenario-based question requires careful reading. From the answer choices provided, which would *not be* a concern for the global team.

With the proliferation of collaborative technology, teams have become stronger and more diverse, as more team members and stakeholders work remotely. It's therefore more commonplace that business analysts and team members have identified and established the optimal approach to engagement over the lifecycle of the initiative. Answer choice B pertains to addressing conflict and contributing to the decision-making process, both of which are cultural considerations and would be of less concern than the other answer choices provided.

Answer Choices A, C, and D: These answer choices pertain more to location and availability and would be areas that require careful attention and planning.

Business Analysis for Practitioners: A Practice Guide, Section 3.3.2.6, "Location and Availability."

PMI Professional in Business Analysis (PMI-PBA)® Examination Content Outline, 2013, "Planning," Task 3.

20. Past projects at Soleil, Aubrey & Hart LLC, a leading management consulting firm, have been tumultuous, resulting in losses in the tens of millions of dollars. In a turnaround effort, the new chief project officer, Garth, recognizes that the organization must focus on key areas to improve the effectiveness of requirements management. As his business analyst, you would suggest:
    a. Skills development, formalization and standardization of processes, garnering the support of senior leadership
    b. Implementing the principles of business analysis and requirements management
    c. Using a business analysis plan, a requirements management plan, and change control
    d. Following the processes and using the tools and techniques from PMI publications

The correct answer is: **A**

As outlined in PMI's *Pulse of the Profession®*, research has shown that to effectively improve business outcomes, organizations need to focus on culture, people, and processes. This includes skills development, formalization and standardization of processes, and garnering the support of senior leadership.

Answer Choice B–D: While these may seem like logical and reasonable answer choices, they don't properly address the question.

PMI's Pulse of the Profession®, Requirements Management: A Core Competency for Project and Program Success, page 4, "Executive Summary."

PMI Professional in Business Analysis (PMI-PBA)® Examination Content Outline, 2013, "Analysis," Task 1.

21. As you verify that the delivered product is fulfilling the outlined requirements, you make note of them, referring to:
    a. A checklist in your team's collaboration portal
    b. The project management plan
    c. The schedule management plan
    d. The time management plan

The correct answer is: **A**

This question tests your knowledge of both domain orientation and tool selection. The business analyst is assessing whether the delivered solution is fulfilling the established requirements, which means we're in the Evaluation domain, performing a quality assessment. As structured documents, checklists can be used to verify that the delivered product is aligned with the customer's requirements. Also known as *checksheets* or *tally sheets,* they are great for simply collecting and tracking data. Once complete, they become project artifacts and are considered organizational process assets.

Answer Choice B: The project management plan is the aggregation of all the subsidiary project plans.
Answer Choice C: The schedule management plan is a subsidiary component of the overall project management plan. This plan outlines the *how* and *when* related to the activities associated with creating, monitoring, and controlling the project schedule.
Answer Choice D: Project time management is a knowledge area and is not a suitable answer choice.
A Guide to the Project Management Body of Knowledge (PMBOK®), Section 8.3.1.3, "Quality Checklists."
PMI Professional in Business Analysis (PMI-PBA)® Examination Content Outline, 2013, "Evaluation," Task 1.

22. You've been assigned to work with Genevieve, the supervisor for the pipefitters at a high-rise office building in midtown Manhattan, to understand why the water pipes are losing pressure. Following a walkthrough, you create an Ishikawa diagram and determine the root cause of the problem:
    a. The journeyman's apprentice assembled the areas of the pipes at which defects were found.
    b. Will necessitate spending money to avoid project failures.
    c. Will result in a claim to the trade union.
    d. Will be categorized as an internal project failure.

The correct answer is: **D**

This scenario-based question tests your knowledge of both the 7QC tools and the cost of quality (COQ). Following the use of an Ishikawa diagram (more commonly referred to as either a fishbone or a cause-and-effect diagram), internal project failures result in rework and scrap. These types of nonconformance costs can be incurred both during the project and after launch as the direct result of failures. Failures found by the project are categorized as internal failure costs.
Answer Choices A and C: These are distracters and don't properly address the question.
Answer Choice B: This answer choice describes conformance costs, which can be categorized as either prevention or appraisal costs.
A Guide to the Project Management Body of Knowledge (PMBOK®), Section 8.1.2.2, Cost of Quality (COQ) *PMI Professional in Business Analysis (PMI-PBA)® Examination Content Outline*, 2013, "Evaluation," Task 2.

23. Saoirse is the business analyst working in the corporate accounting department of a Fortune 100 multinational company. She is assisting with the deployment of Hyacinth P&L, a software package designed to consolidate financial information from accounting systems in 106 countries. As part of a companion project, she'd like to ask her global colleagues for their input on an assorted number of topics. What is the most efficient way to engage her colleagues?
    a. Send them an email containing the questions, requesting that they respond in kind.
    b. Schedule an audio conference at a mutually convenient time for all participants.
    c. Have your assistant Luna prepare a Survey Monkey.
    d. Plan a focus group with all participants within the next week.

The correct answer is: **C**

This is a very wordy question and provides a lot of unnecessary information. On the exam, *remember to focus your answer on the question only and weed out the noise.* In this scenario, the question is, "What's the most effective way to elicit responses on a number of topics from a dispersed group of stakeholders?" Questionnaires and surveys are the most efficient means of polling and aggregating responses from a large population. Please keep in mind, response rates can average between 4% and 5% within organizations.

Answer Choice A: Although email may be easy and seem like a logical response, it is not the most efficient means to poll and aggregate responses from a large population.

Answer Choice B: Audio conferences are by no means effective platforms for engaging colleagues in the described scenario.

Answer Choice D: This would not be an appropriate choice, considering the dispersed nature of the stakeholders.

Business Analysis for Practitioners: A Practice Guide, Section 4.5.5.8, "Questionnaires and Surveys."

PMI Professional in Business Analysis (PMI-PBA)® Examination Content Outline, 2013, "Analysis," Task 1.

24. Talia has witnessed a dramatic change in her organization since they embraced business analysis. They are continuously monitoring capability and capacity, and they are identifying and implementing changes to improve project delivery. It can be said that the organization, its people, and its processes and tools are:
    a. Maturing
    b. Static
    c. Aligned
    d. Focused

The correct answer is: **A**

This scenario describes an organization that is maturing. For organizations to be successful, they need to change, and projects are one way to facilitate that change. Organizations that are skillful at project execution continuously monitor capability and capacity and identify and implement changes to improve project delivery.

Answer Choice B: Static implies that the organization is lacking in progress and change. This would not be an appropriate answer choice to describe the scenario.

Answer Choices C and D: These answer choices may seem logical; however, maturing more accurately describes the organization.

Pulse of the Profession®, Requirements Management: A Core Competency for Project and Program Success, page 4; "Executive Summary."

PMI Professional in Business Analysis (PMI-PBA)® Examination Content Outline, 2013, "Analysis," Task 7.

25. Everett has drafted documents that accurately describe the solution to be built. It can be said that his requirements documents are:
    a. Complete
    b. Precise
    c. Unambiguous
    d. Correct

The correct answer is: **D**

This question tests your definitional knowledge of requirements document characteristics. Processes drive requirements, which need to be translated into high-quality, properly formatted documents, often referred to as *specifications*. One of the nine characteristics is correctness, which is not absolute; this can only be achieved through progressive elaboration and refinement.

Answer Choice A: *Complete* is also a reasonable answer choice, but it implies that there could be missing elements. Answer choice D is the better selection.
Answer Choice B: *Precise* is synonymous with *precision*.
Answer Choice C: *Unambiguous* is synonymous with *clarity*.

Business Analysis for Practitioners: A Practice Guide, Section 4.11.5.1, "Functional Requirements."
PMI Professional in Business Analysis (PMI-PBA)® Examination Content Outline, 2013, "Analysis," Task 6.

26. You work for the Great Brick Company, which for the past 150 years has had a reputation for producing the strongest, most durable bricks on the market. Over the last few months, contractors have reported sporadic cases of bricks cracking and crumbling. You have been asked by Oliver, a senior product manager, for assistance determining the cause of the defect. To begin the analysis, you decide to start with a:
    a. Monte Carlo analysis
    b. Tornado diagram
    c. Pareto chart
    d. Ishikawa diagram

The correct answer is: **D**

In this scenario, to determine the cause of the defect, a business analyst could use a root-cause analysis tool, like an Ishikawa diagram (also known as a cause-and-effect or fishbone diagram). This is one of the 7QC tools. Other potential tools could be 5 Whys, interrelationship diagrams, and process-flow diagrams. Costs associated with rework, scrap, liabilities, warrantees, and lost business are all categorized as nonconformance.

Answer Choice A: A Monte Carlo analysis is a simulation technique based on random sampling, which generates quantitative data to assist with informed decision making.
Answer Choice B: A Tornado diagram is a special type of bar chart, in which the data categories are shown on the y (vertical) axis. The finished chart resembles a tornado, ordering the items with the biggest impact starting at the top.
Answer Choice C: A Pareto chart, named after Italian economist Vilfredo Pareto (1895), is a data representation based on the Pareto Principle, which is commonly referred to as the 80/20 rule. The principle suggests that approximately 20% of the causes account for 80% of the work.

Business Analysis for Practitioners: A Practice Guide, Section 2.4.4.2, "Cause-and-Effect Diagrams."
PMI Professional in Business Analysis (PMI-PBA)® Examination Content Outline, 2013, "Needs Assessment," Task 2.

27. Construction has just completed at your executive sponsor Bobby Phil's new restaurant in downtown Manhattan. During design, a number of requirements were identified, such as sustaining oven temperature, color temperature ranges for all the guest areas, and required refrigerator and freezer temperatures. Now the project team is preparing to train employees for the new restaurant. On what document would the training requirements be identified?
    a. Training and cutover requirements
    b. Business requirements
    c. Nonfunctional requirements
    d. Transition requirements

The correct answer is: **D**

This is a very wordy question, designed to distract exam candidates, which is quite common on the PMI-PBA® Exam. *Remember to focus on exactly what the question is asking*—in this case, it's the last sentence, "On what document would the training requirements be identified?" Transition requirements focus on temporary activities to transition from the *as-is* to the *to-be* states. Tasks typically associated with transition requirements are training and data conversion.

Answer Choice A: This is a made-up answer choice.

Answer Choice B: Business requirements describe the higher-level needs of the enterprise. Typically in the form of a business requirements document (BRD), they describe the purpose/intent of the component and metrics to evaluate its impact on the organization. BRDs contain the component goals, which are linked to the strategic plan. The BRD focuses on the entire enterprise, not on specific organizational levels.

Answer Choice C: Nonfunctional requirements are a subset of solution requirements, which describe the environmental conditions of the component. They are typically characterized as *quality* or *additive* requirements and are described as "must haves."

Requirements Management: A Practice Guide, Section 5.2.2."Define Types of Requirements."

PMI Professional in Business Analysis (PMI-PBA)® Examination Content Outline, 2013, "Analysis," Task 6.

28. Your sponsor, Alessandra, would like to use a tool that is contingent on peer pressure to resolve a conflict. You advise her:
    a. This sounds a little Machiavellian and should not be a part of any sound method for resolving conflict.
    b. We can use the Delphi tool.
    c. The nominal group technique would be an appropriate solution.
    d. Take a step a back and think this through and evaluate other options, because peer pressure is not part of any sound method of resolving conflict.

The correct answer is: **B**

The Delphi is a structured technique used to facilitate informed decision making, designed to allow participants to comment anonymously on a particular subject matter. Commonly used for drafting policies, forecasting, or solutionizing, it relies on a panel of subject matter experts to arrive at a consensus. The panel can consist of individuals both internal and external to an organization and is always moderated by a facilitator. When used properly, all participants remain anonymous, even after the final decision or report is released. The facilitator is commonly an expert and may also hold a particular point of view. When using the Delphi technique, there are no less than two rounds of review; the process typically begins with a questionnaire. The strength of the tool is that it relies on peer pressure to reach consensus.

Answer Choices A and D: These are made-up answer choices.

Answer Choice C: The nominal group technique is optimally used so that all attendees can participate equally. The tool is also especially helpful in cases in which the team is having difficulty deriving potential solutions. It does not rely on peer pressure.

Business Analysis for Practitioners: A Practice Guide, Section 4.15.1, "Delphi."

PMI Professional in Business Analysis (PMI-PBA)® Examination Content Outline, 2013, "Analysis," Task 5.

29. Your sponsor, Callum, is familiar with the terms *quality assurance* and *quality control*. When he suggests that you identify the causes of metal failure in the company's bolts and recommend corrective action, he's referring to:

a. Quality analysis
b. Control charts
c. Quality control
d. Pareto Diagrams

The correct answer is: **C**

On the exam, you can expect a number of questions pertaining to both quality assurance (QA) and quality control (QC). Quality assurance is focused on prevention; with quality control the emphasis is on testing and inspection. Quality control (a) identifies the causes of poor product quality, (b) recommends corrective action, and (c) validates that the deliverables are in line with stakeholder expectations.

Answer Choice A: Quality analysis is a term commonly used in software development; it is also concentrated on inspection, ensuring that procedures are followed as established by the project management office.

Answer Choice B: Control charts are often used in Six Sigma to visually show how a process changes over time. There are three limits that the business analyst will study: the average, the upper control limit, and the lower control limit.

Answer Choice D: A Pareto chart, sometimes called a Pareto diagram, is named after Italian economist Vilfredo Pareto (1895). Data is represented based on an analysis that contends that approximately 20% of the causes account for 80% of the work. When you see "Pareto" on the exam, remember the 80/20 rule.

A Guide to the Project Management Body of Knowledge (PMBOK®), Section 8.3, "Quality Control."

PMI Professional in Business Analysis (PMI-PBA)® Examination Content Outline, 2013, "Evaluation," Task 2.

30. Paladin is a business analyst for Anwen Foods Inc, a world-class manufacturer of kitchen equipment for pizza restaurants. They are planning to expand their products to the Korean barbecue market, and in the process, they would gain a better understanding of how competitors' products are used. What form of observation would best help them to experience how their product would be used and to ascertain the criteria for acceptance in the marketplace?
    a. Active
    b. Simulation
    c. Passive
    d. Participatory

The correct answer is: **D**

This question tests your knowledge of the observation concept, which allows individuals to gain a greater understanding of the situation under real-world conditions. There are four categories of observation: (a) passive, (b) active, (c) participatory, and (d) simulation. With participatory, the observer simulates or recreates the activities from the perspective of the end user—in this scenario, experiencing how the product would be used.

Answer Choice A: When using active observation, the observer would ask questions of the individual performing the process.

Answer Choice B: When using the simulation technique, the observer mimics or recreates the activities from the perspective of the end user. This can also be performed in test or training facilities.

Answer Choice C: When using passive observation, there is absolutely no interruption with the individual performing the process.

Business Analysis for Practitioners: A Practice Guide, Section 4.5.5.6, "Observation."

PMI Professional in Business Analysis (PMI-PBA)® Examination Content Outline, 2013, "Analysis," Task 3.

31. Your sponsor, Zander, requested a meeting with you and the project manager, Carter, to discuss documenting the requirements change management process. As you prepare for the meeting with Zander, you and Carter agree:

   a. As the business analyst, you'll take accountability for documenting the plan, and it will become part of the project management plan.

   b. As the PM, Carter will document the plan as a part of the business analysis plan.

   c. As the business analyst, you'll document the process and communicate it to the stakeholders.

   d. You and Carter will jointly develop the approach and communicate it to all relevant stakeholders.

The correct answer is: **A**

This question is representative of ones you're likely to find on the exam, introducing several distracting words, in this case names of team members. Remember, the requirements change process can either be documented in one of two places. Either in the business analysis plan or included in the change management plan, which is a subsidiary of the project management plan. The most important consideration is that the approach and plan is agreed to by the project manager and business analyst and that it's documented and communicated to stakeholders.

Answer Choice B: If the project manager were to document the plan, it would become a subsidiary of the project management plan, not the business analysis plan.

Answer Choices C & D: These are partially correct answer choices; however, answer choice A is more complete—identifying it will become part of the project management plan.

Business Analysis for Practitioners: A Practice Guide, Section 3.4.14, "Define the Requirements Change Process."

PMI Professional in Business Analysis (PMI-PBA)® Examination Content Outline, 2013, "Planning," Task 4.

32. Miriam has just joined the project team in the lead role responsible for coordinating and monitoring testing as the third round of system integrated testing commences. The team is planning to conduct ten rounds of testing for this highly complex automatous vehicle. As the team progresses to later rounds of system integrated testing, they are questioning why new issues are emerging.

   a. New issues are emerging as new functionality is being tested.

   b. Prior to Miriam's arrival, the test plan was inadequate.

   c. Although the team was conducting testing from the start, they were not charting their results.

   d. Due to the complex nature of the testing, the team needed to purchase a tool to automate the testing, which has uncovered the new issues.

The correct answer is: **C**

One of the biggest challenges with predictive methodologies (e.g., waterfall) is that each phase (i.e., plan, design, build, test, and deploy) must be completed before the next phase begins, and there are generally established exit criteria for each phase. Leading up to the testing phase, organizations have invested a significant amount of resources. The goal is to introduce the concept of evaluation early on in the process. In the case of adaptive methodologies (such as those covered under the agile umbrella), the focus shifts to the principles outlined in the Agile Manifesto, which are intended to deliver value at an increased velocity as compared to the predictive development lifecycle. The two most important aspects of testing are (1) to evaluate and measure early, and (2) to evaluate and measure frequently. Documentation and charting results are key tenets of this process.

Answer Choice A: System integrated testing verifies interoperability and coexistence with other products, applications, and solutions. After the first round, new functionality is typically not introduced, as the focus shifts to refactoring.

Answer Choice B: While the test plan may have been inadequate, answer choice C is a slightly better option.

Answer Choice D: This is a distractor; *remember, on the exam, never buy a tool.*

Business Analysis for Practitioners: A Practice Guide, Section 6.4, "Plan for Evaluation of the Solution."

PMI Professional in Business Analysis (PMI-PBA)® Examination Content Outline, 2013, "Evaluation," Task 1.

33. Alyssa has found that subject matter experts are unsupportive of the business analysis activities detailed within her plan, and there was a misunderstanding regarding their level of involvement for the requirements activities. How could Alyssa have avoided this situation?
    a. She could have involved the subject matter experts when defining the timing and sequence of activities.
    b. She could have involved the broader project team when defining all the activities.
    c. She could have collaborated with the project manager on the approach to business analysis activities.
    d. She could have sought approval of the business analysis plan.

The correct answer is: **D**

Similar to the planning activities associated with project management, the activities for business analysis planning are critical to ensuring that all stakeholders clearly understand the approach and intent for the effective management of product requirements. Prior to putting the business analysis plan into action, it's critical that you have approval from the sponsor and engagement of the key stakeholders, whose resources will be supporting the activities and deliverables. Resource commitment is often a contentious topic and one of concern for non-projectized organizations.

Answer Choice A: Whereas subject matter experts may have been involved with the timing and sequencing of activities, their managers would be the ones to allocate their commitment.

Answer Choice B: This is a distractor.

Answer Choice C: Although it's critical to involve the project manager when developing the approach to business analysis activities, doing so would not be sufficient to avoid the situation described in the scenario.

Business Analysis for Practitioners: A Practice Guide, Section 3.5.6, "Obtain Approval of the Business Analysis Plan."

PMI Professional in Business Analysis (PMI-PBA)® Examination Content Outline, 2013, "Planning," Task 7.

34. After several weeks of effort, you have just completed the business analysis work plan for a highly complex project. Following a review session with the project team and key stakeholders, your project manager, Flynn, has expressed concern about full integration with the project management plan. To simplify matters, you suggest that:
    a. The fine details can be tracked in a separate plan; summary information should be integrated with the overall project management plan.

b. Because a majority of the project is focused on requirements, all the details can be integrated, and as the business analyst, you will maintain the overall plan.

c. Considering the highly complex nature of the project, we can maintain two separate project plans.

d. To properly manage the project, both plans need to be fully integrated.

The correct answer is: **A**

Work plans developed by the business analyst detail the product activities to be accomplished by the performing organization. It's leading practice to have an integrated BA work plan at the summary level, with the other project activities. This can assist with sequencing of activities, leveling of resources, and helping to uncover any gaps or inconsistencies. In doing so, the BA will have accountability for managing and directing the work detailed on their plan, and reporting status to the PM.

Answer Choice B: Only in cases in which organizations have dual PM/BA roles would the business analyst maintain the overall project management plan.

Answer Choice C: It's not uncommon for business analysts and project managers to maintain two separate detailed work plans. When done, the project manager will add summary tasks representing the details on the BA work plan to track and monitor deliverables such that there is mid-level integration.

Answer Choice D: While this may be ideal, it's not very realistic. A fully integrated plan to a granular level would be cumbersome to manage.

Business Analysis for Practitioners: A Practice Guide, Section 3.5.2, "Build the Business Analysis Work Plan."

PMI Professional in Business Analysis (PMI-PBA)® Examination Content Outline, 2013, "Planning," Task 4.

35. The business lead for the human capital thread of your program recommended a number of changes to the Change Control Board (CCB). With minimal supporting documentation, the board approved all changes. Now during system integrated testing, your team is uncovering issues. What could have been done to ensure the CCB knew the full picture?

a. The business lead could have conducted an assessment to understand the relationships and considerations to related development objects.

b. The program manager could have asked the business analysts to provide their recommendation to the Change Control Board in person rather than via email.

c. The Change Control Board should have deferred approving all change requests until after system integrated testing.

d. The Change Control Board should have adopted a policy of no changes unless they were foundational to the organization's achieving its business objectives.

The correct answer is: **A**

At the point of system integrated testing (SIT), the testing is focused on verifying the interoperability and coexistence of the solution with other products and applications. To ensure that the Change Control Board understood the broader picture, the business lead should have conducted an impact assessment in collaboration with the project manager and business analyst. From the business analyst's perspective, they would concentrate their efforts on understanding the relationship between the requested change and the existing requirements, understanding the change from a holistic perspective. In addition, a dependency analysis could be conducted to determine whether a requirement cannot exist without the presence of another. Some tools show these relationships visually in the form of traceability trees.

Answer Choice B: This answer choice doesn't properly address the situation.

Answer Choice C: Changes can be proposed at any point in the project; deferring approval wouldn't address the problem.

Answer Choice D: This an unrealistic answer choice, as changes are inherent to projects.

Business Analysis for Practitioners: A Practice Guide, Section 5.8.3, "Impact Analysis."

PMI Professional in Business Analysis (PMI-PBA)® Examination Content Outline, 2013, "Traceability and Monitoring," Task 5.

36. Your subject matter expert for the order entry team, Preston, is describing how a button should work on a redesigned web page. How should these requirements be classified?
   a. Business
   b. Nonfunctional
   c. On the requirements traceability matrix
   d. Functional

The correct answer is: **D**

Functional requirements are a subset of solution requirements, which describe the behaviors or function of the component—a statement of conformity. Described in Yes/No terms, they articulate specific functions that enable stakeholders to complete their goals and intentions.

Answer Choice A: Business requirements describe the higher-level needs of the enterprise. Typically in the form of a business requirements document (BRD), they describe the purpose/intent of the component and metrics to evaluate its impact on the organization. BRDs contain the component goals that are linked to the strategic plan. The BRD focuses on the entire enterprise, not specific organizational levels.

Answer Choice B: The button requirement would be identified on the tracker, but it would not describe how it should work.

Answer Choice C: Nonfunctional requirements are a subset of solution requirements, which describe the environmental conditions of the component. They are typically characterized as *quality* or *additive* requirements and are described as "must haves."

Requirements Management: A Practice Guide, 5.2.2. "Define Types of Requirements."

PMI Professional in Business Analysis (PMI-PBA)® Examination Content Outline, 2013, "Analysis," Task 6.

37. You are the business analyst collaborating with your project manager, Graham, on a project that, when complete, will affect nearly all the residents in a community in Iowa. During your needs assessment, you recommend completing a RACI. Graham asks, what's a RACI? You explain:
   a. It stands for Revitalization Assistance for Community Improvement, a document the sponsor requested be included in the charter.
   b. It stands for Risk-Adjusted Complication Index, an analysis to determine exactly how many residents will be affected by the project.
   c. It is a type of responsibility assignment matrix.
   d. It stands for responsible, accountable, consulted, informed.

The correct answer is: **C**

RACI charts, also known as *responsibility assignment matrices* (RAMs) and *linear responsibility charts* (LRCs), are incredibly useful tools with a broad range of applicability. **On the exam, you can expect a few questions pertaining to this topic and tool.** The responsibility assignment matrix (RACI) was created during project planning, establishing accountability for requirements signoff. This matrix was included as part of the requirements management plan and business analysis plan. They help to clearly identify who's **r**esponsible for completing the work; the one and only person who is **a**ccountable; experts who can be **c**onsulted; and others who need to be **i**nformed.
Answer Choices A & B: These are made-up answer choices.
Answer Choice D: This answer choice outlines the contents of the RACI.
Business Analysis for Practitioners: A Practice Guide, Section 2.3.1,"Identify Stakeholders."
PMI Professional in Business Analysis (PMI-PBA)® Examination Content Outline, 2013, "Needs Assessment, Task 4."

38. Your college intern, Maverick, inquires as to who can request a significant change to the approved requirements traceability matrix. You share:
    a. Once documented, only the customer can request a significant change to the approved product.
    b. The Change Control Board must review and vote on the change request.
    c. Any stakeholder can request a significant change to the approved product, provided the request is documented.
    d. Any team member can recommend a change, provided the request is documented.

The correct answer is: **C**

As the subject matter experts and technical resources develop specification documents, build models, and execute test cases, changes are inevitable. It's the business analyst's responsibility to manage requested changes in accordance with the change control plan, which was produced in the Planning domain, all while managing the interrelated aspect of product delivery. Remember, on a project, change is inevitable, and on adaptive lifecycle projects, change is welcome and expected. *Key for the exam: while any stakeholder can verbally suggest a change to the product, it must be documented and submitted to the Change Control Board to be considered.*
Answer Choice A: This answer choice is misleading, as any stakeholder can request a change.
Answer Choice B: While this is an accurate statement, it doesn't address who can request a change.
Answer Choice D: This answer choice is also misleading, as it states, "recommend a change."
A Guide to the Project Management Body of Knowledge (PMBOK®), Section 4.5, "Perform Integrated Change Control."
PMI Professional in Business Analysis (PMI-PBA)® Examination Content Outline, 2013, "Traceability and Monitoring," Task 5.

39. Your sponsor, Selena, has offered to assist with your efforts to plan for solution evaluation. From the answer choices below, which should not be included as part of your efforts?
    a. Evaluation criteria, acceptance thresholds, how results will be analyzed and reported
    b. When and how often evaluation will be performed
    c. Special measurement tools not used as part of solution evaluation
    d. Focus groups, observations, surveys, qualitative and quantitative activities

The correct answer is: **C**

Evaluation occurs at multiple points during the project; in the planning phase, the business analyst is identifying these activities, the tools and techniques that will be used, and how the information will be collected and reported. They are essentially the tasks and activities that (a) validate test results, (b) analyze and communicate gaps, (c) fulfill the work with stakeholders to obtain signoff, and (d) conclude with an evaluation of the deployed solution. This question is asking "which should *not* be included." *Remember to read the questions very carefully.*

Answer Choices A, B, and D: These all include aspects that *should* be included as part of the efforts to plan for evaluation.

Business Analysis for Practitioners: A Practice Guide, Section 3.4.15, "Define the Solution Evaluation Process."

PMI Professional in Business Analysis (PMI-PBA)® Examination Content Outline, 2013, "Planning," Task 6.

40. Aryan is an experienced business analyst and is highly skilled in elicitation. Tobias, a colleague leading a focus group that Aryan is attending, is having difficulty extracting key requirement baseline information from attendees. During a brief 15-minute break, Aryan suggests Tobias try:
    a. Using only closed-ended and context-free questions
    b. Primarily using contextual and context-free questions, with the occasional closed-ended question
    c. Concentrating to use only open-ended and context-free questions
    d. Using a combination of open-ended, contextual, context-free, and closed-ended questions

The correct answer is: **D**

Elicitation sessions require the use of a number of skills to extract the most relevant and appropriate information. The Practice Guide lists a number of soft skills, including: (a) active listening, (b) empathy, (c) body language, (d) question selection, (e) question sequencing, and (f) influencing. In this scenario, by incorporating a variety of question styles (open-ended, closed-ended, contextual, and context-free), the business analyst is able to accomplish the intended goals and objectives of the focus group.

Answer Choice A: Closed-ended questions, also known as *forced choice*, are best used for confirmation. While context-free questions are appropriate in most all situations, this is not the correct answer choice because of the limitations associated with closed-ended questions.

Answer Choice B: This answer choice lists the three types of questions: (a) contextual, (b) context-free, (c) and closed-ended. Despite neglecting to mention open-ended questions, the phrasing of the answer choice is also not appropriate, as there should be a balance.

Answer Choice C: This answer choice is partially correct; however, it is missing contextual and closed-ended questions.

Business Analysis for Practitioners: A Practice Guide, 4.5.2.1, "Types of Questions."

PMI Professional in Business Analysis (PMI-PBA)® Examination Content Outline, 2013, "Analysis," Task 5.

41. Collaborating with Rhett, your project manager, you have logically decomposed the total scope of work related to the product in a hierarchical manner. What was the output of your effort?
    a. Process model
    b. Work breakdown structure
    c. Capability table
    d. Salience diagram

The correct answer is: **B**

The work breakdown structure (WBS) is a fundamental project management deliverable; it is a logical decomposition, represented in the form of a hierarchy, representing the total scope of the project. It is often accompanied by a WBS dictionary, which is a detailed description of each WBS package.

Answer Choice A: Process models describe business processes; they can take the form of process flows, use cases, or user stories.

Answer Choice C: A capability table outlines the resources needed to solve problem or capitalize on an opportunity.

Answer Choice D: The stakeholder salience diagramming method is used to assist team members when the need arises to prioritize competing stakeholder requests (Wood, 1997). This is an advanced topic; *on the exam, it is unlikely you will encounter questions pertaining to the salience matrix, categories, or diagrams.* Questions will focus on the intent, which is to categorize stakeholders based on the factors of power, urgency, and legitimacy.

Business Analysis for Practitioners: A Practice Guide, Section 5.2.1, "What is Traceability?"

PMI Professional in Business Analysis (PMI-PBA)® Examination Content Outline, 2013, "Traceability and Monitoring," Task 1.

42. The VP of human resources, Septimus, convenes a meeting with Adalyn, Buster, Geneva, Madoc, Adrienne, and Cade, intending to discipline them for playing cards at their desk. What is the appropriate response?
    a. Septimus must refer to the corporate policy for playing poker during working hours.
    b. Adrienne and Cade suggest they were playing Iterative Manjaro, a card gamification technique to arrive at consensus.
    c. Geneva and Madoc offer that the cards weren't based on suits, but rather on the Fibonacci sequence, and were being used to estimate the project.
    d. Adalyn and Buster refer Septimus to the team's Scrum master for clarification, justification, and support of their activities.

The correct answer is: **C**

This question tests your knowledge of agile tools for estimating. Planning poker is a consensus-based technique used by agile teams to estimate the product backlog. It's based on a modified Fibonacci sequence: 0, 1, 2, 3, 5, 8, 13, 20, 40 and 100. You'll note that up to 13, each number is the sum of the two preceding numbers. For example, 5 is 2 plus 3.

Answer Choices A, B, & D: These are all made-up responses.

PMI Professional in Business Analysis (PMI-PBA)® Examination Content Outline, "Knowledge and Skills," #15, Estimation Tools and Techniques.

PMI Professional in Business Analysis (PMI-PBA)® Examination Content Outline, 2013, "Analysis," Task 4.

43. You are the business analyst for Xgen, a global pharmaceutical company supporting the ERP upgrade project. You are four months from go-live and have spent $20 million USD to date. The estimated cost to complete the upgrade is $525,000 USD, all associated with training, as all the development is complete. Your company has just merged with a multinational conglomerate, which completed a similar project last year. You suggest to the project sponsor and CFO:

a. The project should continue as planned; once live we can begin depreciating the capital.
b. We should complete a needs assessment using the results to update the business case and project charter.
c. Using virtual classrooms, training can be fast-tracked, thus delivering the project ahead of schedule and under budget, allowing the team to devote time to the merger.
d. The CFO should meet with the Technology Subcommittee of the Board of Directors to request approval to hire a consulting firm to provide their assistance with the merger.

The correct answer is: **B**

This is a complicated question and requires thoughtful attention. Although the company has invested significant sums of money, it may no longer be in their best interest to continue with the project. Needs assessments are conducted when external factors influence or impact the project. In this example, a merger has just occurred, and although the project is nearing the finish line, it may no longer be viable, as the organization may want to consider the multinational conglomerate ERP instead—a needs assessment will provide the quantifiable basis for a decision.

Answer Choice A: While this may seem like a logical answer choice, it may not be in the best interest of the company. A needs assessment could help make this determination.

Answer Choice C: This is a distractor and doesn't properly focus the team's attention.

Answer Choice D: *On the exam, remember you should never buy a tool or hire a consultant.* Other answer choices will generally provide a sounder approach to a situation.

Business Analysis for Practitioners: A Practice Guide, Section 2.2, "Why Perform Needs Assessments."

PMI Professional in Business Analysis (PMI-PBA)® Examination Content Outline, 2013, "Needs Assessment," Task 1.

44. While building the traceability matrix, Ada has found a number of development objects that seem to be related. In the process of conducting a dependency analysis, she documented:
    a. Implementation, subset, and benefit dependencies
    b. Value and implementation dependencies
    c. Benefit, subset, and value dependencies
    d. Implementation dependencies and related objects

The correct answer is: **A**

When creating the traceability matrix, it's common to find related and dependent objects. A dependency analysis is a technique that will determine whether the requirement could exist without the presence of another. Appropriately categorizing and tracing dependencies will improve stakeholder expectations and the success of the project.

Answer Choices B & C: These answers choices are partially correct; the examples of dependent relationships identified in the Practice Guide are subset, implementation, benefit, and value.

Answer Choice D: This answer choice is partially correct—related objects is not a valid answer.

Business Analysis for Practitioners: A Practice Guide, Section 5.3, "Relationships and Dependencies."

PMI Professional in Business Analysis (PMI-PBA)® Examination Content Outline, 2013, "Traceability and Monitoring," Task 3.

45. You work for a clothing manufacturer that is looking to increase both market share and time to market, considering the recent delays in ocean freight. Your product manager, Ophelia, suggests exploiting 3D printers and selling the pattern code for the clothing, rather than the clothing itself.

You think back to business school—is this a "Blue Sky," "Blue Ocean," or "Red Ocean" concept? Either way, to assess the viability, you decide to start by:

a. Meeting with all internal stakeholders to conduct a rationalized group technique (RGT) session and outline the opportunity.

b. Conducting a study of the potential opportunity to determine the viability

c. Developing a business case with Ophelia and presenting it your sponsor for consideration

d. Conducting a proof of concept using the Monte Carlo technique

The correct answer is: B

This is a very wordy question, and it introduces many unnecessary facts. The question is looking to establish the viability of a potential opportunity. To recommend a viable and appropriate solution, the business analyst would conduct an opportunity analysis focused on launching the new product.

Answer Choices A & D: These are a made-up answer choices.

Answer Choice C: While this might seem like a reasonable answer, prior to developing the business case, the business analyst would first need to determine the value proposition.

Business Analysis for Practitioners: A Practice Guide, Section 2.4.4, "Perform Root Cause Analysis on the Situation."

PMI Professional in Business Analysis (PMI-PBA)® Examination Content Outline, 2013, "Needs Assessment," Task 2.

46. Crispin is helping you build and maintain the stakeholder register for a highly regulated generic pharmaceutical. As the list is quickly growing, how can the register be organized to simplify the stakeholder communication and engagement activities?

a. Crispin should first identify the target audience, then create an interest table.

b. Crispin should create an interest table, followed by a power/interest grid.

c. You should consult with the project office to determine how they would prefer the stakeholder register to be organized.

d. Crispin could add designations and group stakeholders to simplify engagement.

The correct answer is: D

To effectively manage stakeholders, especially on large initiatives, adding designations and grouping them can simplify both the communication and engagement activities. While the Practice Guide offers a few suggestions, it's ultimately the business analyst's responsibility to determine the most relevant designations.

Answer Choices A & B: A stakeholder interest table (power/interest grid) maps categories for engagement, such as unknown, blocker, neutral, supporter, and champion. While this would be helpful, it would not help in the organization of stakeholders.

Answer Choice C: This answer choice is a distractor; while it may seem logical, as a business analyst, you should be familiar with the methodologies as established by the project office.

Business Analysis for Practitioners: A Practice Guide, Section 3.3.3, "Techniques for Grouping or Analyzing Stakeholders."

PMI Professional in Business Analysis (PMI-PBA)® Examination Content Outline, 2013, "Planning," Task 3.

47. In your role as a business analyst for Khaleesi Consulting LLC, you are talking with Jacob about the last iteration of software development. The team has just completed testing, and they are quite satisfied with the results. Who has accountability to document the results?

a. Jacob should suggest that you document the results and post them to the team's collaboration site; this is a business analyst's responsibility.

b. As the project manager, Jacob offers to take the lead and document that stakeholders are satisfied with the results.

c. You suggest that Mary, the project coordinator and communication lead, update the information in the team's collaboration site.

d. Jacob suggests that the subject matter experts who performed the testing have accountability for documenting and posting the results to the team's collaboration site.

The correct answer is: A

It is the responsibility of the business analyst to thoroughly document test results, decisions, outcomes, and issues at the end of each phase, sprint, iteration, or test cycle. This also includes customer perception and satisfaction with the delivered solution and creates the basis for knowledge transfer and supporting documentation for lessons learned.

Answer Choices B–D: These answer choices are all distractors. While you may welcome the help, absent a test lead, it is the responsibility of the business analyst to thoroughly document the outcomes.

Requirements Management: A Practice Guide, page 54, Section 9.2.1, "Document, Project, or Phase Closure Activities."
PMI Professional in Business Analysis (PMI-PBA)® Examination Content Outline, 2013, "Evaluation," Task 1.

48. Following a test cycle, Camille is working to understand the relationship that may exist between two variables. What type of analysis is Camille conducting?
a. Quantitative analysis
b. Data capture and logging analysis
c. Regression analysis
d. Organizational analysis

The correct answer is: C

During testing, to validate whether the delivered solution is aligned with stakeholder requirements, business analysts may rely upon analytical techniques to uncover any gaps or issues. In addition to regression analysis, the Practice Guide lists three other techniques: (a) benchmarking, (b) gap analysis, and (c) trend analysis.

Answer Choice A: A quantitative analysis produces results that are measurable and actionable; while in the broadest sense this could help to understand relationships, answer choice C is a better option.

Answer Choice B: This is a made-up answer choice.

Answer Choice D: An organizational analysis can be focused on both internal and external elements and is used to evaluate its resources and potential. A common technique is a SWOT analysis, which focuses on strengths and weaknesses (internal) and opportunities and threats (external).

Requirements Management: A Practice Guide, Section 9.3.2, "Analytical Techniques."
PMI Professional in Business Analysis (PMI-PBA)® Examination Content Outline, 2013, "Evaluation," Task 1.

49. You are working with Yvaine and Fritz to create a presentation to leadership on the key elements of the service- and operational-level agreements. When building this presentation, you want to ensure that:

a. Each slide has one topic, with six bullet points and a maximum of six words per bullet (1-6-6 Rule).

b. Each slide is engaging, with text minimized.

c. Each slide has one topic, with five bullet points and a maximum of five words per bullet (1-5-5 Rule).

d. You consult with your communications department for guidelines on presentations.

The correct answer is: **B**

When building presentations, you need to make them engaging and relevant; otherwise, you'll lose your audience. The best approach to building an engaging presentation is to minimize text and bullets and speak to the audience rather than read from a text-packed slide deck.

Answer Choices A & C: While these may seem like logical choices, they are distractors and made-up answer choices.

Answer Choice D: This is a reasonable choice; however, as an experienced business analyst, you should have an understanding of the organization's guidelines on presentations.

PMI Professional in Business Analysis (PMI-PBA)® Examination Content Outline, "Knowledge and Skills," #6, Communication Skills.

PMI Professional in Business Analysis (PMI-PBA)® Examination Content Outline, 2013, "Analysis," Task 8.

50. Rhys believes his requirements documents are complete and is preparing to present them to the Project Steering Committee for review. Before doing so, what should he ensure that they all contain?

a. All necessary business requirements

b. Requirements identified that produce measurable outcomes, plus all necessary requirements

c. All necessary requirements

d. Labels and references to all figures; requirements identified that produce all necessary outcomes; responses identified for each input; and all necessary requirements.

The correct answer is: **D**

Specification completeness is typically determined by the organization's project management office and is dependent on the delivery methodology. However, it's generally accepted that to be considered complete they must meet the following criteria: All aspects of the requirement are documented; inputs and outputs are identified; the document contains labels and references for all figures, tables, and diagrams.

Answer Choices A–C: These answer choices are all distractors; the question is focused on what the documents should contain, which is universal to the requirement types.

Business Analysis for Practitioners: A Practice Guide, 4.1.5.1, "Functional Requirements."

PMI Professional in Business Analysis (PMI-PBA)® Examination Content Outline, 2013, "Analysis," Task 7.

51. The project team has a very elaborate SharePoint site to post team documents and to collaborate, although it's referenced infrequently for requirements traceability information, because your spreadsheet has become the single source of truth. In addition, maintaining the SharePoint site is very time consuming, and it's become stale over the last few months. The Change Control Board (CCB) has just authorized the addition of new attributes that will further categorize business value. At a minimum, what should you do?

a. Update your spreadsheet with the new attributes, plus all the associated projects as directed by the CCB.

b. Update your spreadsheet with the new attributes, plus the associated projects as directed by the CCB, and communicate this update to your project team.

c. Update your spreadsheet with the new attributes, plus the associated projects as directed by the CCB, and communicate this update to your stakeholders.

d. Update your spreadsheet and the team SharePoint site with the new attributes, plus the associated projects as directed by the CCB, and communicate this update to your stakeholders.

The correct answer is: D

Because the team site has not been shut down, the attributes and information must be maintained on both systems; otherwise, stakeholders will question the validity of both tools.

Answer Choice A: This answer choice is partially correct. However, it neglects to communicate the update with stakeholders and fails to update the team site, which remains in use.

Answer Choice B: This answer choice is partially correct. However, communication must extend beyond the project team and must update the team site, which remains in use.

Answer Choice C: This answer choice is partially correct. However, it fails to update the team site, which remains in use.

Business Analysis for Practitioners: A Practice Guide, Section 5.2.3.1, "Requirements Attributes."

PMI Professional in Business Analysis (PMI-PBA)® Examination Content Outline, 2013, "Traceability and Monitoring," Task 4.

52. Using actual data from a prior similar project, Kelsey is attempting to validate estimates provided to her by lead subject matter experts. What estimation technique is she using?
    a. Parametric
    b. Expert
    c. Relational
    d. Analogous

The correct answer is: D

The analogous estimation technique uses data from prior projects of similar size to estimate work effort and cost when there is minimal information available. Estimating is an iterative process, and over time, these estimates are further refined.

Answer Choice A: Parametric estimating considers the relationships among variables for estimating cost and duration.

Answer Choices B & C: These are made-up estimating techniques.

A Guide to the Project Management Body of Knowledge (PMBOK®), Section 6.5.2.2, "Analogous Estimating."

Business Analysis for Practitioners: A Practice Guide, Section 3.5.2.6, "Estimate the Work."

PMI Professional in Business Analysis (PMI-PBA)® Examination Content Outline, 2013, "Planning," Task 4.

53. Your sponsor, Elian, and the VP of new product development, Fawn, would like to establish metrics to help evaluate whether the delivered solution is achieving its intended goals. What can be used to quantitatively evaluate the solution?
    a. Project management information system
    b. Key performance indicators

c. Metrics and acceptance criteria

d. Approved requirements documentation

The correct answer is: **B**

Key performance indicators (KPIs) are metrics that can be used objectively to evaluate whether a solution is achieving its intended goals and objectives.

Answer Choice A: The project management information system (PMIS) is a repository for storing project artifacts, collaborating, and knowledge management.

Answer Choice C: This is good answer choice; however, answer choice B is more specific providing key performance indicators as the tool to quantitatively evaluate the solution. KPIs are metrics that can be used to evaluate if the delivered solution is achieving its intended goals.

Answer Choice D: This is a distractor; while very important, the approved requirements documentation would not contain metrics.

Business Analysis for Practitioners: A Practice Guide, Section 6.5.2, "Consider Key Performance Indicators."

PMI Professional in Business Analysis (PMI-PBA)® Examination Content Outline, 2013, "Planning," Task 6.

54. Your organization has decided to undergo a business transformation to improve efficiency and eliminate waste. As part of this initiative, they are considering implementing optical character recognition (OCR) technology in their accounts payable department. Unfortunately, due to organizational turnover, no one truly understands the complete invoice/receipt-to-pay process. What steps can you take to learn more about the process?

a. Create process models documenting the as-is state; conduct interviews with the AP department; review any existing documentation.

b. Observe the AP clerks; conduct interviews with stakeholders; develop documentation outlining the to-be state; create process models documenting the as-is state.

c. Observe the AP clerks; conduct interviews with stakeholders; review any existing documentation; create process models documenting the as-is state.

d. Observe the AP clerks; conduct interviews with the AP department; review any existing documentation; create process models documenting the as-is state.

The correct answer is: **C**

In this scenario, the objective is to fully analyze the business opportunity as previously documented in the situation statement. To learn more about the process, the business analyst could observe the AP clerks, conduct interviews with stakeholders, review any existing documentation, and create process models documenting the as-is state. The resulting deliverables are focused on documenting the expected benefits.

Answer Choice A: This answer choice is partially correct. However, it's missing the technique of observation.

Answer Choice B: This answer choice is partially correct. However, in this scenario the business analyst would not be developing documentation outlining the to-be state, but rather would be creating process models documenting the as-is state.

Answer Choice D: This answer choice is partially correct. However, interviews shouldn't be limited to the accounts payable department.

Business Analysis for Practitioners: A Practice Guide, Section 2.3.2, "Investigate the Problem or Opportunity."

PMI Professional in Business Analysis (PMI-PBA)® Examination Content Outline, 2013, "Needs Assessment," Task 2.

55. Ace is an expert in business analysis and knows that feasibility is a key characteristic of a properly written requirements document. From the list provided below, what is one factor that can determine the feasibility of a proposed solution?
    a. Operational feasibility.
    b. Technology feasibility.
    c. Time and cost feasibility.
    d. Feasibility can best be determined by assessing a variety of factors.

The correct answer is: **D**

Feasibility is a key characteristic of properly written requirement documents. Solution or product feasibility can only be assessed by taking into account a variety of elements. The Practice Guide describes four categories: (a) operational, (b) technology/system, (c) cost-effectiveness, and (d) time. On the exam, remember feasibility can best be determined by assessing a variety of factors.
Answer Choice A: An operational feasibility assessment will determine if the delivered solution will be used and supported by all stakeholders.
Answer Choice B: A technology feasibility assessment will evaluate if the proposed solution is viable based on current-day knowledge, tools, and equipment.
Answer Choice C: A time feasibility assessment will determine if the proposed solution can be delivered within the timeline of the project, whereas a cost feasibility assessment will determine if the proposed solution can be delivered and supported considering the budgetary requirements of the organization.
Business Analysis for Practitioners: A Practice Guide, 4.11.5.1, "Functional Requirements."
PMI Professional in Business Analysis (PMI-PBA)® Examination Content Outline, 2013, "Analysis," Task 7.

56. Working with your project manager, Brooklyn, you have just delivered your fourth project in 36 months using a waterfall delivery method. You've now been asked to lead a highly visible project based on scrum with nearly the same project team. In terms of change control, the team has agreed:
    a. That Brooklyn will document the process in the change management plan
    b. That they will adapt a flexible approach to change control, because the team anticipates that requirements will evolve as the project progresses
    c. To use the same process and templates as with all their other projects
    d. That they will defer to the project management office for guidance on the process

The correct answer is: **A**

This question leads in with unnecessary information to confuse you and slow down your exam cadence. The question is asking, "What is the change control methodology that is used on agile projects?" Although agile and other adaptive lifecycle projects plan for and anticipate change, the process must first be documented in either the project manager's change management plan or the business analyst's plan. In addition to having a clearly documented plan, they must also have the approval of the governing body.
Answer Choice B: While it's true that agile projects must be flexible and anticipate change, the process must be documented.
Answer Choice C: This is a distractor, as the process and methodologies are not congruent with waterfall and agile projects.

Answer Choice D: For an experienced business analyst, this would not be a logical or appropriate choice, as they should be familiar with the processes, tools, techniques, and methodologies.

Business Analysis for Practitioners: A Practice Guide, Section 3.4.14, "Define the Requirements Change Process."

PMI Professional in Business Analysis (PMI-PBA)® Examination Content Outline, 2013, "Planning," Task 4.

57. Sadie has a meeting later in the week to discuss solution requirements for a new line of organic thread at the textile mill. Prior to the meeting, where can Sadie look to gain insight into the organization and prior initiatives?
    a. The business analysis plan
    b. The project management information system
    c. The business analysis plan, requirements management plan, and plan for change control
    d. The requirements management plan

The correct answer is: B

Prior to meeting with stakeholders, reviewing existing documentation is an excellent approach for gaining insight into the environment. Information gathered can form the basis for productive conversations. In this scenario, by referring to the project management system—or even the configuration management system (CMS)—the business analyst could garner useful information to enhance their understanding of the current state. This elicitation technique is known as *documentation analysis*.

Answer Choices A, C, & D: These are all partially correct answer choices, since these plans would provide insight into the organization. However, they would not provide any information about past initiatives. This could only be obtained from the project management information system (PMIS) or the configuration management system (CMS).

Business Analysis for Practitioners: A Practice Guide, Section 4.5.5.2, "Documentation Analysis."

PMI Professional in Business Analysis (PMI-PBA)® Examination Content Outline, 2013, "Analysis," Task 1.

58. Your project manager, Ciaran, has suggested that the team consider using a tool based on two axes representing opposing viewpoints and interests, essentially creating a table with four cells to enable informed decision making. What tool is Ciaran suggesting the team use?
    a. An options analysis
    b. A decision table
    c. A quadrant analysis
    d. A strategic table

The correct answer is: C

This question tests your knowledge, understanding, and practical use of a commonly used matrix tool. A quadrant analysis, also referred to as a 2x2 matrix, is a flexible tool that allows teams to categorize viewpoints and interests in a grid comprising four cells. Not showing an example and providing extraneous context adds to the complexity of this question.

Answer Choice A: An options analysis technique is used to evaluate an investment decision, often used when creating a business case.

Answer Choice B: A decision table is one of the three models used to administer polices.

Answer Choice D: This is a made-up answer choice.

PMI-PBA® Examination Content Outline, "Knowledge and Skills," #15, "Estimation Tool and Techniques."

PMI Professional in Business Analysis (PMI-PBA)® Examination Content Outline, 2013, "Analysis," Task 4.

59. During a focus group at the law firm of Adelaide Nolan LLC, the team is having difficulty arriving at a decision between two options. To help the team evaluate a Yes/No decision based on the option pair, Jane, your subject matter expert in torts, suggests that they should consider using:
    a. A mediator who can objectively assess each position and provide a recommendation
    b. The Business Process Modeling Language (BPML) to optimize the process
    c. A method that will enable the team to reach consensus, whereby each participant will document the advantages and disadvantages of their position
    d. The requirements modeling language to visually show the solution requirements linked to the objectives and goals of the organization

The correct answer is: C

Although didactic interaction (DI) is the least common of the four group decision-making techniques that are likely to appear on the PMI-PBA® Exam, it's a very useful technique and one that should be in every business analyst's arsenal. As brainstorming, the nominal group technique, and the Delphi technique are all situationally appropriate, the same is true for didactic interaction. DI is best used when the team is evaluating a Yes/No decision or a pair-based selection (Option 1/ Option 2). For example, you're planning a once-in-a-lifetime family vacation; should you go to Europe or Mexico? When using this tool, each team or participant must document the advantages and disadvantages of their position.
Answer Choice A: This is a distractor. As a business analyst, you should have a more profound approach to addressing the scenario.
Answer Choice B: Modeling languages consist of Business Process Modeling Notation (BPMN), Requirements Modeling Language (RML), System Modeling Language (SysML), and Unified Modeling Language (UML). This answer choice is not situationally appropriate.
Answer Choice D: This is a made-up answer choice.
Business Analysis for Practitioners: A Practice Guide, 4.15, "Resolve Requirements-Related Conflicts."
PMI Professional in Business Analysis (PMI-PBA)® Examination Content Outline, 2013, "Analysis," Task 3.

60. Rebecca is preparing a project charter for a new line of dehydrated fruit. She has all the information from the needs assessment that you conducted, but she would also like to include information pertaining to costs and benefits, as well as to explore many of the aspects related to the recommendation. What should you do?
    a. Work with a financial analyst to prepare NPV, IRR, and ROI assessments.
    b. Provide Rebecca with the business drivers, economic viability, and success criteria for the recommendation.
    c. Provide Rebecca with an impact analysis.
    d. Work with Rebecca to outline the business goals and objectives.

The correct answer is: B

This scenario-based question is testing your knowledge of business cases. Business drivers, economic viability, and success drivers are the foundations of a sound cost–benefit analysis. Once created, the business case will be updated over the course of the project and used to support effective decision making; furthermore, it serves as a solid reminder of why the initiative was approved.

Answer Choices A & D: These are partially correct answer choices; however, the analysis would extend beyond valuation measures and simply outlining the business goals and objectives.

Answer Choice C: This is a distractor; an impact analysis is typically associated with change requests.

Business Analysis for Practitioners: A Practice Guide, Section 2.6, "Assemble the Business Case."

PMI Professional in Business Analysis (PMI-PBA)® Examination Content Outline, 2013, "Needs Assessment," Task 1.

61. Imogen, the VP of strategic marketing and customer engagement, would like to strengthen the relationship with customers and identify opportunities to improve the company's website. Wanting to focus externally, she commissions you to create:
    a. An ecosystem map
    b. A customer ecosystem map
    c. A value engineer map
    d. A user journey map

The correct answer is: **D**

When analyzing processes, user (aka *customer*) journey maps are used to visually articulate the overall story from the user's perspective. Focusing externally, they include elements such as personas, timeline, reactions, actions within the system, and methods of interaction. They can optionally include supporting personas. They are actionable, in that they provide a mechanism for business analysts to broaden their understanding of user behaviors by incorporating storytelling and visualization.

Answer Choice A: An ecosystem map is a type of scope model used for visually articulating all the relevant systems, their relationships, and their associated interfaces.

Answer Choice B: A customer ecosystem map is a made-up tool.

Answer Choice C: A value engineer map is a made-up tool.

Examination Content Outline Knowledge and Skills, #26, "Problem Solving and Opportunity Identification Tools and Techniques."

PMI Professional in Business Analysis (PMI-PBA)® Examination Content Outline, 2013, "Analysis," Task 2.

62. Your sponsor, Simba, is new to the concept of business analysis. Although you've explained that requirements are the sole justification for the existence of a project, he's asked for additional context. From the answer choices below, which would you not provide?
    a. Business analysis reduces project risks.
    b. Business analysis sets expectations with stakeholders as to activities that will be performed.
    c. Business analysis activities will be planned up front in sufficient detail for the duration of the project.
    d. An overabundance of business analysis planning can negatively impact the project.

The correct answer is: **C**

Be careful as you read the questions—in this case, "What would you *not* provide?" Projects almost always evolve over time, and as business analysts, it is critical that we support the evolving needs of the organization. While projects are based on the business case, it may be in the organization's best interest to plan everything up front. The primary goal is to ensure that stakeholders support the work effort and that they have a clear understanding of their roles and deliverables; this is achieved by thinking critically about the product/solution and the overall project.

Answer Choices A, B, & D: Each of these answer choices provides a reasonable level of justification for the incorporation of business analysis efforts on projects. When implemented properly, it can lead to a reduction in project risks, and it sets expectations with stakeholders as to activities that will be performed. As you consider these activities, please keep in mind, *an overabundance of business analysis planning can negatively impact the project.*

Business Analysis for Practitioners: A Practice Guide, Section 3.2.1, "Rationale."

PMI Professional in Business Analysis (PMI-PBA)® Examination Content Outline, 2013, "Planning," Task 1.

63. Aurelia would like to create several archetypes so that the project team can design an effective software solution. She would like to stay away from the typical outline commonly found in use cases. Which of the below could you recommend for a project in a hospital?
    a. General practitioner doctor, general surgeon, specialist surgeon
    b. Administrative support, clinical staff
    c. "Abigail Kian," "Eliana Bennett," "Savannah Knox"
    d. Overhead, revenue generating

The correct answer is: **C**

This question tests your knowledge in two areas: the first is synonyms used in the industry, the second is application of the associated technique. Personas define an archetype—a type of person who would use a system. They represent fictitious people who are based on the team's knowledge of real users. In practice, you should never use someone's actual name in a meeting or when designing a system. Actors represent roles, and are associated with use cases.

Answer Choices A, B, & D: Each of these answer choices is incorrect—they list titles and functions, which are not personas. The names "Abigail Kian," "Eliana Bennett," and "Savannah Knox" represent personas that would be used in use cases.

Examination Content Outline Knowledge and Skills, #35, "Stakeholder Analysis."

PMI Professional in Business Analysis (PMI-PBA)® Examination Content Outline, 2013, "Analysis," Task 6.

64. Your organization is deploying a new, state-of-the-art website. Working with Penelope, an SME from marketing, you begin a process of outlining resources who will implement and support the system, along with those who will use and benefit from the website. You are:
    a. Identifying the transition and operations team along with customers, so that you can properly plan a go-live event.
    b. Identifying system users who will support and use the system.
    c. Identifying team members; as go-live approaches, all resources will be asked to provide additional support to ensure a smooth cutover.
    d. Identifying individuals and groups who may be affected or perceive themselves to be effected by the launch of the website.

The correct answer is: **D**

In this scenario-based question, after removing the unnecessary information, you are simply being asked, "What are stakeholders?" Stakeholders are all individuals and groups who are affected or who perceive they will be affected by your initiative. Project managers and business analysts should collaborate on the identification and continual update of this information.

Answer Choices A–C: These are all distractor answer choices. While they are providing types of stakeholders, they are not addressing the true nature of the question.

Business Analysis for Practitioners: A Practice Guide, Section 2.3.1, "Identifying Stakeholders."

PMI Professional in Business Analysis (PMI-PBA)® Examination Content Outline, 2013, "Needs Assessment, Task 4."

65. Your consultant, George, has inquired if further review or approval is required as the payroll team submits their product documentation. To what artifact can you refer George?
   a. Product or solution management plan
   b. Scope management plan
   c. Change control plan
   d. Requirements management plan

The correct answer is: **D**

In organizations with dual project manager/business analyst roles, the requirements management plan and business analysis plan are often produced by the same person, whereas in larger organizations, these documents will be prepared by different individuals, often a project manager and business analyst who are collaborating. The requirements management plan will cover elements of both the project and product—identifying stakeholders and their roles, establishing the framework for communications, and articulating the guidelines for managing requirements—whereas the business analysis plan focuses solely on the activities and deliverables related to the efforts of business analysis. PMI is a proponent of plans; *exam questions will focus on the contents, purpose, and intent of these project artifacts.*

Answer Choice A: This is a made-up project artifact.

Answer Choice B: The scope management plan is a subsidiary of the overall project management plan. The plan elaborates the details of how project scope will be defined, developed, monitored, controlled, and verified.

Answer Choice C: The change management plan is a subsidiary of the overall project management plan. This plan establishes the ground rules and roles and responsibilities for how stakeholders will propose changes, how these changes will be documented, and when proposed changes will be reviewed and decisions communicated.

Business Analysis for Practitioners: A Practice Guide, Section 3.4.2, "What to Include in the Business Analysis Plan."

PMI Professional in Business Analysis (PMI-PBA)® Examination Content Outline, 2013, "Traceability and Monitoring," Task 2.

66. Your subject matter experts, Zane and David, who are both certified public accountants, have a disagreement as to how the core financial management system should be structured. After lengthy discussions, it is determined that they would use the data warehouse for reporting. This is an example of what conflict management technique?
   a. Compromising
   b. Smoothing
   c. Agreeing
   d. Collaborating

The correct answer is: **D**

On the exam, you can expect questions on any of the five conflict management techniques. These techniques are: (a) withdrawing or avoidance, (b) smoothing or accommodating, (c) compromising or reconciliation, (d) forcing or directing, and (e) collaborating or problem solving. Conflict often presents opportunities for dialogue; while inevitable on projects, when addressed appropriately, stakeholders will appreciate your candor and effectiveness. Collaborating is an example of a win–win conflict management technique and is the preferred method of addressing conflicts and disagreements.

Answer Choice A: Compromising or reconciliation to some extent satisfies both parties by bringing about a faster result, as both parties have something to gain.

Answer Choice B: When using the smoothing technique, one party tends to concede more to the other, attempting to be accommodative.

Answer Choice C: This is a made-up conflict management technique.

PMI Professional in Business Analysis (PMI-PBA)® Examination Content Outline, 2013, Knowledge and Skills, "Conflict Management."

PMI Professional in Business Analysis (PMI-PBA)® Examination Content Outline, 2013, "Traceability and Monitoring," Task 3.

67. Prior to meeting with stakeholders to estimate the work effort for a project, Bailey, an experienced business analyst, decides to review planning information from prior projects of similar size. What elicitation techniques could she rely upon?
    a. Document analysis
    b. Alternative analysis
    c. Interviews
    d. Focus group

The correct answer is: **A**

While building the work plan, business analysts can employ a number of techniques to refine task and activity work efforts. Document analysis is an elicitation technique in which material from projects with similar characteristics is reviewed, to gain better understanding of the prior environment. This is most commonly performed before meeting with stakeholders, to form the basis for conversations.

Answer Choice B: Alternative analysis is one of the scheduling tools and techniques; it examines various ways to deliver the solution while preserving the iron triangle of quality as constrained by time, cost, and schedule.

Answer Choices C & D: Although interviews and focus groups are also elicitation techniques, like document analysis, they would not provide the required level of detail in this scenario.

Business Analysis for Practitioners: A Practice Guide, Section 3.5.2.6, "Estimate the Work."

PMI Professional in Business Analysis (PMI-PBA)® Examination Content Outline, 2013, "Planning," Task 3.

68. Following a very long and sometimes challenging requirements confirmation process, you are now asking for sign-off. You anticipated that this would be a routine process, but subject matter experts Eleni, Fia, Behati, and Delta are unable to provide an opinion to Griffith, the senior VP of consumer plastics. Why are they unable to provide an opinion?

a. While they signed off on the requirements, they were not involved in the confirmation sessions.

b. They were not listed on the RACI matrix.

c. They were not involved in any of the requirements sessions.

d. They only participated in didactic interaction sessions.

The correct answer is: **C**

This question introduces a number of personas in an attempt to add unnecessary complexity to a relatively easy question. Although requirement documents can be very well written and of high quality, they can meet resistance if key stakeholders have not been involved in their development or had a chance to review them ahead of time. *For the exam, and in practice, please remember the nine characteristics of high-quality requirements documentation: unambiguous, precise, consistent, correct, complete, measurable, feasible, traceable, and testable.* To ensure that requirements documentation satisfies these characteristics, the business analyst can use checklists to validate that documents meet these criteria.

Answer Choice A: There is nothing in the question to confirm or negate whether they signed off on the requirements or their level of participation during the confirmation sessions.

Answer Choice B: This is a distractor answer choice and doesn't completely address the question. RACI charts, also known as *responsibility assignment matrices* (RAMs) and *linear responsibility charts* (LRCs), establish accountability for requirements signoff. They help to clearly identify who's responsible, the one and only person who is accountable, experts who can be consulted, and others who need to be informed.

Answer Choice D: There is nothing in the question that indicates that they participated in a group decision-making session trying to reach consensus on a Yes/No or Option 1/Option 2 decision.

Business Analysis for Practitioners: A Practice Guide, Section 4.14, "Approval Sessions."

PMI Professional in Business Analysis (PMI-PBA)® Examination Content Outline, 2013, "Analysis," Task 5.

69. In the development of a business case, you're working with Daisy, an experienced financial analyst, on the cost–benefit summation. Which financial valuation method addresses the time to recover a project investment?

a. Net present value (NPV)

b. Return on investment (ROI)

c. Internal rate of return (IRR)

d. Payback period (PBP)

The correct answer is: **D**

While this question adds background to the scenario, it's focusing on your understanding of each of the valuation techniques. The results of this analysis will form the basis for the rough order-of-magnitude estimates of costs and benefits. The payback period (PBP) is the amount of time required for the organization to recover the cost of the project, typically represented in months. Projects with longer payback periods represent increased risk to the organization.

Answer Choice A: The net present value (NPV) technique estimates the future value of the project's expected benefits.

Answer Choice B: Return on investment (ROI) is the benefit of a project divided by the cost of the project. Many organizations establish ROI hurdle rates before a project can be considered.

Answer Choice C: The internal rate of return (IRR) is the projected annual yield of the project.

Business Analysis for Practitioners: A Practice Guide, Section 2.5.6.1, "Payback Period (PBP)."

PMI Professional in Business Analysis (PMI-PBA)® Examination Content Outline, 2013, "Needs Assessment," Task 2.

70. You're presenting a business case to your executive sponsor, along with your financial analyst, Olive, highlighting the financial valuation methods used to justify the investment. Your sponsor, Colton, asked that you expand the analysis to include the initial and ongoing costs, along with the projected annual yield of the investment. Upon returning with Olive, you decide to prepare:
    a. An internal rate of return (IRR) analysis
    b. A return on investment (ROI) analysis
    c. A present value versus future value (PV vs. FV) assessment
    d. A net present value (NPV) analysis

The correct answer is: **A**

This scenario is testing your understanding of each of the valuation techniques. The results of this analysis will form the basis for the rough order-of-magnitude estimates of costs and benefits. An internal rate of return (IRR) analysis is performed when organizations are considering CAPEX (capital expenditure) projects to determine the profitability of projects. When the NPV for the project is set to zero, IRR is the discount rate.

Answer Choice B: Return on investment (ROI) is the benefit of the project divided by the cost of the project. Many organizations establish ROI hurdle rates before a project can be considered.

Answer Choice C: Present value (PV) and future value (FV) measure how the value of money changes over time.

Answer Choice D: The net present value (NPV) technique estimates the future value of the project's expected benefits.

Business Analysis for Practitioners: A Practice Guide, Section 2.5.6.3, "Internal Rate of Return (IRR)."

PMI Professional in Business Analysis (PMI-PBA)® Examination Content Outline, 2013, "Needs Assessment," Task 2.

71. Julius, a college intern, has inquired as to the differences between the requirements management plan and the business analysis plan. How would you best answer his question?
    a. The requirements management plan is a subsidiary plan of the overall project management plan, whereas the business analysis plan is a complementary artifact.
    b. The requirements management plan covers the project, whereas the business analysis plan is focused on the effort associated with business analysis.
    c. Both documents are complementary artifacts.
    d. The requirements management plan covers both the project and product, whereas the business analysis plan is focused on the effort associated with business analysis.

The correct answer is: **D**

The business analysis plan and requirements management plan provide the overarching direction throughout the project; they formally establish the *how* and *when* for all solution-development activities. The requirements management plan covers the overall project and the product, and it describes how requirements overall will be elicited, analyzed, documented, and managed. The business analysis plan covers the activities to be conducted, deliverables to be produced, and key requirement process decisions (e.g., prioritization approach, documentation strategies validation, communication, approvals, and change management).

Answer Choice A: This answer choice is partially correct; the requirements management plan is a subsidiary plan of the overall project management plan. However, the business analysis plan is more than a complementary artifact.

Answer Choice B: Also a partially correct answer choice. The requirements management plan does cover the project, and the business analysis plan is focused on the effort associated with business analysis. However, the requirements management plan also includes product/solution-related elements.

Answer Choice C: This is an incorrect answer choice.

Business Analysis for Practitioners: A Practice Guide, Section 3.4.1."Business Analysis Plan vs. Requirements Management Plan."

PMI Professional in Business Analysis (PMI-PBA)® Examination Content Outline, 2013, "Planning," Task 3.

72. As part of the United States Antarctic Program, Callie is stationed at the Amundsen–Scott South Pole Station as a research scientist. Your team is in the process of building a software application that will improve the accuracy of weather modeling, with a significant contribution to geosciences. As you are based in the United States, and there is little opportunity to conduct live interviews to review evaluation metrics and acceptance criteria, what method of elicitation might you consider?
    a. Unstructured
    b. Asynchronous
    c. Structured
    d. Synchronous

The correct answer is: **B**

This scenario-based question introduces a lot of context, which can be confusing during the actual exam. ***Remember to focus on what the question is truly asking;*** in terms of interviewing techniques, recall that there are two categories of interview techniques and two methods. The asynchronous interviewing technique is ideal when interviews cannot be conducted in real time (synchronous). When using the asynchronous interviewing technique, the interview questions are scripted, allowing the interviewee to note their responses either by sending an email, posting to a wiki, or recording in a video collaboration system. In some instances, asynchronous interviews are preferred by project teams, so that they can both revisit interviewee responses and accommodate team members' schedules and availability.

Answer Choice A: When using the unstructured method, the interviewer comes prepared with a list of questions. However, the interviewee is definitely only asked the first question. The interview progresses on its own, relying on the skill of the business analyst to guide the conversation.

Answer Choice C: In structured interviews, the interviewer has a defined set of questions, and the elicitation is conducted with the intent of asking all the questions of the list.

Answer Choice D: Synchronous interviews are conducted in real time, either in person or via audio or video conference. With the proliferation of collaboration tools and globalization of our workforce, it's quite common for synchronous interviews to be held in formats other than in person. However, if provided the opportunity, in-person interviews would be the preferred method.

Business Analysis for Practitioners: A Practice Guide, Section 4.5.5.5, "Interviews."

PMI Professional in Business Analysis (PMI-PBA)® Examination Content Outline, 2013, "Analysis," Task 8.

73. Your team comprises very skilled individuals from 30 countries, all of whom work remotely, except for a semiannual company meeting. As the business analyst, while performing due diligence to ensure that the requirements remain aligned to the evolving needs of the organization, you want to focus on one key element:
    a. Addressing feelings of isolation
    b. Establishing tools for effective decision making
    c. Ensuring that the appropriate technology is in place
    d. Establishing parameters for effective and consistent communication

The correct answer is: **D**

This is a complex question, and selection of the correct answer should follow a process of elimination. The question can be restated as, ". . . to ensure that the requirements remain aligned to the evolving needs of the organization, what can you do?" Keep in mind, as the business analyst, what can you control? Virtual teams allow organizations to expand their talent pool and incorporate viewpoints and perspectives from a wider geographic dispersion. The most important factor to consider when working with virtual teams is establishing parameters and guidelines for effective and consistent communication.
Answer Choice A: Addressing feelings of isolation should absolutely be addressed; however, it is outside the responsibility of the business analyst. This would be the responsibility of the line manager.
Answer Choice B: The project management office would be accountable for the establishment and training on tools that enable effective decision making.
Answer Choice C: The IT teams would have accountability to ensure that the appropriate technology is in place.
PMI-PBA® Examination Content Outline, Knowledge and Skills, # 6, "Virtual Teams."
A Guide to the Project Management Body of Knowledge, Section 9.2.2.4, "Virtual Teams."
PMI Professional in Business Analysis (PMI-PBA)® Examination Content Outline, 2013, "Analysis," Task 7.

74. Natalia works as a business analyst for a highly regulated hedge fund, Kinsley & Associates LLC. She's profoundly aware of the need to link elements of the solution to the organization's policies and procedures. To do so, what model could Natalia use?
    a. Interrelationship diagram
    b. State table
    c. Decision table
    d. Entity relationship diagram

The correct answer is: **C**

Rule models, which consist of business rule catalogs, decision trees, and tables, are used to assist with classifying and verifying the business processes, procedures, and frameworks that the initiative must support.
Answer Choice A: Interrelationship diagrams, one of the tools used in scope modeling, aid in visualizing complex problems and relationships. When using the tool, is quite common to find factors that influence each other. In cases in which there is more than one influencing factor, the team needs to determine which factor is stronger, and note only that one.

Answer Choice B: A state table is a form of data model, which can be used to illustrate information within the process. It can be used to model the valid states of an object and any transitions between states over their lifecycle.

Answer Choice D: An entity relationship diagram is also a form of data model; it is used to uncover objects that are related or that may have dependencies with other objects.

Business Analysis for Practitioners: A Practice Guide, Section 4.10.3, "Categories of Models."

PMI Professional in Business Analysis (PMI-PBA)® Examination Content Outline, 2013, "Analysis," Task 6.

75. In an effort to leverage the significant investment in an enterprise software application, uncover pain points, and optimize the business processes, your CFO, Lydia Maxwell, requests that you perform what type of analysis?
   a. S.A.V.E. assessment
   b. Value engineering assessment
   c. Optimum value assessment
   d. Transformative value assessment

The correct answer is: **B**

This question tests your knowledge of the problem-solving and opportunity-identification tools and techniques used in the Analysis domain. The only viable answer choice is value engineering assessment. Value engineering (VE) is based on quantitative measures and is focused on improving business value in existing investments. VE focuses on improving the viability and usefulness of the solution while reducing overall costs.

Answer Choices A, C, and D: These are all made-up answer choices.

Examination Content Outline Knowledge and Skills, #26, "Problem Solving and Opportunity Identification Tools and Techniques."

PMI Professional in Business Analysis (PMI-PBA)® Examination Content Outline, 2013, "Analysis," Task 2.

76. Azriel, your summer intern, is helping you plan the requirements approval and confirmation sessions. As you create the slide deck, you realize:
   a. These should be two distinct meetings, with approval occurring only after the solution is confirmed.
   b. In the interest of time, both topics can be discussed concurrently.
   c. The slides should be created following the 1-6-6 rule, each slide has one topic, with six bullet points and a maximum of six words per bullet.
   d. Your sponsor should start the session with a warm opening, setting the stage.

The correct answer is: **A**

This scenario-based question challenges your understanding of solution approval sessions. While it may seem logical that approval and confirmation sessions are related, in fact they are two completely separate subjects. *On the exam, please remember, approval of solution requirements can only occur once the approach to address the problem or opportunity has been confirmed.*

Answer Choices B, C, & D: These are all made-up answer choices designed to have you second guess your instinct for the correct answer choice.

Business Analysis for Practitioners: A Practice Guide, Section 4.14, "Approval Sessions."

PMI Professional in Business Analysis (PMI-PBA)® Examination Content Outline, 2013, "Analysis," Task 5.

77. In your business case, you've included an analysis outlining the initial and ongoing costs, the projected annual yield of the investment, and an assessment addressing the time to recover the project investment. Prior to returning to meet with your executive sponsor, Hugo, Kate, your financial analyst, suggests that you include a section comparing the amount of the investment to the future value of the expected benefits. You are discussing a:
    a. Net present value (NPV) analysis
    b. Return on investment (ROI) analysis
    c. Payback period (PBP) analysis
    d. Internal rate of return (IRR) analysis

The correct answer is: **A**

In the question, there is a lengthy back story, crafted with the intent of slowing your cadence, requiring a skilled business analyst to drive to the root of the question. The question is, "In using what valuation technique do you compare the amount of investment versus the future value of the expected benefits?" From the answer choices, the only valid choice is Net Present Value (NPV) analysis. The NPV analysis technique is used for CAPEX projects to calculate the total cash flow (both in and out), which is directly attributed to the initiative over a period of time. The analysis is used to determine the profitability of the capital investment.

Answer Choice B: Return on investment (ROI) is the benefit of the project divided by the cost of the project. Many organizations establish ROI hurdle rates before a project can be considered.

Answer Choice C: The payback period (PBP) is the amount of time required for the organization to recover the cost of the project, typically represented in months. Projects with longer payback periods represent increased risk to an organization.

Answer Choice D: The internal rate of return (IRR) is the projected annual yield of the project.

Business Analysis for Practitioners: A Practice Guide, Section 2.5.6.4, "Net Present Value (NPV)."

PMI Professional in Business Analysis (PMI-PBA)® Examination Content Outline, 2013, "Needs Assessment," Task 2.

78. You are a business analyst for Daenerys Electronics Co, a sole-source supplier to Wolf Automotive Ltd, a hypercar company catering to the ultra-wealthy. Wolf has decided to move production from Frankfurt to Düsseldorf, Germany, some 232.2 km via the A3. Your factory is located in Hachenburg, Germany, geographically about halfway been the two factories, but there is an increase in travel time. How should the business analyst for Wolf address Daenerys Electronics' interests?
    a. As a project requirement, ensure that all the processes and conditions are met for a successful transition.
    b. As a component of nonfunctional requirements, ensure that there is no disruption to the just-in-time order process, by suggesting a buffer to the travel time.
    c. As part of stakeholder requirements, address the quantifiable interests of Daenerys.
    d. In the transition requirements plan, ensure that Daenerys is aware of the new ship-to address, and the relevant change in travel time.

The correct answer is: **C**

This scenario-based question introduces a lot of unnecessary information. With this type of question, first establish roles and the overall situation: Daenerys Electronics is a supplier to Wolf

Automotive Ltd. Then rephrase the question. "As the business analyst for Wolf Automotive Ltd., how should you address Daenerys Electronics' interest as you plan the move of production?" From the answer choices, the only option in this scenario is choice C. Stakeholder requirements provide the foundation for solution requirements and address the quantifiable interests of all interested parties either impacted or perceived to be impacted by the initiative.

Answer Choice A: Project requirements are the responsibility of the project manager, and the context is not situationally appropriate.

Answer Choice B: Nonfunctional requirements are based on quality and behavior attributes; they can cover items such as availability, capacity, continuity, and performance. Tolerance ranges can include worst-case value (minimum acceptable value), target value (most-likely value), and wished-for value (best-case value).

Answer Choice D: Although updating the ship-to address is a critical task, it doesn't appropriately address the scenario.

Requirements Management: A Practice Guide, Section 5.2.2, "Define Requirements Types."

PMI Professional in Business Analysis (PMI-PBA)® Examination Content Outline, 2013, "Analysis," Task 6.

79. Naomi is responsible for maintaining the requirements traceability matrix. At the end of each week, she would like to update the project team as to status and progress related to the development objects. How would this be most effectively accomplished?
    a. Documentation management
    b. Status reports
    c. Communications management
    d. Integrated change control

The correct answer is: **B**

When read carefully, this question is relatively simple and challenges your knowledge of each of the provided answer choices. Overall, business analysts can expect to spend up to 80% of their time communicating. In the context of their day, this is quite significant; one way to ensure that a consistent, redundant message is delivered is via status reports. Weekly status reports are a great tool for informing the team how they are tracking to plan and advise of any issues, risks, or management asks.

Answer Choice A: This is a distractor; while documentation management is an important tool, it alone would not be enough to effectively update the project team with regard to status and progress of development objects.

Answer Choice C: This is also a distractor; project communications management is a knowledge area that includes processes such as (a) planning communications management, (b) managing communications, and (c) controlling communications.

Answer Choice D: Integrated change control is a process tasked with evaluating and considering changes to scope. While important to the overall process, it would not be the means or the forum to update the project team with regard to status and progress of development objects.

PMI-PBA® Examination Content Outline, "Knowledge and Skills, Reporting Tools and Techniques."

PMI Professional in Business Analysis (PMI-PBA)® Examination Content Outline, 2013, "Traceability and Monitoring," Task 4.

80. Ava is in the process of updating the configuration management systems with items that directly impact project work. From the list below, what should not be included?
    a. Building material from China will be delayed due to an inclement weather event.
    b. The HR team will not complete functional specifications on time due to competing priorities.
    c. The non-production application server sporadically reboots.
    d. A request is made to accelerate the timeline to secure a reduction in materials cost.

The correct answer is: **D**

Although a briefly worded question, it needs to be read very carefully—"What should *not* be included in the updating of the configuration management systems." To start, you'll need to have an understanding of the configuration management system (CMS) as a change control tool. The proper management of issues and risks is core to this task and an important facet of the documentation within the CMS. Risks are events that, should they occur, could have a positive or negative impact on the project. Issues are events that are presently impacting the project. Answer choice D is an opportunity or a positive risk that the team needs to evaluate.

Answer Choices A–C: These are all incorrect answers and do not appropriately address the scenario.

A Guide to the Project Management Body of Knowledge (PMBOK®), "Risk Management," pg. 310.

PMI Professional in Business Analysis (PMI-PBA)® Examination Content Outline, 2013, "Traceability and Monitoring," Task 4.

81. Your senior business analyst, Toulouse, is creating an entity relationship diagram to visually represent objects and their relationships. While building the diagram, you remind him to notate the ordinality. He agrees and proceeds to update the diagram by:
    a. Adding composite attributes to the strong and weak entities
    b. Adding derived attributes to the multivalued relationships
    c. Indicating the minimum number of times an instance in one entity can be associated with instances in a related entity
    d. Drawing the maximum number of times an instance in one entity can be associated with instances in a related entity

The correct answer is: **C**

This is an advanced question that challenges your understanding of entity relationship diagrams. In most cases, if you haven't used the tool, this type of question will be a struggle. When creating an entity relationship diagram or business data diagram, ordinality is used to uncover mandatory and optional relationships. It represents the minimum number of times an instance in one entity can be associated with instances in a related entity. Objects can include concepts, individuals, or whatever the business analyst considers important. It is often modeled using a crow's foot notation.

Answer Choice A: The terms *strong* and *weak* are used, respectively, to describe the entity set with the primary key and those which lack sufficient attributes from the primary key.

Answer Choice B: This is a made-up answer choice combining the terms *derived* and *multivalued*. A derived attribute is an attribute based on another, whereas attributes with more than one value are considered multivalued.

Answer Choice D: This is the definition of cardinality: the maximum number of times an instance in one entity can be associated with instances in a related entity. Conversely, ordinality is the minimum number of times an instance in one entity can be associated with an instance in the related entity.

Requirements Management: A Practice Guide, Section 4.10.10.1, "Entity Relationship Diagram."

PMI Professional in Business Analysis (PMI-PBA)® Examination Content Outline, 2013, "Analysis," Task 2.

82. You're collaborating with Reese, an experienced project manager, on the business analysis plan. With regard to the section on change management, which of the following plans addresses who will provide information to stakeholders following a meeting of the Change Control Board?
    a. Human resource management plan
    b. Stakeholder management plan
    c. Resource management plan
    d. Communications management plan

The correct answer is: **D**

This question tests your knowledge of the core project management plans, which can sometimes be confusing, as they are similar in name. In this scenario, communications management plan, which is a subsidiary plan of the overall project management plan, outlines *who* is accountable, *methods* for communication, and *timing* of the communication. Key words for plans are *how*, *when*, and *by whom*. When communicating, remember the six key questions to address: who, what, when, where, why, and how.

Answer Choice A: The human resource management plan outlines the project roles, responsibilities, reporting relationships, and management of staff over the duration of the project.

Answer Choice B: The stakeholder management plan establishes how stakeholders will be effectively engaged over the duration of the project.

Answer Choice C: This is the abbreviated nomenclature for the human resource management plan described in answer choice A.

Business Analysis for Practitioners: A Practice Guide, Section 5.8.3.4, "Impact on Project Management."

PMI Professional in Business Analysis (PMI-PBA)® Examination Content Outline, 2013, "Planning," Task 4.

83. You work for a paper mill and have been asked to fill the role of PM/BA on a new self-sealing envelope project. Your lead subject matter expert, Milos, seems to be introducing requirements that are not in line with the paper mill's business needs. What should you do?
    a. Review the requirements with your sponsor
    b. Track the requirements on a matrix
    c. Ask Milos to provide the business justification
    d. As the project manager/business analyst, defer to Milos as the subject matter expert

The correct answer is: **B**

The requirements traceability matrix is a critical project artifact that tracks requirements from their origin through Evaluation. Among other items, it can be used to track needs, opportunities, goals, or objectives. If it's found that something cannot be traced, the business analyst would follow the escalation procedure outlined in the business analysis plan.

Answer Choice A: A review with the project sponsor could come after the items were added to the requirements traceability matrix, provided that this was the escalation procedure outlined in the business analysis plan. The first step would be to validate your assumptions with the SME by tracing and tracking the requirements to the goals and objectives of the business.

Answer Choice C: While this may seem like a logical answer choice, it could result in unwelcomed conflict. The better approach is to add the items to the tracker, then work with the SME to trace and track the requirements to the goals and objectives of the business.

Answer Choice D: This is a correct response; however, answer B is the better choice, because it establishes a tangible outcome, adding the items to the requirements traceability matrix.

A Guide to the Project Management Body of Knowledge, 5.2.3.2, "Requirements Traceability Matrix."

PMI Professional in Business Analysis (PMI-PBA)® Examination Content Outline, 2013, "Planning," Task 2.

84. In the process of reviewing requirements documents, Caroline uncovers several related objects with her subject matter experts. What is one way to visualize the order and reliance of objects?
    a. Caroline and her subject matter experts can build a process flow diagram.
    b. Caroline can create a dependency graph.
    c. Caroline and her subject matter experts can use a requirements traceability matrix.
    d. The team can use an Ishikawa/fishbone diagram.

The correct answer is: **B**

Although presented as a brief scenario, the question challenges your understanding of the four answer choices—specifically, the tools and techniques for process analysis. To visualize the order and dependency of requirements, business analysts can create a dependency graph, which will also show the order of operation.

Answer Choice A: Had the answer choice been process models, this could be a viable choice; however, process flow diagrams are used to describe elements of a solution, process, or project. They are powerful tools for visualizing both current (as-is) and future (to-be) states, often referred to as *swim-lane diagrams.* This would not be the best technique for visualizing the order and reliance of objects.

Answer Choice C: The requirements traceability matrix is used for monitoring and controlling product scope by linking requirements to an organization's goals and objectives.

Answer Choice D: Cause-and-effect diagrams, a variation of which is a fishbone diagram (aka *Ishikawa diagram*) are used to investigate the high-level causes of occurring problems.

Examination Content Outline, Knowledge and Skills, #27, "Process Analysis Tools and Techniques."

Business Analysis for Practitioners: A Practice Guide, Section 5.3, "Relationships and Dependencies."

PMI Professional in Business Analysis (PMI-PBA)® Examination Content Outline, 2013, "Analysis," Task 2.

85. Your executive sponsor, Enzo, asked that you collaborate with Cillian, a veteran project manager, on a key document that will outline the need for action, the root causes, and the main contributors of the problems, along with the rank order of the recommendations. Cillian suggests that Enzo is referring to a:
    a. Business case
    b. Situation statement
    c. Project charter
    d. Capability framework

The correct answer is: **A**

This question introduces a lot of unnecessary information and further tests your understanding of the key elements of the value proposition. These, in total, are essentials of the business case, which contains the justification and cost–benefit analysis for the project. It serves as a reminder of why the initiative was approved and will be revisited and updated over the duration of the initiative.

Answer Choice B: Situation statements present a complete understanding of the problem to be addressed or the opportunity to be pursued, along with the contributing effects and the overall impacts. They provide enough detail to establish the problem (or opportunity) of "x," the effect of "y," with the impact of "z."

Answer Choice C: The charter is a key artifact that formally recognizes and authorizes the project and provides the project manager with the authority to execute the project.

Answer Choice D: A capability framework is a simple outline of an organization's capabilities, which can comprise human, financial, or technological components.

Business Analysis for Practitioners: A Practice Guide, Section 2.6, "Assemble the Business Case."

PMI Professional in Business Analysis (PMI-PBA)® Examination Content Outline, 2013, "Needs Assessment," Task 1.

86. Your data conversion lead, Wolfgang, has just reviewed the requirements traceability matrix and identified all aspects to consider, as per the organizational records and retention policy. Where should these requirements be identified?
    a. Business requirements document
    b. Stakeholder requirements document
    c. Transition requirements document
    d. Functional requirements document

The correct answer is: **C**

This question tests your knowledge across two continuums, the first being requirement types, the second, the applicability of an organizational records-and-retention policy in a given scenario. Transition requirements focus on temporary activities to move from the as-is to the to-be states. Tasks typically associated with transition requirements are training and data conversion. In this case, the records-and-retention policy would be an input to the data conversion requirements.

Answer Choice A: Business requirements describe the higher-level needs of the enterprise; they describe the purpose/intent of the component and metrics to evaluate its impact on the organization. They contain the component goals, which are linked to the strategic plan. The business requirements document (BRD) focuses on the entire enterprise, not specific organizational levels.

Answer Choice B: Stakeholder requirements describe the needs of a stakeholder or group of stakeholders and how they will benefit and interact with the component.

Answer Choice D: Functional requirements are a subset of solution requirements, which describe the behaviors or function of the component—a statement of conformity (described in Yes/No terms).

Requirements Management: A Practice Guide, Section 5.2.2. "Define Types of Requirements."

PMI Professional in Business Analysis (PMI-PBA)® Examination Content Outline, 2013, "Analysis," Task 6.

87. Claire works for ZooZle, a complex internet search and news company. While documenting solution constraints, what might be something she needs to keep in mind?

a. Access to the search engine and news may be limited based on the country of origin.
b. The success of the solution is based on a future event.
c. All product team members may not stay to witness the delivered solution.
d. The iron triangle of quality as constrained by time, cost, and scope.

The correct answer is: **A**

In this scenario, the question is challenging you to analyze each of the answer choices to determine which is the most applicable in terms of solution constraints. To begin, you must first understand that constraints are restraining factors, which can be tied to either the (a) product/solution or (b) the project, and which can impede the execution of a project. Solution constraints can encompass culture, geography, policies, and regulations. Today, it's not uncommon for governments to restrict access to both news and internet search; ergo, answer choice A is the most appropriate selection.
Answer Choices B & C: These are made-up answer choices.
Answer Choice D: Project constraints can consist of time, cost, scope, and quality.
Business Analysis for Practitioners: A Practice Guide, Section 4.11.4.2, "Documenting Constraints."
PMI Professional in Business Analysis (PMI-PBA)® Examination Content Outline, 2013, "Analysis," Task 2.

88. The agile software development methodology is new to your organization, and you are very excited to be part of the team. You've been asked to work with Oscar, a user-interface expert, on the actions for each button that is to be part of the order-entry screen. When selecting the template to track your requirements, you choose:
a. The W3C Standards template from the Open Web Platform, as this will ensure cross-platform compatibility
b. A template from your project management office that tracks the behavior of the screen and buttons
c. The agile burndown and burnup template
d. A value stream map, as one of the guiding principles is to reduce customer wait time and eliminate non-value-added time

The correct answer is: **B**

In this scenario, the business analyst would need to think through each of the answer choices to determine the most appropriate means of selecting a template to track requirements. Templates ensure consistency and are designed so that business analysts and project managers are aware of both the required and the optional information to deliver a solution. In many organizations, templates are administered by the project management office.
Answer Choice A: This is a made-up answer choice.
Answer Choice C: In agile, burndown charts show how much work is remaining in a given project; burnup charts show the inverse—how much work has been completed. Neither of these tools would help to track requirements.
Answer Choice D: Value stream maps identify non-value-added time, commonly referred to as waste, within a process.
Business Analysis for Practitioners: A Practice Guide, Section 1.7.2, "Requirements Types."
PMI Professional in Business Analysis (PMI-PBA)® Examination Content Outline, 2013, "Needs Assessment," Task 1.

89. During an interview, your executive sponsor, Adelaide, is providing very detailed responses but does not seem to answer the questions in a refined, direct manner. In an effort to reframe

the interview, you've requested that Adelaide respond based on validation. What result will this yield?

a. Responses will be used as lead-ins to follow-up questions.

b. It will produce responses based on forced-choice answers.

c. Responses will be based only on the current topic of proposed solutions.

d. Adelaide will quantitatively validate or substantiate her responses.

The correct answer is: B

This is a complicated question, challenging your understanding of both interviewing techniques and types of questions. In interviewing stakeholders, there are a variety of questions that can lead to more thought-out and comprehensive responses. There are four categories of questions: (a) open-ended, (b) closed-ended, (c) contextual, and (d) context-free. Closed-ended questions elicit responses based on limited answer choices, either forced or limited choice or confirmation.

Answer Choice A: Context-free questions are typically used as lead-ins.

Answer Choices C & D: These are made-up answer choices.

Business Analysis for Practitioners: A Practice Guide, Section 4.5.2.1, "Types of Questions."

PMI Professional in Business Analysis (PMI-PBA)® Examination Content Outline, 2013, "Analysis," Task 1.

90. You've been invited to meet with the CEO of Blythe Consulting LLC, along with several other senior leaders of the product development team. The team has just completed the baseline of the requirements, and Brian Blythe begins the meeting by reviewing the project justification. As the business analyst, where should you document this information?

a. In the meeting minutes

b. As part of the solution scope statement

c. In the business case

d. In the business requirements document

The correct answer is: D

This question challenges you to think through where you are in the process and what is the most appropriate response based on the scenario. With the requirements baselined, the focus shifts to writing the specification documents. In this case, the CEO is covering aspects that should be included as part of the business requirements document. These include the high-level essentials and the rationale for the investment. Business requirements describe the higher-level needs of the enterprise and focus on the entire enterprise, not specific organizational levels.

Answer Choice A: This may seem like a logical answer choice, and while some relevant information will be captured in the meeting minutes, you would not document the component goals or establish the linkage to the strategic plan.

Answer Choice B: The solution scope statement establishes the boundary for the initiative covering elements such as acceptance criteria, assumptions and constraints, deliverables, exclusions, and scope description.

Answer Choice C: The business case is the cost–benefit analysis for the project and contains the value proposition for the solution. Although updated over the life the project, there is nothing in this scenario that suggests the business case should be updated.

Requirements Management: A Practice Guide, Section 5.2.2, "Define Requirements Types."

PMI Professional in Business Analysis (PMI-PBA)® Examination Content Outline, 2013, "Analysis," Task 6.

91. You are the subject matter expert representing the distribution logistics center of a textile company. Your business analyst, Nathaniel, is facilitating sessions with technicians Austin and Scarlett on the required training and associated operational changes for a proposed relocation. What type of requirements is Nathaniel eliciting?
   a. Technical requirements, as the relocation will require moving several servers and phone systems
   b. Operational requirements, as this is more than an IT initiative—the entire business unit is relocating
   c. Transition requirements, as the logistics center considers the capabilities needed to migrate to the future state
   d. Foundational requirements; essentially, the business process drives the training for the proposed operational changes

The correct answer is: C

As you approach this question, you recall that there are five types of requirements applicable to business analysts: (a) business requirements, (b) stakeholder requirements, (c) solution require-ments that are either (c1) functional requirements or (c2) nonfunctional requirements; and (d) tran-sition requirements. The Practice Guide also discusses project, program, and quality requirements specific to project management. Transition requirements are temporary in nature and are used to facilitate the passage from current (as-is) to future (to-be) state (most commonly associated with training and data conversion). From the answer choices, C is the only valid option.
Answer Choices A, B, & D: Technical, operational, and foundational requirements are not valid requirement types.
Business Analysis for Practitioners: A Practice Guide, Section 1.7.2, "Requirement Types."
PMI Professional in Business Analysis (PMI-PBA)® Examination Content Outline, 2013, "Needs Assessment," Task 1.

92. At the request of the operations business lead, the Change Control Board (CCB) approved a $250,000 USD workaround, based on information from the requirements documentation. Once the development team started to review the details of the workaround, their estimate nearly tripled. At an emergency meeting of the CCB, your sponsor, Charles, detailed the issues with the change approval process, asked for the team's suggestions on how to improve, and inquired whether there was any part that was working well. Why?
   a. Charles wanted to prevent further cases of improper estimating.
   b. The business units were exploiting a weakness in the change approval process.
   c. Charles found the process to be too light and wanted to add additional rigor.
   d. Charles was conducting a retrospective.

The correct answer is: D

Projects by nature are dynamic, and to be successful, teams need to adapt and adjust course in rapid fashion. To properly address this question, you need to understand the definition of *retrospective* and know that use of the technique is not limited to the end of sprint iterations. Retrospectives are opportunities to: (a) understand what is working well and should be continued; (b) evaluate what has not worked well; (c) determine the necessary actions the team needs to take to improve the process going forward.

Answer Choices A, B, & C: These answer choices are all distractors and are not focused on the actual question.

PMI-PBA° Examination Content Outline, Knowledge and Skills, #19, "Lessons Learned & Retrospectives."

PMI Professional in Business Analysis (PMI-PBA)° Examination Content Outline, 2013, "Traceability and Monitoring," Task 5.

93. You work for an IT service organization that was recently certified as ISO 9004 compliant. As a result, your director, Tuesday, would like to update service quality, performance, and response guidelines for both new and existing customers. What artifact describes these metrics?
    a. Service-level agreements
    b. ISO 9004 requirements
    c. Quality management plan
    d. Requirements management plan

The correct answer is: **A**

This is an advanced question, and it presents topics that are very likely to appear on the exam. There is some distracting information, so your focus should be on, "The document contains service quality, performance, and response guidelines for both new and existing customers." Once you identify that actual question, there is only one possible answer choice. You'll recall, there are two types of agreements to manage stakeholder expectations. The first, service-level agreements (SLAs), are contracts between service organizations and customers that establish performance metrics and response times. Once established, these metrics can continually be validated during Evaluation. The second, operational-level agreements, are contracts within the service organization. **Answer Choice B:** ISO 9004 is a quality management standard, not a contractual obligation or requirement. **Answer Choice C:** The quality management plan establishes the quality requirements and standards for which the project will be measured. **Answer Choice D:** The requirements management plan will cover elements of both the project and the product. It provides the overarching direction throughout the project and formally establishes the *how* and *when* for all solution development activities.

A Guide to the Project Management Body of Knowledge, Section 4.1.1.3, "Agreements."

PMI Professional in Business Analysis (PMI-PBA)° Examination Content Outline, 2013, "Planning," Task 6.

94. Elise, a senior project manager who is also filling the role of business analyst, is collaborating with Atticus, the subject matter expert from the mayor's office, on a document that will outline the compliance requirements for a new procurement application. What document are they working on?
    a. Solution requirements
    b. Quality requirements
    c. Business requirements
    d. Stakeholder requirements

The correct answer is: **B**

It's common for organizations to have the same resource filling the role of project manager and business analyst. In these situations, the resource needs to ensure that they are not overly focusing on one aspect versus another, because both proper project management and proper business analysis are critical to ensuring that the project fulfills its value proposition. As you approach this question, you'll recall that there are five requirement types applicable to business analysis: (a) business; (b) stakeholder; (c) solution, which can be either (c1) functional or (c2) nonfunctional; and (d) requirements. The Practice Guide also discusses project, program, and quality requirements specific to project management. In this question, the focus is on compliance requirements. The quality requirements document establishes the measures and metrics to ensure project completion, while addressing compliance requirements and standards.

Answer Choice A: Solution requirements describe the characteristics of the component (i.e., features, functions), which are further decomposed to functional and nonfunctional requirements.

Answer Choice C: Business requirements describe the higher-level needs of the enterprise.

Answer Choice D: Stakeholder requirements describe the needs of a stakeholder or group of stakeholders and how they will benefit from and interact with the component.

Requirements Management: A Practice Guide, Section 5.2.2., "Define Types of Requirements."

PMI Professional in Business Analysis (PMI-PBA)® Examination Content Outline, 2013, "Analysis," Task 1.

95. Following a meeting with senior leadership, your executive sponsor, Brennan, shared privately that the company was being acquired by another firm. He expressed concern that this action could have significant implications for your product and didn't want to spend money unnecessarily on the project. As a result of the meeting, you decide to:
    a. Collaborate with the project manager on ways to inform the team and place the project on hold pending further guidance from your sponsor
    b. Review the project management plan
    c. Continue the project until the information is made public
    d. Convene a meeting of the Change Control Board to discuss the implications

The correct answer is: **B**

Change can affect the project from many directions; in the broadest sense, they can originate from the top down (leadership) or bottom up (functional or project team member). This is an example of change originating from the Board of Directors, or a top-down change. The first step in this scenario would be to review the change management plan, which is a subsidiary of the project management plan. This project artifact will outline the process for integrated change control. Any changes to the project, including termination or hiatus, would need to follow the process outlined in the plan.

Answer Choice A: While this may seem like a logical response, it's (a) not appropriate to discuss this with anyone else—the information was shared privately; and (b) the change management plan is the first place to look, as it will establish the proper guidelines.

Answer Choice C: This may also seem like a logical choice; however, the executive sponsor advised that they didn't want to spend money unnecessarily on the project.

Answer Choice D: Until the project management plan is reviewed, this would be premature.

Business Analysis for Practitioners: A Practice Guide, Section 3.4.14, "Define the Requirements Change Process."

PMI Professional in Business Analysis (PMI-PBA)® Examination Content Outline, 2013, "Planning," Task 4.

96. You're facilitating a workshop for Alexandra and several other key stakeholders at QQQ Manufacturing Incorporated. You've just completed outlining all the key stakeholder needs and are now

trying to determine the critical characteristics for a new product. What technique are you most likely using?

a. Quality function deployment
b. Voice of the customer
c. Quality assurance
d. Joint design/development

The correct answer is: **A**

This is an advanced question, and one that is very likely to appear on the exam. After removing the commentary, the question is asking, "What technique would you use to determine the critical characteristics for a new product?" From the answer choices, the only choice is quality function deployment (QFD), a planned method of defining stakeholder needs and translating them into factional plans.

Answer Choice B: The voice of the customer exercise is preceded by defining the customer in a QFD exercise. The voice of the customer technique removes noise and isolates each requirement by identifying: (a) what is the customer saying? (b) why are they saying it? (c) what are their aversions? and (d) what are their actual requirements?

Answer Choice C: Quality assurance is focused on preventing defects through planning and by inspecting work-in-progress (WIP).

Answer Choice D: This is made-up answer choice.

A Guide to the Project Management Body of Knowledge (PMBOK®), Section 5.2.2.3, "Facilitated Workshops."

PMI Professional in Business Analysis (PMI-PBA)® Examination Content Outline, 2013, "Evaluation," Task 2.

97. Ziggy is preparing to present the business analysis plan to key stakeholders for final approval following several rounds of meetings, during which the attendees continually questioned the rationale and were having difficulty justifying the resource commitments. What could Ziggy have done to prepare stakeholders prior to the initial approval meeting?

a. Sent out an advance copy to all invited attendees requesting their support
b. Sent out an advance copy to all invited attendees requesting they send you any questions in advance
c. Met with the project manager to review a draft copy of the plan
d. Met with his sponsor to review a draft copy of the plan, along with a few bullet points outlining the justification

The correct answer is: **D**

This question draws upon on a number of your skills as a business analyst, notably meeting facilitation techniques and cultural awareness. Because project plans can be perceived as being bureaucratic, stakeholders are more inclined to support the business analysis plan if it has strong support from the sponsor. Clearly documenting the rationale will also help to justify the resource commitment. In this scenario, had the business analyst met with the sponsor ahead of time to review the plan, both would have been better prepared for the approval meeting.

Answer Choices A & B: While sending out an advance copy to all stakeholders for review ahead of the meeting is a leading practice, neither are best answer choices, because they don't engender support from the sponsor.

Answer Choice C: While this is also a leading practice, a meeting with project manager would not be sufficient in and of itself to address the concern at hand.

Business Analysis for Practitioners: A Practice Guide, Section 3.5.4, "Document the Rationale for the Business Analysis Approach."

Examination Content Outline, Knowledge and Skills, #24, "Political and Cultural Awareness."

PMI Professional in Business Analysis (PMI-PBA)® Examination Content Outline, 2013, "Planning," Task 4.

98. Caius is working to categorize the models that will be used while the team is drafting specification documents. Entity relationship diagrams, data dictionaries, and state tables are all considered:
    a. Models
    b. State models
    c. Data models
    d. Process data models

The correct answer is: **C**

This question tests your understanding and knowledge pertaining to the categorization of entity relationship diagrams, data-flow diagrams, data dictionaries, state tables, and state diagrams. These are all data models, which can be used to illustrate information within a process. Their purpose is to enable the investigation and documentation pertaining to the lifecycle of information, within both the processes and the related applications.

Answer Choice A: This answer choice is unspecific.

Answer Choice B & D: These are both made-up answer choices.

Business Analysis for Practitioners: A Practice Guide, Section 4.10.3, "Categories of Models."

PMI Professional in Business Analysis (PMI-PBA)® Examination Content Outline, 2013, "Analysis," Task 6.

99. Working with Ezekiel, your executive sponsor and VP of distribution and logistics for a pet food company, you've both agreed there are two options for addressing the opportunity. Ezekiel suggests using a method whereby team members can vote to determine the most suitable or preferred option. You suggest:
    a. A matrix in which options are weighted and ranked, with criteria that align with objectives set forth in the needs assessment
    b. Planning poker, allowing stakeholders to vote and score options
    c. A needs assessment
    d. A scored capability table in which each team member's vote is weighted in proportion to their role in the organization

The correct answer is: **A**

This is a very easy question, once you remove all the distractor answer choices. Following the completion of requirements prioritization, solutions can be quantitatively evaluated with the stakeholders and scored on two dimensions:

(a) First, establishing the expectation score, based on a simple scale: 0 = does not meet; 1 = partially meets; and 2 = fully meets expectations. This establishes the degree to which the solution meets stakeholders' expectations.

(b) This is followed by a Kano score—the degree to which stakeholders were satisfied with the items that (1) partially or (2) fully meet expectations. Kano is measured on five dimensions: 1 = dissatisfied; 2 = disappointed; 3 = neutral; 4 = satisfied; and 5 = delighted.

The scoring outputs of the evaluation are used to further refine the analysis using techniques such as pair-wise analysis, weighted ranking, and multi-criteria scoring. With the qualitative and quantitative analysis complete, the business analyst will use the complete analysis as an input to the business case. Weighted ranking is an objective technique to determine the preferred option, based on its alignment to the organizational imperatives.

Answer Choice B: This is a made-up answer combination, as planning poker does not allow stakeholders to vote and score options. Rather, it's a consensus-based technique used by agile teams to estimate the product backlog. Planning poker is based on a modified Fibonacci sequence: 0, 1, 2, 3, 5, 8, 13, 20, 40, and 100. You'll note, each number up to 20 is the sum of the two preceding numbers. For example, 5 is 2 plus 3.

Answer Choice C: This is a distractor and doesn't properly focus the team's attention.

Answer Choice D: This is a made-up answer choice.

Business Analysis for Practitioners: A Practice Guide, Section 2.5.5.1 Weighted Ranking.

PMI Professional in Business Analysis (PMI-PBA)® Examination Content Outline, 2013, "Needs Assessment," Task 2.

100. At what appears to be an interval of every 15 days, your product owner reprioritizes the work to be addressed in the next iteration. What do they hope to achieve?
    a. They are performing a reprioritizing analysis.
    b. They are performing a backlog reprioritizing analysis.
    c. They are grooming the backlog to manage scope.
    d. They are preparing a sprint analysis to manage scope.

The correct answer is: **C**

This question tests your understanding and experience with the agile delivery methodology. There are several key concepts within this question: (a) sprint, 15-day iteration; (b) product owner, who is responsible for prioritizing the backlog and defining stories; (c) backlog, which is a prioritized list of user stories and features. To manage scope and ensure that the backlog contains appropriate items that are of value to the organization, the product manager will groom the backlog ahead of each sprint.

Answer Choices A, B, & D: These are all made-up answer choices.

Business Analysis for Practitioners: A Practice Guide, Section 5.6.1, "Benefits of Using the Traceability Matrix to Monitor Requirements."

PMI Professional in Business Analysis (PMI-PBA)® Examination Content Outline, 2013, "Traceability and Monitoring," Task 3.

101. Aria, the subject matter expert for a new line of autonomous vehicles, is impressed with all the features and highlighted functionality. However, after the vehicle started, the windows began to roll up and down by themselves and the radio's volume was acting erratically. When she attempted to turn the car off, the sunroof opened. As the SME, is Aria more concerned about grade or quality?
    a. Quality: all the defects render the product useless.
    b. Grade: all the defects render the product useless.
    c. Quality and grade: all the defects render the product useless.
    d. Neither: Aria is concerned about precision and accuracy.

The correct answer is: **A**

This is a difficult question and tests your understanding of quality management. You'll recall there are differences between *quality* and *grade*. Quality is determined by how closely the product satisfies the identified requirements. Grade is commonly associated with features in the delivered solution. In this scenario, a feature-rich product or solution (high grade) with many bugs and poor documentation would be useless due to the poor quality.

Answer Choice B: Grade is a measure of features, in some cases low grade may be acceptable, even preferred over low quality.

Answer Choice C: This is an incorrect statement; low grade does not render the product useless.

Answer Choice D: This answer choice is incorrect; precision is a degree of *correctness*, whereas accuracy is a measure of *exactness*.

A Guide to the Project Management Body of Knowledge (PMBOK®), "Introduction to Quality Management," page 228.

PMI Professional in Business Analysis (PMI-PBA)® Examination Content Outline, 2013, "Evaluation," Task 1.

102. Pippin is the business analyst on a Kanban project. What tool can she use to report the remaining items to be addressed?
    a. Burndown chart
    b. Estimate at completion
    c. Estimate to complete
    d. Variance report

The correct answer is: **A**

For those business analysts with agile experience, this is a relatively easy question. The emphasis of Kanban projects is on deliverable status, displayed prominently on Kanban boards (aka *information radiators*). Common categories are (a) ideas/backlog, (b) to do, (c) doing (work-in-progress, WIP), and (d) done. Concentrating on project delivery, burndown charts are used with projects that follow an adaptive lifecycle and list the remaining items to be completed. Conversely, burnup charts show how much work has been completed.

Answer Choice B: Estimate at completion is the forecasted cost of the final delivered solution, per the requirements documentation.

Answer Choice C: Estimate to complete are the anticipated funds required to finish the solution, per the requirements documentation.

Answer Choice D: Whereas a variance report will show planned versus actual, it would not be an appropriate mechanism to report the remaining items to be addressed.

Business Analysis for Practitioners: A Practice Guide, Section 6.5.3.1, "Project Metrics as Input to the Evaluation of the Solution."

PMI Professional in Business Analysis (PMI-PBA)® Examination Content Outline, 2013, "Planning," Task 6.

103. Amelie, a recent college graduate, is assisting you in identifying the cardinality and multiplicity of relationships. What is a common manner of displaying these relationships?
    a. Entity relationship diagram
    b. Crow's foot notation
    c. Context diagram
    d. Wireframe

The correct answer is: **B**

This can be a difficult question if you are not familiar with data modeling techniques to illustrate information within processes. Entity relationship diagrams (ERDs), or business data diagrams, illustrate business objects and their mutual relationships. Cardinality and multiplicity are both aspects modeled within the ERD; a common means of doing so is via the crow's foot notation method. Cardinality is the maximum number of times an instance in one entity can be associated with instances in a related entity. Conversely, ordinality is the minimum number of times an instance in one entity can be associated with an instance in the related entity. Multiplicity is the minimum and maximum permitted members in the set, which can include (a) one to one, (b) one to many, (c) many to one, (d) and many to many.

Answer Choice A: The crow's foot notation is a graphical symbol to indicate the cardinality and multiplicity within the entity relationship diagram.

Answer Choice C: Context diagrams are created to show all the direct system and human interfaces within a solution.

Answer Choice D: Wireframes are outlines, blue prints, or schematics that illustrate the general look and feel of a proposed solution.

Business Analysis for Practitioners: A Practice Guide, Section 4.10.10.1, "Entity Relationship Diagrams."

PMI Professional in Business Analysis (PMI-PBA)® Examination Content Outline, 2013, "Analysis," Task 6.

104. Connor, the chief engineer for a microprocessor that will be used in a new line of autonomous bicycles in New York City, documents a problem during the last round of testing prior to go-live. Connor is very outspoken, and many team members defer to his opinion. As the business analyst, you:

a. Interview each of the stakeholders involved with testing and document their results.

b. Repeat the system and integrated testing cycles.

c. Facilitate a MultiVoting session.

d. Consult with the project manager and escalate the concern to the project sponsor.

The correct answer is: **C**

On the exam, there will be a number of questions pertaining to the techniques and methods to enable and support group decision making. It's important to remember that PMI is a proponent of taking action, and working in groups (even virtually) is highly favored to bring items to resolution. Although there are four methods which are likely to appear on the exam (brainstorming, nominal group technique, Delphi technique, and didactic interaction), from the answer choices, MultiVoting is the only option. Synonymous with the nominal group technique, this method is used to quantitatively produce consensus for a particular discussion topic, essentially refining results generated during brainstorming sessions. By using this method, each participant has an opportunity to participate equally.

Answer Choice A: While conducting an interview may seem like a reasonable approach, it will not help the team reach consensus. This can only be achieved via a group decision making process.

Answer Choice B: Repeating the test would more than likely produce the same the result for Conner; you are not addressing the concern—that others didn't raise a similar concern.

Answer Choice D: This answer choice is a distractor.

Business Analysis for Practitioners: A Practice Guide, Section 6.8, "Facilitate the Go/No-Go Decision."

PMI Professional in Business Analysis (PMI-PBA)® Examination Content Outline, 2013, "Evaluation," Task 3.

105. You're collaborating with your project manager, Caspian, on creating an illustration that supports the prioritization of solution elements based on their value to the business. Impressed by your well-thought-out model, your sponsor, Cecilia, inquired as to the name of the model. You advise:
    a. Business Objective Model
    b. Business Rule Model
    c. Business Value Model
    d. Business SWOT Model

The correct answer is: **A**

Business objective and goal models are diagrams that visualize the relationship between solution elements and the value to the organization. The illustrations help to support prioritization efforts to ensure that the elements that provide the greatest value have the highest prioritization.

Answer Choice B: Rule models comprise business rule catalogues, decision trees, and decision tables.

Answer Choice C: Value model is a made-up answer choice.

Answer Choice D: SWOT is mnemonic acronym to assess strategy, goals, and objectives, the letters stand for: **s**trengths, **w**eaknesses, **o**pportunities, and **t**hreats. *On the exam, remember, strengths and weaknesses are internal to the organization, whereas opportunities and threats are external.*

Business Analysis for Practitioners: A Practice Guide, Section 4.10.7.1, "Goal Model and Business Objective Model."

PMI Professional in Business Analysis (PMI-PBA)® Examination Content Outline, 2013, "Analysis," Task 2.

106. It would appear that the team has reached an impasse and is struggling to reach a consensus on the best approach to address a defect in the manufacturing process. Although each proposed solution has merit, the team can only select one. Ciaran, a trained business analyst, proposes the nominal group technique. To begin, he explains that the first step is to:
    a. Rank and score the concepts, then quantitatively agree upon a solution.
    b. List only designs that are relevant, so the team can rank and score the ideas.
    c. Quantitatively agree upon a solution, based only on ideas that are relevant.
    d. Identify all the possible theories, even listing ones that may appear to be unrealistic.

The correct answer is: **D**

The key word in the first sentence is *consensus*—this should immediately trigger you to think group decision making techniques. You'll recall there are *there are four methods that are likely to appear on the exam (brainstorming, nominal group technique, Delphi technique, and didactic interaction).* Originally developed by Andre Delbecq and Andrew H. Van de Ven, the nominal group technique (also referred to as *NGT* or *MultiVoting*) is used to quantitatively produce consensus for a particular discussion topic, essentially refining results generated during brainstorming sessions. Because brainstorming sessions can often result in ideas that are not feasible or realistic, or are simply too blue sky for the moment, NGT provides a means to narrow down the proposed solution set to something that participants can agree upon. Furthermore, it was designed such that the points of view of all participants are taken equally into consideration.

Answer Choice A: Ranking and scoring is the last step in the process.

Answer Choices B & C: These are distractors; the idea is to list all possible ideas, even those that may seem unrealistic.

Business Analysis for Practitioners: A Practice Guide, Section 4.15.2, "MultiVoting."

PMI Professional in Business Analysis (PMI-PBA)® Examination Content Outline, 2013, "Analysis," Task 5.

107. While giving a presentation, Brody states, "Although risks can become issues, not all issues started as risks. And assumptions are . . ."
    a. Practices that should be avoided, because they are not valid or warranted.
    b. True, real, or certain without required proof.
    c. For the most part, may or may not hold true over the life of the project.
    d. Should not be documented, because they are baseless.

The correct answer is: **B**

This is an easy question, and is simply the definition of *assumptions*. Similar to issues, risks, and decisions, assumptions also need to documented and tracked over the lifecycle of the solution, from inception through evaluation. They are factors that are considered to be true, real, or certain without proof or demonstration.

Answer Choices A & C: These are made-up answer choices.

Answer Choice D: Because assumptions can become risks and issues, they need to be documented, so that they can be proactively addressed.

Business Analysis for Practitioners: A Practice Guide, Section 4.11.4.1, "Documenting Assumptions."

PMI Professional in Business Analysis (PMI-PBA)® Examination Content Outline, 2013, "Analysis," Task 1.

108. Your project is part of a larger program that is significantly ahead of schedule. In the interest of the customer, Navy Kate LLC, your lead developer, Mabel, decided to build a new module that the customer had mentioned during the needs assessment, even though the module never made it to the approved requirements traceability matrix. She worked on this module for 90 days, with the project coordinator's knowledge, before the other components caught up for system testing. Because the project coordinator was aware of the work effort and reported the status each week to the PMO, was this acceptable?
    a. Yes, the developer built the module in the interest of the customer, and status was reported to the PMO.
    b. No, this is considered scope creep.
    c. Yes, because the developer received permission from the customer.
    d. No, this is considered gold plating.

The correct answer is: **B**

There are two concepts introduced in the answer choices that are often confused: *scope creep* and *gold plating*. The requirements baseline establishes the boundaries for the deliverables associated with the product. Any work effort outside of this baseline must be approved by the Change Control Board. The scenario provided is an example of scope creep. Although scope creep can happen for a number of reasons, the most common are (a) customer interference, (b) poorly defined scope, (c) inadequate change control, and (e) poor communication within the team.

Answer Choices A & C: Mabel's work effort was not appropriate. Although the module was built in the interest of the customer and status was reported to the PMO, the scenario does not mention that a change request was approved to add the work to the baseline.

Answer Choice D: Gold plating implies additional features were added to the product, often done by team members to demonstrate their skills and abilities.

Business Analysis for Practitioners: A Practice Guide, Section 5.6.1, "Benefits of Using the Traceability Matrix to Monitor Requirements."

PMI Professional in Business Analysis (PMI-PBA)® Examination Content Outline, 2013, "Traceability and Monitoring," Task 5.

109. Your organization is considering a major investment in redesigning its manufacturing line. Your sponsors, Emilia and Caroline, have requested that you conduct an assessment to determine the organization's capability, capacity, willingness, and commitment to the project. What type of analysis will you perform?
    a. Organizational readiness assessment
    b. Capability and capacity assessment
    c. Quadrant analysis assessment
    d. Organizational factors analysis

The correct answer is: **A**

This is a definition-based question, representative of ones that are very likely to appear on the exam. Presented are four answer choices that are reasonable; if you are unsure of the answer, the best approach is through process of elimination. Organizational readiness assessments are used to ensure that the organization is prepared to undertake the initiative. They provide for an understanding of the current state by identifying enablers and constraints and the key goals and objectives as the organization prepares to transform to the desired state.

Answer Choice B: *Capability* refers to the team's ability to complete the work, whereas *capacity* addresses factors such as (a) resourcing, (b) financial, and (c) equipment.

Answer Choice C: *Quadrant analysis* is used for plotting and understanding data elements. Although there is no limit to the number of grids, the most common is a 2x2 matrix. This flexible tool allows teams to categorize viewpoints and interests in a grid format.

Answer Choice D: This is a made-up answer choice.

PMI Professional in Business Analysis (PMI-PBA)® Examination Content Outline, Knowledge and Skills, #22, "Organizational Assessment."

PMI Professional in Business Analysis (PMI-PBA)® Examination Content Outline, 2013, "Analysis," Task 1.

110. Alistair, the lead for your time-and-attendance system, hung a whiteboard outside his cubical, with columns representing the project phases. What was the intent?
    a. To establish a platform on which to base discussions
    b. To illustrate the requirements lifecycle
    c. To provide the basis for elicitation
    d. To provide alternative analysis for satisfying the business need

The correct answer is: **A**

This is an example of an information radiator, a concept that is very likely to appear on the exam, and a leading practice for communicating with the organization, not just team members. Communication is paramount in projects, and this is a simple way of sharing information and establishing a platform on which to base discussions. An example of an information radiator is a Kanban board, which is often used in projects with adaptive lifecycles to visually display work in progress (WIP). By using sticky notes, items can easily be adjusted. Common Kanban categories are (a) ideas/backlog, (b) to do, (c) doing (work-in-progress, WIP), and (d) done. However, in this scenario, the idea was adopted for waterfall and iterative delivery.

Answer Choice B & D: These are incorrect answer choices and distractors.

Answer Choice C: *On the exam, remember words have specific meaning.* In this case rather than *elicitation,* the better word is *discussion.*

Business Analysis for Practitioners: A Practice Guide, Section 5.8.3, "Impact Analysis."

PMI Professional in Business Analysis (PMI-PBA)® Examination Content Outline, 2013, "Traceability and Monitoring," Task 4.

111. Aurelia is working with stakeholders to determine which project requirements will be deferred to a future phase of her modernization initiative at the bakery plant. The team cannot agree on the decision-making process, their overall authority, or how project requirements are to be prioritized. To what document can they refer?
    a. Requirements management plan
    b. Requirements matrix
    c. Project plan
    d. Business analysis plan

The correct answer is: **A**

If you answered D, you need to read this question a little more carefully; it asks *how project requirements are to be prioritized.* The requirements management plan will cover elements of both the project and product, outlining the decision-making process, team members' overall authority, and how project requirements are to be prioritized. Furthermore, it establishes the framework for communications and provides the guidelines for managing requirements, whereas the business analysis plan focuses solely on the activities and deliverables related to the efforts of business analysis. From a project perspective, the team would consult the requirements management plan.

Answer Choice B: This is a made-up answer choice; *on the exam, be very careful of wording—it does not state requirements traceability matrix.*

Answer Choice C: The project plan is a work plan and would not provide this information.

Answer Choice D: If the question focused on the solution or product, the team would look to the business analysis plan.

PMI Professional in Business Analysis (PMI-PBA)® Examination Content Outline, Knowledge and Skills, #13, "Elements of the Requirements Management Plan."

PMI Professional in Business Analysis (PMI-PBA)® Examination Content Outline, 2013, "Analysis," Task 3.

112. Wesley works for a consumer electronics company that is focused on building innovative games. Because of the constantly changing dynamic of the industry, Wesley is keenly aware that requirements can lose relevance overnight. To estimate the effort required to perform due diligence as to the relevance of the solution, where would Wesley not look?

a. A competitor's press release
b. The business analysis plan
c. Lower-level components of the work breakdown system
d. The project management information system (PMIS)

The correct answer is: **B**

The obscure nature of wording on PMI exams can add to a further level of complexity. In this scenario, it's asking, "Where would Wesley *not* look?" This means that three answer choices state where Wesley *would* look, so via a process of elimination and grouping by similarities, you're left with only one choice. The business analysis plan is the *how* and *when;* it establishes the overarching direction throughout the project. This would *not* be a source of information to determine the relevance of the solution.

Answer Choice A: He could look to published information in a press release to determine product relevance.

Answer Choice C: By aggregating lower-level data from the WBS, the business analyst could begin to determine the effort required to perform due diligence as to the relevance of the solution.

Answer Choice D: Historical information in the project management information system (PMIS) is an excellent source of information, in addition to the configuration management system.

A Guide to the Project Management Body of Knowledge, Section 6.4.2, "Estimate Activity Resources: Tools and Techniques."

PMI Professional in Business Analysis (PMI-PBA)® Examination Content Outline, Knowledge and Skills, #34, "Scheduling Techniques."

PMI Professional in Business Analysis (PMI-PBA)® Examination Content Outline, 2013, "Analysis," Task 7.

113. Sam is the chief project officer of a consulting company that is aspiring to achieve greatness and knows that to deliver solutions on time, within budget, and aligned to stakeholder expectations, they need to be really good at business analysis. To be good at business analysis, what else should they have expertise in?
    a. Requirements management
    b. Project delivery methodologies
    c. Organizational change control
    d. Resource management (human, financial, equipment)

The correct answer is: **A**

According to PMI's research, the number 2 reason projects fail is poor requirements management (RM). Requirements management is focused on the key tenets of collecting, documenting, analyzing, tracing, prioritizing, and agreeing on requirements, while ensuring rigorous change control and communication methods are in place.

Answer Choice B: From the perspective of the project manager, they should have expertise in project delivery methodologies; the business analyst should have a strong working understanding of agile, iterative, incremental, lean, and waterfall.

Answer Choice C: Although organizational change control is important, it is typically the responsibility of the OCM lead, not the business analyst.

Answer Choice D: Resource management (human, financial, equipment) is generally the responsibility of the project manager.

PMI's Pulse of the Profession®, "Requirements Management: A Core Competency for Project and Program Success," page 2, Introduction.

PMI Professional in Business Analysis (PMI-PBA®) Examination Content Outline, 2013, "Analysis," Task 1.

114. Your executive sponsor, Malachi, asked for your help preparing a document that will formally authorize the project to colonize Mars. During your conversation, he referenced statements of work (SOWs), government regulations, lessons learned, and the business case. What project document was Malachi referring to?
    a. Project scope statement
    b. Project charter
    c. Project situation statement
    d. Project recommendation

The correct answer is: **B**

This question contains noise, which can be quite distracting during the exam, which is why it's critical that both in practice and professionally, you have a solid understanding of all the project artifacts. ***Most importantly, read the questions carefully.*** Following the approval of the business case (project valuation, cost–benefit analysts), the project can be chartered. The project charter is the document that formally authorizes the project and establishes the boundaries. It contains sections such as (a) approval requirements, (b) assumptions/constraints, (c) summary budget, (d) high-level description, (e) measurable objectives, (f) justification (or purpose), (g) summary milestones, (h) high-level requirements, (i) high-level risks, (j) stakeholders, (k) sponsor, (l) project manager (outlining responsibility and authority), and (m) success criteria for the project.

Answer Choice A: Although the project scope statement contains (a) assumptions/constraints, (b) acceptance criteria, (d) deliverables, (e) exclusions, and (f) scope description, it does not formally authorize the project.

Answer Choice C: The situation statement is one of the deliverables from the Needs Assessment domain. It presents a complete understanding of the problem to be addressed or the opportunity to be pursued, along with the contributing effects and the overall impacts. It does not formally authorize the project.

Answer Choice D: The recommendation is a section of the business case.

A Guide to the Project Management Body of Knowledge (PMBOK®), Section 4.1 "Develop Project Charter."

PMI Professional in Business Analysis (PMI-PBA)® Examination Content Outline, 2013, "Needs Assessment," Task 5.

115. Melissa, an expert in interface design, is creating a model to demonstrate the precise user interactions within a global financial consolidation system. The development team prefers this model, because it places the requirement statements associated with each element on the screen. What model does Melissa's development team prefer?
    a. On-screen contextual model
    b. Display-action-response model
    c. Wireframe model
    d. High-fidelity prototype model

The correct answer is: **B**

This question tests your working understanding of interface models, which can consist of (a) report tables, (b) system interface tables, (c) user interface flows, (d) wireframes, and (e) display-action-response models. Often used in combination with wireframes and screen mockups, display-action-response models are very useful in articulating requirements for user interfaces. The model is commonly used when there are large number of stakeholders or highly complicated interfaces, or when the development team is unfamiliar with the solution.

Answer Choice A: This is a made-up answer choice.

Answer Choice C: Wireframes are outlines, blue prints, or schematics that illustrate the general look and feel of a proposed solution. They are often very flexible and highly adaptable for use in various situations. Commonly used for newsletters and website design, wireframes show the flow of specific logic and business functions, depicting all user touchpoints.

Answer Choice D: High-fidelity prototypes are a functioning model of the final deliverable.

Business Analysis for Practitioners: A Practice Guide, Section 4.10.11.4, "Wireframes and Display-Action-Response."

PMI Professional in Business Analysis (PMI-PBA)® Examination Content Outline, 2013, "Analysis," Task 2.

116. Your sponsor, Augustus, has just signed off on the Harriet–Keira project, a transformational initiative that has modernized a previously mothballed manufacturing facility. Prior to shutting down the existing facility, in your role as business analyst, you need to:
    a. Notify the city that work is shifting as part an existing labor agreement.
    b. Conduct a cost–benefit realization analysis.
    c. Collaborate with the project manager to document the steps required to move from the old to the new facility.
    d. Ensure that there are adequate resources to conduct a risk assessment the day the facility opens.

The correct answer is: **C**

Projects by nature are about change, moving from one state to another. Although this scenario provides sufficient contextual information, most of it is distracting. From the answer choices provided, there is only one choice appropriate to the role of business analyst: "Collaborate with the project manager to document the steps required to move from the old to new facility." As a collaborative effort, the project manager and business analyst will create the transition plan, which provides the team with the necessary information to migrate from the as-is to the to-be state.

Answer Choices A, B, and D: These are all distractor answer choices. *During the exam, remember to focus on what the question is asking and how that relates best to the answer choices.*

Requirements Management: A Practice Guide, Section 9.1.1, "Transition Plan."

PMI Professional in Business Analysis (PMI-PBA)® Examination Content Outline, 2013, Evaluation, Task 3.

117. You are the business analyst for a hotel loyalty program; Violet, your subject matter expert for guest reservations, is describing requirements for family travelers and requirements for those that travel on business. What is she describing?
    a. A subset
    b. An implementation relationship
    c. A value dependency
    d. A rationale dependency

The correct answer is: **A**

This question tests your knowledge and understanding of dependencies and relationships. When completing the requirements traceability matrix, it's quite common to find elements that are related to each other. By using the dependency analysis technique, these relationships can be examined and documented. It's leading practice to group these dependent elements for appropriate oversight over the duration of the initiative. The Practice Guide describes the four most common dependencies and relationships: (a) subset, (b) implementation, (c) benefit, and (d) value. The scenario is describing a subset relationship—both family travelers and business travelers are a subset of guests.

Answer Choice B: Implementation suggests that before one element can be fulfilled another must be implemented.

Answer Choice C: For a given element to be of value, an enabler component would have to be present.

Answer Choice D: This is a made-up answer choice.

Business Analysis for Practitioners: A Practice Guide, Section, "5.3.1 Subset."

PMI Professional in Business Analysis (PMI-PBA)® Examination Content Outline, 2013, "Traceability and Monitoring," Task 4.

118. You're the office manager for a construction company that is six months into a five-planned-year development of a large retirement community. Levi, the district manager for your primary supply company, stops by the site office and offers a substantial discount on material if you can commit to a regular cadence of orders. You determine:

    a. With four and a half years to go, this is amazing—the money saved can be shared with the team as a bonus.

    b. A needs assessment should be conducted to analyze the opportunity.

    c. The discount will amount to over $3 million USD when analyzing the proposal in your job-costing software.

    d. The results from a Monte Carlo analysis suggest that construction will slow at certain points in time, and you will not be able to maintain the order thresholds and cadence.

The correct answer is: **B**

This scenario challenges the practical instinctive approach against the PMI defined methodology. Although you might want to rush in and commit to a regular cadence of orders, the first step should be to thoroughly and thoughtfully consider the scenario. This would begin by conducting a needs assessment, to fully assess the business problem or opportunities. A needs assessment can be mandated by a sponsoring organization, recommended by a business analyst, or, in this case, formally requested by the office manager (a stakeholder). Although typically conducted prior to start of a project or program, external factors influenced by the market or regulations can necessitate a needs assessment.

Answer Choices A & C: These answer choices are worded to distract and confuse you from the correct course of action: conducting a needs assessment.

Answer Choice D: This is a made-up answer, introducing the Monte Carlo technique as a distractor. A Monte Carlo analysis is a simulation technique based on random sampling, which generates quantitative data to assist with informed decision making.

Business Analysis for Practitioners: A Practice Guide, 2.2, "Why Perform Needs Assessments?"

PMI Professional in Business Analysis (PMI-PBA)® Examination Content Outline, 2013, "Needs Assessment," Task 1.

119. As the project manager for a newly implemented software solution, you are about to commence the close activities. Before doing so, your sponsor, Emmett, has inquired whether the delivered solution is providing business value. In turn, you ask Faye, the project business analyst, to:
    a. Review the results from the nominal group technique session along with the project signoff documents.
    b. Review the project's acceptance criteria.
    c. Measure and validate the solution referring to SLAs and OLAs.
    d. Evaluate the delivered solution and conduct a net promoter survey.

The correct answer is: **D**

This is an advanced question on a tool that is very likely to appear on the exam. This task focus on evaluating the delivered solution and performance over time. The net promoter survey determines customer's satisfaction and loyalty. The net promotor score (NPS) is a metric system designed by Fred Reichheld (Reichheld, 2003) to gauge customer loyalty and brand satisfaction. It can be based on a variety of scales; however, the most common is Zero to Ten, based on a single open question, for which the response is Zero, "Not at all," to Ten, "Extremely." Common qualifiers are "Satisfied" and "Likely."

Answer Choice A: This is a distractor; the scenario does not call for a technique involving group decisions.

Answer Choice B: Acceptance criteria is very broad, and although categorized as a validation tool and technique, it would not be sufficient to quantitatively determine if the delivered solution was providing business value. Acceptance criteria is most commonly used during both testing and solution signoff.

Answer Choice C: Service-level agreements (SLA's) are contracts between service organizations and customers that establish performance metrics and response times. Once established, these metrics can continually be validated during Evaluation. The second, operational-level agreements (OLAs), are contracts within the service organization.

Business Analysis for Practitioners: A Practice Guide, Section 6.2, to review the "Purpose of Solution Evaluation."

PMI Professional in Business Analysis (PMI-PBA)® Examination Content Outline, 2013, Evaluation, Task 4.

120. While assisting you with writing project documentation, your intern from a local community college, Dante, inquired as to the principle behind solution requirements. How would you best respond to his question?
    a. They encompass all the requirements as identified by the stakeholders.
    b. They describe the features and functions to fulfill the stakeholder requirements.
    c. Solution requirements address the foundational needs of the organization.
    d. Solution requirements address the KPIs to determine a successful outcome.

The correct answer is: **B**

Solution requirements describe all the characteristics of the component identified by the stakeholders; *for the exam remember this includes characteristics such as features and functions.* Solution requirements are further decomposed into functional (utility) and nonfunctional (warrantee) requirements. They cover both the discrete items that fulfill a business need and the environmental/quality attributes often associated with service levels.

Answer Choice A: This answer choice is too general. Within the responsibility of the business analyst, remember there are five categories of requirements: business, stakeholder, functional, nonfunctional, and transition.

Answer Choice C: Requirements that address the higher-level needs of the organization are known as business requirements.

Answer Choice D: This is a made-up answer choice.

Requirements Management: A Practice Guide, Section 5.2.2, "Define Types of Requirements."

PMI Professional in Business Analysis (PMI-PBA)® Examination Content Outline, 2013, "Analysis," Task 6.

121. You work for Aoibhinn Medical Supply, a manufacturer of equipment for first responders and paramedics. Because your customers have expressed concern with your stretchers, your company is undertaking a complete redesign. In an effort to actively engage with your stakeholders and understand how they are using your products, you offer to participate in a disaster simulation. During this simulation exercise, you seek clarification of their activities, ask what they like about your current product line, and ask how the product can be improved. You are participating in what form of observation?

    a. Simulation
    b. Active
    c. Passive
    d. Participatory

The correct answer is: **B**

This question tests your knowledge of the observation concept, which allows individuals to gain a greater understanding of the situation under real-world conditions. There are four categories of observation: (a) passive, (b) active, (c) participatory, and (d) simulation). When using the active observation technique, the observer actively engages with stakeholders, at times may interrupt the process for clarification, and asks for opinions and feedback in real-time.

Answer Choice A: When using the simulation technique, the observer mimics or recreates the activities from the perspective of the end user.

Answer Choice C: When using passive observation, there is absolutely no interruption with the individual performing the process.

Answer Choice D: When conducting participatory observation, the observer would jointly perform the activities with the individual performing the process.

Business Analysis for Practitioners: A Practice Guide, Section 4.5.5.6, "Observation."

PMI Professional in Business Analysis (PMI-PBA)® Examination Content Outline, 2013, "Analysis," Task 2.

122. Collaborating with your project manager, Greyson, you've defined the business need and solution scope and ensured that the product is aligned with the goals and objectives of the business. What should you do next?

    a. Understand stakeholders' interests so that the requirements can be baselined.
    b. Use valuation techniques such as payback period (PBP), return on investment (ROI), internal rate of return (IRR), or net present value (NPV) as inputs to the business case.
    c. Develop the solution scope statement based on either the RCA or the opportunity analysis.
    d. Develop an interrelationship analysis so that stakeholders can visualize the proposed solution scope.

The correct answer is: **A**

This question tests your knowledge of the sequence of tasks in within the Needs Assessment domain. *Exam questions that ask, "What would you do next?" or "To this point, what have you completed?" are common and can be tricky.* To overcome these questions, it's important that you have a clear understanding of the sequence of domain tasks. In this scenario, once the business need and solution scope are defined and it's been validated that the solution is aligned with the goals and objectives of the organization, the business analyst would begin the process of establishing a baseline for the prioritization of requirements. The requirements baseline is a measurement for future comparison, and is the last task in this domain.

Answer Choice B: The use of valuation techniques would occur during Task 2.

Answer Choice C: The solution scope statement is created as output of Task 1.

Answer Choice D: Interrelationship diagrams are used for visualizing complex problems and relationships. When using the tool, is quite common to find factors that influence each other. In the context of the ECO, this would occur during Task 2, the collection and analysis of information.

PMI Professional in Business Analysis (PMI-PBA)® Examination Content Outline, sequence of the "Needs Assessment Domain Tasks."

PMI Professional in Business Analysis (PMI-PBA)® Examination Content Outline, 2013, "Needs Assessment," Task 5.

123. In planning for the implementation of a new enterprise software application, Will, the director of IT, would like to outline the skills, competencies, and requirements for each of his staff, because their responsibilities will be changing both during the initiative and at go-live. As the business analyst assigned to support the IT team, what can you suggest?

   a. The output should detail the expectations of each staff member, all required training, and any suggested certifications; these can also be used for any future job postings.

   b. He should refer to the Skills Framework for the Information Age.

   c. The output should detail how staff will be managed, appraised, and trained.

   d. We should contact our human resource business partner to assist with assessment.

The correct answer is: **A**

In most cases, projects introduce change to organizations. In some cases, the change may necessitate hiring additional staff or modifying the roles and responsibilities of current staff. In either of these cases, a job analysis is used to outline the position requirements and core competencies. The tool is also typically used when posting positions and serves as the basis for annual performance appraisals. They will often contain common vernacular for the skills and competencies for the role.

Answer Choice B: On the exam, beware of answer choices that refer to other certifying bodies. Although this seems like a logical answer, it is not specific enough to address the question.

Answer Choice C: The human resources management plan will address how staff will be managed, appraised, and trained. It will not address staff skills, competencies, and requirements.

Answer Choice D: As a business analyst, it shouldn't be necessary to defer to your HR business partner; you should be familiar with the tools at your disposal.

Business Analysis for Practitioners: A Practice Guide, Section 3.3.3.1, "Job Analysis."

PMI Professional in Business Analysis (PMI-PBA)® Examination Content Outline, 2013, "Planning," Task 3.

124. From your experience on iterative projects, you recognize the value of maintaining a spreadsheet to manage requirements. You've been tasked with supporting a project that is using an adaptive lifecycle. When team members submit user stories, how might you track them?
    a. Using a requirements tractability matrix
    b. Adding them to the epic log
    c. Tracking them in the user story matrix
    d. Adding them to a backlog

> The correct answer is: **D**
>
> This question challenges your understanding of adaptive lifecycle projects as compared to iterative projects. ***On the exam, remember, projects that follow the adaptive lifecycle can include agile methods such as scrum, extreme programming, Kanban, and rapid application development.*** They are often associated with change-driven methods, where it's anticipated that the end product is going to evolve over the duration of the project. When using the adaptive lifecycle approach, the backlog is used to maintain the list of requirements. Ahead of each sprint, a selection of items from the backlog are selected and prioritized.
>
> **Answer Choice A:** The requirements tractability matrix is associated with predictive and iterative lifecycle delivery methodologies.
>
> **Answer Choice B:** This is a made-up answer choice. There is no log to track epics. Epics are large user stories that require further decomposition.
>
> **Answer Choice C:** This is a made-up answer choice; the correct term is backlog, not user story matrix.
>
> *Business Analysis for Practitioners: A Practice Guide,* Section 5.5.3, "Maintaining the Product Backlog."
>
> *PMI Professional in Business Analysis (PMI-PBA)® Examination Content Outline,* 2013, "Traceability and Monitoring," Task 2.

125. Following your second sprint, your product manager, Innes, inquired as to the team's performance. In terms of metrics, what would you provide him?
    a. Variance
    b. Vertical velocity variance
    c. Mean variance velocity
    d. Velocity

> The correct answer is: **D**
>
> On all projects, not just those that follow the adaptive delivery methodology, measurements and metrics are essential to determine the current state relative to the goal. Concentrating on delivery, for projects that follow the adaptive lifecycle, velocity reports the backlog items completed during a sprint or delivery interval.
>
> **Answer Choices A, B, & C:** These are all made-up answer choices.
>
> *Business Analysis for Practitioners: A Practice Guide,* Section 6.5.3.1, "Project Metrics as Input to the Evaluation of the Solution."
>
> *PMI Professional in Business Analysis (PMI-PBA)® Examination Content Outline,* 2013, "Planning," Task 6.

126. Your Change Control Board recently voted on several requests; three were approved, two were denied, and one was deferred pending further information. Upon leaving the room with Finn, your project manager, you both agreed:
    a. In your role as business analyst, you would only communicate the approved and denied requests. Pending the outcome of the next meeting, you would share the decision on the deferred request.
    b. Finn, as the project manager, would communicate the decisions on all requests to all stakeholders.
    c. In your role as business analyst, you would communicate the decisions regarding all requests to all stakeholders.
    d. Finn, as the project manager, would only communicate the decision regarding the deferred request, as representatives for the five requests were present.

The correct answer is: **C**

This question challenges your understanding of the integrated change control process, adding an additional level of complexity by weaving in statistics, which are presented in an inconsistent format. Regardless of the outcome, it is the business analyst's responsibility to communicate and record the decisions from all Change Control Board meetings. This should be done in a consistent manner that enables the sharing of knowledge and for future reference.

Answer Choice A: This is a distractor answer choice. Although as the business analyst, you would have accountability for communication, it wouldn't be limited to only those requests that were approved or denied.

Answer Choices B & D: Depending on your organization, these may be true statements. However, *on the exam you'll need to put aside some experiences and think like a PMI business analyst.* In this scenario, it's the business analyst's responsibility to communicate the outcome on all decisions regarding integrated change control.

Business Analysis for Practitioners: A Practice Guide, Section 5.8.3.5, "Recommending a Course of Action."

PMI Professional in Business Analysis (PMI-PBA)® Examination Content Outline, 2013, "Traceability and Monitoring," Task 5.

127. Your primary stakeholders are having a difficult time understanding the problems and the complex relationships among all the components. To simplify the relationships, your sponsor, Aveline, suggests that you:
    a. Conduct a 5 Whys analysis to help all stakeholders gain a better understanding of the problem and complex relationships.
    b. Develop a cause-and-effect diagram to trace the problem to complex relationships among all the components.
    c. Present the variables in the form of an interrelationship diagram.
    d. Create an opportunity analysis diagram to simplify the relationships.

The correct answer is: **C**

This question challenges your experience and understating of root-cause analysis techniques and scope models. Interrelationship diagrams are a form of cause-and-effect diagram, used in circumstances in which the business analyst would like to illustrate complex problems with nearly unmanageable relationships among interrelated components. They are optimally used for visualizing complex problems and relationships. When using the tool, is quite common to find factors that influence each other.

Answer Choice A: The 5 Whys technique probes using rephrased and often repeated questions to gain a full understanding of the situation. It would not be appropriate to understand the complex relationship among components.

Answer Choice B: This is good answer choice; however, it is too general. Answer choice C is more specific, offering to present the variables in the form of an interrelationship diagram.

Answer Choice D: This is a made-up answer choice.

Business Analysis for Practitioners: A Practice Guide 2.4.4.2, "Cause-and-Effect Diagrams."

PMI Professional in Business Analysis (PMI-PBA)® Examination Content Outline, 2013, "Needs Assessment," Task 2.

128. A key team member, Mira, has recommended that the team conduct a brand-satisfaction and loyalty survey as the team celebrates the one-year anniversary of the product launch. How will participants be grouped?
    a. Into three categories: promoters, satisfied, and detractors
    b. Into two categories: those who responded and those who did not
    c. Into three categories: brand loyal, promoters, and neutral
    d. Into three categories: neutral, satisfied, and passive

The correct answer is: **A**

This question tests your knowledge and understanding of the net promotor score (NPS), a metric system designed by Fred Reichheld (Reichheld, 2003) to gauge customer loyalty and brand satisfaction. Although it can be based on a variety of scales, the most common is Zero to Ten, tied to a single open question, the response to which is Zero, "Not at all," to Ten, "Extremely"; common qualifiers are "Satisfied" and "Likely."

Once respondents complete the questionnaire, scores are tallied and grouped as follows:

Grouping	Score	Description
Promoters	9–10	Respondents who are brand loyal and would recommend your product
Passive	7–8	Respondents who are satisfied, but would consider other products and brands
Detractors	0–6	Respondents who are dissatisfied with your product or brand

Answer Choice B: This is a made-up answer choice.

Answer Choice C: This answer choice is partially correct—promoters are brand loyal. Unfortunately, it's missing passive and detractors.

Answer Choice D: This answer choice is partially correct. Passive are respondents who are satisfied. Unfortunately, it's missing promoters and detractors.

Requirements Management: A Practice Guide, Section 9.1.2, "Final Customer Acceptance."

PMI Professional in Business Analysis (PMI-PBA)® Examination Content Outline, 2013, Evaluation, Task 4

129. Your sponsor, Esmeralda, has requested assistance in preparing an artifact that will contain a list of stakeholders, milestones, assumptions, constraints, and success criteria. You will be assisting with preparation of the:
    a. Solution scope statement
    b. Value proposition statement
    c. Charter
    d. Product solution and value justification

The correct answer is: **C**

The charter is a key project artifact that formally authorizes the project. It will be issued by the sponsor, based on value proposition outlined in the business case, to address the problem or opportunity outlined in the scope statement. The charter will also outline, at a high level, the project requirements, the risks, and the justification for the investment. It will also outline measurable objectives and establish the responsibility and authority of the project manager.

Answer Choice A: The solution scope statement establishes the boundary for the initiative, covering elements such as acceptance criteria, assumptions and constraints, deliverables, exclusions, and scope description. Although there is some overlap, the charter is the only document that contains a list of stakeholders, milestones, assumptions, constraints, and success criteria.

Answer Choice B: A value proposition statement outlines how you intend to address a customer's needs or improve upon the current state.

Answer Choice D: This is a made-up answer choice.

A Guide to the Project Management Body of Knowledge (PMBOK®), Section 4.1.3.1, "Project Charter."

PMI Professional in Business Analysis (PMI-PBA)® Examination Content Outline, 2013, "Needs Assessment," Task 5.

130. You work in a fairly stable industry, and projects are generally completed on time and within budget. In working with your sponsor, Nala, she'd like to gauge project capriciousness over the planned lifecycle. What is one measurement that can be considered?
    a. Assessing the items presented at change control
    b. Monitoring the status of requirements
    c. Evaluating the defects and resolution time
    d. Measuring the efficiency of the project

The correct answer is: **A**

This is a difficult scenario if you don't know the meaning of *capriciousness*. Remember, on the exam there will be 20 pretest questions, for which your answers (correct or wrong) will not count toward the final score. In all cases, you can't tell which are pretest questions or those that count toward your score, so it is best to approach difficult questions in a pragmatic manner. Step one: identify what the examiner is asking: "What is one measurement that can be used over the planned lifecycle to gauge *X*."

Answer Choice B: Monitoring the status of requirements will communicate where they are in the lifecycle: pending functional specifications, pending technical specifications, in development, in testing, et cetera. This is a possible answer, but how does it relate to capriciousness?

Answer Choice C: Evaluating the defects and resolution time is only applicable during Evaluation, not the planned lifecycle. This would not be the best answer choice.

Answer Choice D: Efficiency is a measure of the output versus the input. Although not a project management earned-value method, the closest comparison is: for every dollar spent, how much directly contributed to the final solution? This would not be the best answer choice.

Based on this analysis, we've narrowed it down to two answer choices: A and B. Capriciousness is an adjective of impulsiveness, unpredictability, and variability. Therefore, monitoring the items presented at change control can be an indicator of project volatility.

Business Analysis for Practitioners: A Practice Guide, Section 6.5.3.1, "Project Metrics as Input to the Evaluation of the Solution."

PMI Professional in Business Analysis (PMI-PBA)® Examination Content Outline, 2013, "Planning," Task 6.

131. Maeve is working on a questionnaire that she intends to distribute to a panel of experts in hopes of soliciting their feedback based on the last round of testing. What tool is Maeve using to protect the anonymity of respondents?
    a. The net promotor tool
    b. Planguage
    c. Entry and exit criteria for expert opinion on leading practice
    d. Delphi

The correct answer is: **D**

There are several tools and techniques that are sure to find their way onto the exam, and the Delphi technique is one of them. Why? Because most business analysts have not experienced the value in using the tool, and it's an easy way to trip up candidates. When you see the words "panel of experts," one of the first things that should come to mind is Delphi. The role of the business analyst is to remain impartial and objective over the life of the project. The Delphi technique is one way the business analyst can arbitrate a discussion and help the team arrive at an acceptable solution. It's a structured process that is used to facilitate informed decision making, designed to allow participants to comment anonymously on a particular subject matter. When planning to use the Delphi technique, the facilitator establishes clear, predefined completion parameters (e.g., number of rounds, mean or median score). One benefit of using this technique is that it's typically staffed by a panel of experts. Remember, it is one approach to Evaluation, one which relies on an anonymous panel of experts to arrive at a consensus through independent voting and review.

Answer Choice A: Net promotor tool is a made-up answer choice. ***Remember, on the exam all words have relevance.*** If the correct term is *net promotor score*, don't select *tool*; this is a giveaway that it's probably not the best answer choice.

Answer Choice B: Tom Gilb created the concept of a planning language (*Planguage*) to address ambiguous and incomplete nonfunctional requirements. The tool uses defined identifiers (e.g., tags), to qualify and quantify the quality elements of requirements.

Answer Choice C: Although this may seem like a logical answer, it's a distractor.

Requirements Management: A Practice Guide, Section 8.3.1, "Solicit Inputs."

PMI Professional in Business Analysis (PMI-PBA)® Examination Content Outline, 2013, "Evaluation," Task 3.

132. At lunch, you're conversing with Aspen, the project manager assigned to your initiative, about the dynamics of the enterprise project. It seems that from month to month, your sponsor, Eero, is slightly altering scope to appease various business units. What is one benefit of this approach?
    a. Reduction in risk
    b. Stakeholder satisfaction
    c. Faster depreciation
    d. Dynamic scheduling

The correct answer is: **A**

When approaching this question, you may find it necessary to select from the best of the answer choices, thereby isolating those that are neutral. Ask yourself, how would each option be related to "altering scope to appease various business units"? When planning a project, the business analyst and project manager need to determine the optimal lifecycle delivery methodology. With large, complex, or enterprise projects, iterative and incremental lifecycles can contribute to a reduction in risk, versus waterfall delivery methodologies, as knowledge is transferred from previous iterations. Although a reduction in risk may seem like a reasonable answer choice, remember to read all options, and then make a selection via a process of elimination.

Answer Choice B: To support a dynamic business environment, stakeholders may appreciate the project manager adjusting scope. However, with no mention of an impact analysis or integrated change control, this would not be a good answer choice.

Answer Choice C: With regard to depreciation, the organization could possibly benefit from scope changes. However, there is no mention of a financial analyst, and the project manager couldn't make this determination on their own.

Answer Choice D: Dynamic scheduling is an algorithm-based technique, in which priorities are calculated during the project delivery phase. This would not be the best answer choice.

A Guide to the Project Management Body of Knowledge (PMBOK®), Section 2.4.2.3, "Iterative and Incremental Lifecycles."

PMI Professional in Business Analysis (PMI-PBA)® Examination Content Outline, 2013, "Planning," Task 3.

133. Your CIO, Zenobia, is negotiating a contract with a solution provider to implement a large-scale software application. Including contingency and management reserves, the project budget is $4,000,000 USD. However, the solution provider is estimating about twice the budget to deliver the work. Without sacrificing scope or quality and still meeting the deadline, Zenobia requests the vendor to present:

   a. A work plan elaborated through the first phase of the project
   b. A quote using off-shore resources
   c. A comparative quote based on adaptive and iterative lifecycles.
   d. An incentive-based contract.

The correct answer is: B

Common characteristics of nearly all projects is that they lack adequate time and resources (human resources, financial, or technology) to deliver the solution as planned. By requesting a quote that uses off-shore resources, the CIO is conducting an alternative analysis. When estimating activities, negotiating contracts, or evaluating make/buy options, it's common practice to explore alternatives. This evaluation can result in a different skill mix or resource allocation, while not sacrificing the quality of the solution.

Answer Choice A: The technique of progressively elaborating the project work plan, in which work is detailed to varying levels based on project phase, is known as *rolling wave planning*. This wouldn't help address the cost concerns—it would only prolong the financial concerns.

Answer Choice C: Requesting comparative quotes based on both the adaptive and iterative lifecycles is a reasonable approach. However, with no change to the iron triangle (quality as constrained by scope, schedule, and cost), there would be negligible cost savings.

Answer Choice D: This is a distractor. Although many types of contracts offer incentives, and a fixed-price incentive fee contract would be preferable to the customer, in this scenario the seller would never agree to the $4,000,000 USD fixed-price incentive fee contract, considering that they are estimating $8,000,000 USD.

A Guide to the Project Management Body of Knowledge (PMBOK®), Section 6.4.2.2, "Alternative Analysis."

PMI Professional in Business Analysis (PMI-PBA)® Examination Content Outline, 2013, "Planning," Task 3.

134. Your sponsor, Ansley, has scheduled a meeting for you to present the outline of scope baseline to the Executive Steering Committee for the next generation of high-speed passenger rail cars that will be used in the change control process. What will you present?
    a. An explanation and tiered decomposition of the scope, including limitations, and a supporting dictionary
    b. The project management information system and the configuration management system
    c. The requirements traceability matrix, the business analysis plan, the BA work plan, and the change management plan
    d. The change management plan, the requirements management plan, and the requirements traceability matrix

The correct answer is: **A**

A baseline is a measurement for future comparison; every artifact should be archived and properly notated for version control for analysis over the duration of the project. The scope baseline is a significant project deliverable and key artifact used in the change control process. It is not one single document, rather it includes the scope statement, the work breakdown structure, and the WBS dictionary. All three elements are required to satisfy the presentation request for the Executive Steering Committee.

Answer Choice B: Either the project management information system or the configuration management system could serve as the repositories for this information.

Answer Choices C & D: These are both distractors; neither satisfies any part of the request to review an outline of the scope baseline. Both answer choices are too granular for the question.

A Guide to the Project Management Body of Knowledge (PMBOK®), Section 5.4.3.1, " Scope Baseline."

PMI Professional in Business Analysis (PMI-PBA)® Examination Content Outline, 2013, "Planning," Task 4.

135. Declan, an intern working on your project, has inquired as to the key processes associated with solution evaluation. You candidly tell Declan, "This isn't my first time at the rodeo." They are:
    a. Tracing and tracking all requirements in the team wiki
    b. Posting the results to the team wiki, selecting techniques that will be used during the evaluation process, and hosting a demonstration for the key stakeholders
    c. Hosting a demonstration following each phase or sprint
    d. Validating frequently and thoroughly

The correct answer is: **B**

This is a question of moderate complexity, and not unlike ones you may encounter on the exam. There is some level of ambiguity, and you need to relate the material beyond the Practice Guides. To approach this question correctly, you need to have a comfort level with the key processes associated with Evaluation: planning, which begins at the onset of the project; validation, to ensure that the solution meets stakeholder requirements; documentation, for auditing and knowledge management; and communication, to ensure that all stakeholders are kept informed. Answer choice C most closely aligns to these key processes.

Answer Choice A: This is a partially correct answer; it addresses the documentation and communication aspects but does not fully address planning or validation.

Answer Choice C: Although demonstrations are very valuable and enable communication, this answer choice does not address the question.

Answer Choice D: This is a distractor and not a complete answer to outline the key processes associated with Evaluation.

Requirements Management: A Practice Guide, Section 8.2, "Solution Evaluation Activities."

PMI Professional in Business Analysis (PMI-PBA)® Examination Content Outline, 2013, "Evaluation," Task 1.

136. Your executive sponsor, Constance, has asked for your help with an assessment; to date only the approach to the initiative has been approved. What should you begin working on?
    a. The solution scope statement
    b. The project scope statement
    c. The project charter
    d. A cost–benefit analysis

The correct answer is: **D**

To answer this question correctly, you need to first understand where you are in the process. The scenario states: ". . . to date only the approach to the initiative has been approved"; this means we are in the Needs Assessment domain. Next, the question is asking "for your help with an assessment." Once the organization understands the initiative and the approach, the next step is to work on the business case, which is a cost–benefit analysis contrasting options, concluding with a recommendation. Once the business case is accepted, the project team can approach the other artifacts.

Answer Choice A: The solution scope statement establishes the boundary for the initiative, covering elements such as (a) acceptance criteria, (b) assumptions and constraints, (d) deliverables, (e) exclusions, and (f) scope description; it is not considered an assessment.

Answer Choice B: The project scope statement contains (a) assumptions/constraints, (b) acceptance criteria, (d) deliverables, (e) exclusions, (f) scope description; it is not considered an assessment.

Answer Choice C: The project charter is the document formally authorizes the project, and establishes the boundaries. It contains sections such as (a) approval requirements, (b) assumptions/constraints, (c) summary budget, (d) high-level description, (e) measurable objectives, (f) justification (or purpose), (g) summary milestones, (h) high-level requirements, (i) high-level risks, (j) stakeholders, (k) sponsor, (l) project manager (outlining responsibility and authority), and (m) success criteria for the project; it is not considered an assessment.

Business Analysis for Practitioners: A Practice Guide, Section 2.6, "Assemble the Business Case."

PMI Professional in Business Analysis (PMI-PBA)® Examination Content Outline, 2013, "Needs Assessment," Task 1.

137. After creating the work breakdown structure (WBS), Oswin is estimating the overall project duration by totaling the estimates of all the lower-level components. What estimating technique is Oswin using?
    a. Three-point
    b. Parametric
    c. Analogous
    d. Bottom-up

The correct answer is: **D**

Estimating is a vital competency for business analysts, and this question tests your knowledge of the four techniques that can be used to estimate activity durations. The bottom-up estimating technique totals the estimates of all lower-level components within the work breakdown structure to approximate project duration. Because estimating is an iterative process, over time, these estimates become further refined.

Answer Choice A: Three-point estimating relies on three factors to establish an approximation for the activities duration: most likely (M), optimistic (O), and pessimistic (P).

Answer Choice B: Parametric estimating considers the relationships among variables for estimating cost and duration.

Answer Choice C: The analogous estimation technique uses data from prior projects of similar size to estimate work effort and cost when there is minimal information available. Because estimating is an iterative process, over time, these estimates are further refined.

A Guide to the Project Management Body of Knowledge (PMBOK®), Section 6.4.2.4, "Bottom-Up Estimating."

Business Analysis for Practitioners: A Practice Guide, Section 3.5.2.6, "Estimate the Work."

PMI Professional in Business Analysis (PMI-PBA)® Examination Content Outline, 2013, "Planning," Task 4.

138. Collaborating with your project manager, Marcus, you've completed a thorough and complex stakeholder identification process. Your sponsor, Garrett, is quite impressed and suggests:
    a. A regular review of the quadrant analysis.
    b. His name can be removed, because Marcus will be leading the project once approved.
    c. The stakeholder register not be shared with others, because it contains confidential information and should not be updated beyond this point.
    d. The stakeholder register be published to the company intranet to foster collaboration among team members.

The correct answer is: **A**

This is an obscure question, one that challenges your rationalization skills with regard to the stakeholder identification process. A quadrant analysis, also referred to as a *2x2 matrix*, is a flexible tool allowing teams to categorize viewpoints and interests in a grid comprising four cells. In terms of stakeholder identification, this can take the form of a power/interest grid. Over the duration of a project, stakeholder interest, influence, impact, and power can change and evolve. Once created, the stakeholder register and associated matrices must be regularly updated and maintained.

Answer Choice B: The sponsor's name would never be removed from the stakeholder artifacts.

Answer Choice C: This is a distractor, although it does contain a valid point of providing proper care for the register to ensure that the content is adequately secured.

Answer Choice D: This is an incorrect answer. Although sharing the register can foster collaboration, the company's intranet is a very broad suggestion. Had the answer choice suggested the *team* page on the intranet, this answer choice could warrant additional consideration.

Business Analysis for Practitioners: A Practice Guide, 2.3.1,"Identify Stakeholders."

A Guide to the Project Management Body of Knowledge (PMBOK®), Section 13.1.2.1, "Stakeholder Analysis."

PMI Professional in Business Analysis (PMI-PBA)® Examination Content Outline, 2013, "Needs Assessment, Task 4."

139. Your project director, Beatrice, completed a project RACI and noted that Soren, the project manager, was responsible for managing quality requirements and scope. Is this correct?

a. No, it is a collaborative effort between the business analyst and the project manager to manage project scope.
b. No, the management of project scope is the responsibility of the business analyst.
c. Yes, Soren is responsible for managing project scope.
d. No, the project stakeholders have accountability to manage and control scope.

The correct answer is: **C**

This is a challenging scenario-based question that requires careful attention to both the question and the answer choices. *First, on the exam, as you approach roles and responsibility questions, please remember that RACI charts can also be known as responsibility assignment matrices (RAMs) or linear responsibility charts (LRCs).* They establish accountability for requirements signoff, and they clearly identify who's responsible; the one and only person who is accountable; experts who can be consulted; and others who need to be informed. Second, although the question mentions "managing quality requirements," this is not in any of the answer choices; the focus should therefore be on "managing scope." In organizations that split the role of project manager and business analyst, the business analyst is responsible for managing *product* scope, and the project manager is responsible for managing *project* scope. Answer choice C is the only selection worded correctly.

Answer Choice A: When the roles are split, the business analyst and project manager do not share responsibility for managing project scope.

Answer Choice B: This is an incorrect answer choice; had the scenario stated that the roles were combined, this could warrant further consideration.

Answer Choice D: This is a distractor and does not address the question in sufficient specificity.

Business Analysis for Practitioners: A Practice Guide, Section 1.7.2, "Requirement Types."

Business Analysis for Practitioners: A Practice Guide, Section 5.2.2, "Benefits of Tracing Requirements."

PMI Professional in Business Analysis (PMI-PBA)® Examination Content Outline, 2013, "Traceability and Monitoring," Task 3.

---

140. Lucy, the product manager for Lucky Rabbit Food Inc, expressed concern that the delivered product might not meet stakeholder expectations. She requests that you hire a consulting firm that has not been associated with the project to:
    a. Conduct an independent verification and validation of the rabbit toys
    b. Conduct day-in-the-life testing and observe rabbits playing with the toys
    c. Use a fishbone diagram to determine why rabbits of certain breeds are not playing with the toys
    d. Conduct a Delphi method vote

The correct answer is: **A**

Ensuring the delivered solution meets the stakeholders' requirements is a critical task for business analysts. Prior to solution signoff, an organization may request an independent verification and validation (IV&V) assessment to ensure that the delivered solution meets the stakeholders' requirements, is aligned with the requirements documentation, and fulfills the value proposition.

Answer Choice B: Day-in-the-life (DITL) testing is a semiformal activity based on typical usage, performed by a team member with in-depth business knowledge. The results obtained from DITL enable both validation and evaluation, confirming whether a product or solution provides the functionality for a typical day of usage for the role, based on the test scenario. This would not be the best answer choice, because the scenario requested someone who had not been involved with the project.

Answer Choice C: Fishbone diagrams (aka *Ishikawa diagrams*) are a form of cause-and-effect diagram used to investigate the high-level causes of occurring problems. This would not be the best answer choice for an evaluation technique.

Answer Choice D: The Delphi method is a structured technique that relies on a panel of subject matter experts to arrive at a consensus. This technique is frequently used for drafting policies, forecasting, or solutionizing. This would not be the best answer choice for an evaluation technique.

Requirements Management: A Practice Guide, Chapter 8, page 49.

PMI Professional in Business Analysis (PMI-PBA)® Examination Content Outline, 2013, "Evaluation," Task 3.

141. In a meeting with your sponsor, Jeff, he asks that you lead a brainstorming session to identify the reasons product management submitted a proposal to redesign the ketchup bottles that have been used for over 100 years. Your session will focus on:
    a. Transition requirements
    b. Solution requirements
    c. The requirements of product management
    d. Business requirements

The correct answer is: **D**

On the exam, you can expect a number of questions testing your understanding of the various types of requirements. In this scenario, the focus is on business requirements, which describe the higher-level needs of the enterprise. Typically in the form of a business requirements document (BRD), they describe the purpose/intent of the component and metrics to evaluate their impact on the organization. BRDs contain the component goals, which are then linked to the strategic plan. The BRD focuses on the entire enterprise, not specific organizational levels. *Both in practice and on the exam, remember that business requirements focus on the "what" in terms of requirements.* They are not further elaborated or decomposed into solution requirements; rather, they focus on identifying the value proposition to address a problem or opportunity.

Answer Choice A: Transition requirements describe the temporary capabilities that must exist for the component to transition from a current to a desired state. They cannot be fully defined until the as-is and to-be solutions are completely defined.

Answer Choice B: Solution requirements describe the characteristics of the component (i.e., features, functions), which are further decomposed to functional and nonfunctional requirements. These are the utility and warrantee requirements for the solution.

Answer Choice C: This is a made-up type of requirement.

Business Analysis for Practitioners: A Practice Guide, Section 1.7.2, "Requirement Types."

PMI Professional in Business Analysis (PMI-PBA)® Examination Content Outline, 2013, "Needs Assessment," Task 1.

142. Braelyn, your subject matter expert, is having difficulty estimating the work effort for a number of activities. In addition, depending on the software developer, there could be additional activities that are simply not known at this time. As the business analyst, how should you note this on the work plan?
    a. Add a reserve for both items at the activity level, and update the work plan as more information becomes available.

b. Collaborate with your project manager and use a consistent approach across the project.

c. Add a contingency reserve for the known activities and a management reserve for those activities that are unknown at this time.

d. Use a management reserve at the activity level for estimating both the work effort and the unknown activities that are not known at this time, then update the work plan as more information becomes available.

The correct answer is: **C**

Scheduling and coordinating activities within the business analysis work plan is an essential task to ensure that everything fits within the allotted time and to validate that nothing was overlooked. This question tests your hands-on experience with building and maintaining work plans, along with the concepts of *reserves*. While planning business analysis activities within projects, it's quite common that not all information is known at the onset. To establish a baseline (measurement for future comparison), duration estimates may include buffers or contingency reserves for known activities where it's difficult to precisely estimate. For unknown activities, management reserves can be used to allocate time that may influence a project.

Answer Choice A: To properly address both known and unknown activities, the business analyst would use a combination of contingency and management reserves. This answer choice is incomplete.

Answer Choice B: This is a distractor and not the correct answer choice.

Answer Choice D: Although management reserves is partially correct, the explanation offered is incorrect.

A Guide to the Project Management Body of Knowledge (PMBOK®), Section 6.5.2.6, "Reserve Analysis."

PMI Professional in Business Analysis (PMI-PBA)® Examination Content Outline, 2013, "Planning," Task 3.

---

143. You are conducting a focus group to build out the requirements traceability matrix. While doing so, you are having difficulty tracing a few items to business goals and objectives. Your sponsor, Jacob, is not concerned and suggests that they are all relevant. What should you do?

a. Add a short textual description, stating "requirement provided by sponsor."

b. Allow the project manager to manage these items, as they pertain to project management.

c. In some circumstances, it is acceptable not to track requirements to business goals and objectives.

d. Ask Jackson, your associate, to conduct an analysis to determine the relevance of the requirement.

The correct answer is: **D**

While building out the requirements traceability matrix, it's common to find elements that are not anchored to the business goals and objectives. The best way to approach these elements is to conduct an analysis to determine the relevance of the requirement to the product and organization. All requirements must be traced and tracked to business goals and objectives to help control product scope, ensure that the delivered solution meets stakeholder expectations, and adds value. In this case, the business analyst needs to trust their judgment and not accept the sponsor's casual comment.

Answer Choice A: This answer choice is a distractor; although the sponsor may have offered the comment, it is not an appropriate response for a business analyst.

Answer Choice B: It's the business analyst, not the project manager, who has accountability for tracing and tracking requirements.

Answer Choice C: This is an incorrect answer choice; all requirements must be traced and tracked to business goals and objectives.

Business Analysis for Practitioners: A Practice Guide, Section 5.2.2, "Benefits of Tracing Requirements."

PMI Professional in Business Analysis (PMI-PBA)® Examination Content Outline, 2013, "Traceability and Monitoring," Task 2.

144. Agnes, the VP of the project management office (PMO), requests that you complete a feasibility study for the proposed recommendation. Why?
    a. A feasibility study will help determine the extent to which the proposed solution addresses the identified opportunity or problem in an effort to ensure that the correct project is implemented.
    b. A feasibility study will help determine the extent to which the proposed solution addresses the identified opportunity.
    c. A feasibility study will outline potential alternatives.
    d. A feasibility study will help determine the extent to which the proposed solution addresses the identified opportunity or problem in an effort to ensure that the correct solution is implemented.

The correct answer is: D

Because there is little information provided in this scenario, the first step in approaching this question is to establish where we are in the process. Upon careful review, we can determine that we're in the Needs Assessment domain, because we are going to complete a "feasibility study for a proposed recommendation." There are five types of feasibility studies that explore the viability of the proposed initiatives and often consider the influence of **p**olitical, **e**conomic, **s**ocial, **t**echnological, **e**nvironmental, and **l**egal (PESTEL) factors and help determine the extent to which the proposed solution addresses the identified opportunity or problem, in an effort to ensure that the correct solution is implemented.

Answer Choice A: This answer choice closely resembles answer choice D and is very close to being correct. However, inserting the word *project* verses solution or product invalidates this answer choice.

Answer Choice B: Feasibility studies often consider (a) operational, (b) technology/system, (c) cost-effectiveness, and (d) time constraints associated with each option. The results produced would not help to determine the extent to which the proposed solution addresses the identified opportunity.

Answer Choice C: Feasibility studies do not outline potential alternatives.

Business Analysis for Practitioners: A Practice Guide, Section 2.5, "Recommend Action to Address Business Needs."

PMI Professional in Business Analysis (PMI-PBA)® Examination Content Outline, 2013, "Needs Assessment," Task 2.

145. Carson is creating screen mockups to classify page elements and their associated functions. As the mockup is decomposed to user interface elements, how are they described?
    a. As unique development objects
    b. As individual requirement objects
    c. From the standpoint of displays and behaviors
    d. From the perspective of a wireframe and relationship to requirement

The correct answer is: **C**

This question is testing your understanding of and experience with Interface models. Screen mock-ups (or wireframes) are used to classify page elements and their associated functions. Once outlined, the mockup is decomposed to user interface elements from the perspective of displays and behaviors.

Answer Choices A, B, & D: These are all made up and incorrect answer choices.

Business Analysis for Practitioners: A Practice Guide, Section 4.10.11.4, "Wireframes and Display-Action-Response."
PMI Professional in Business Analysis (PMI-PBA)® Examination Content Outline, 2013, "Analysis," Task 2.

146. During user acceptance testing, your subject matter expert, Phoebe, signs off that there are no defects with the program and the documentation is acceptable. However, she is not satisfied with the software application as delivered, commenting that it is missing some features. As a result, you:
    a. Agree, the software does not fully meet the business requirements.
    b. Suggest Phoebe review the business requirements documents.
    c. Advise that the software meets the business requirements; she is concerned with the low quality.
    d. Advise that the software meets the business requirements; she is concerned with the low grade.

The correct answer is: **D**

On the exam, you can expect a number of questions pertaining to both *quality* and *grade*. This question challenges your understanding of both concepts. Based on the information provided in the scenario, the solution meets the higher-level needs of the organization, and Phoebe is voicing concern with regard to the low grade: "It is missing some features." A low-grade software solution is one that contains a limited number of features and functions. It may often be preferable to the conversely feature-rich software solution (high-grade), which could have number of problems and software bugs.

Answer Choice A: This is an incorrect choice; there is nothing to suggest the delivered solution does not meet the business requirements.

Answer Choice B: This is a distractor and not the correct answer choice.

Answer Choice C: This is not the correct answer choice. Low quality would suggest there were defects with the delivered solution.

A Guide to the Project Management Body of Knowledge (PMBOK®), "Introduction to Quality Management," page 228.
PMI Professional in Business Analysis (PMI-PBA)® Examination Content Outline, 2013, "Evaluation," Task 1.

147. Trinity, VP of human resources and head of your Executive Steering Committee, asked that, in collaboration with Ivor, your project manager, you present artifacts that can influence scope management for the new human capital management system. What will you present?
    a. Scope management plan, requirements management plan, quality management plan
    b. Work breakdown structure and work breakdown structure dictionary
    c. Scope management plan, scope baseline, and requirements traceability matrix
    d. On-boarding, transfer, promotion, and off-boarding procedures; travel reimbursement policies and detailed information on the applicant portal

The correct answer is: **D**

This is a difficult question and tests your understanding and knowledge pertaining to artifacts that can influence scope management. The first step when approaching these types of questions is to identify any commonalities or outliers among the answer choices, then, through process of elimination, attempt to determine the best answer choice for the scenario. Answer Choice D lists organizational process assets (on-boarding, transfer, promotion, and off-boarding procedures; travel reimbursement policies and detailed information on the applicant portal) which can significantly influence the project and product scope management processes. Most notable in this category are organizational policies and procedures and detailed lessons learned from prior initiatives.

Answer Choices A–C: These are all project-related artifacts and would be created based on the solution scope and overall structure of the project.

A Guide to the Project Management Body of Knowledge (PMBOK®), Section 5.1.1.4, "Organizational Process Assets"; and Section 2.1.4

PMI Professional in Business Analysis (PMI-PBA)® Examination Content Outline, 2013, "Planning," Task 3.

148. You are facilitating a session with product management, engineering, and manufacturing to understand the high-level causes for problems with the floating-point calculation in a newly designed microprocessor. You begin the session by drawing a fishbone diagram and place the problem to be addressed:
    a. At the head of the fish, which is facing right
    b. At the head of the fish, which is facing left
    c. Along the spine of the fish
    d. At the head of the fish, facing either left or right

The correct answer is: **D**

Fishbone diagrams are visualization tools for categorizing the potential causes of a problem and uncovering their root cause. Also known as *Ishikawa diagrams*, when using this type of cause-and-effect diagram, the problem is represented at the head of the visualization, which can either face left or right.

Answer Choices A & B: As the problem is represented at the head of the visualization, it's the preference of the facilitator as to whether the problem (head of the fish) should face left or right.

Answer Choice C: The categories are often used to group causes along the spine.

Business Analysis for Practitioners: A Practice Guide, Section 2.4.4.2, "Cause-and-Effect Diagrams."

PMI Professional in Business Analysis (PMI-PBA)® Examination Content Outline, 2013, "Needs Assessment," Task 2.

149. Team members Beckett and Eliza have outlined the steps to (a) ensure that quality standards are effectively used during the project, (b) document how the project will demonstrate adherence to standards, and (c) recommend any necessary changes. At the same time, team members Sophia and Christian are focused on communications management. What elements are Beckett and Eliza not concerned with:
    a. Developing the communications management plan, ensuring that all stakeholders receive timely and relevant information about the project
    b. Performing quality assurance, quality management planning, and controlling quality
    c. Performing quality management
    d. Controlling quality

The correct answer is: **A**

This is a long, wordy question, which, if approached in haste, could be answered incorrectly. The last sentence asks, "What elements are Beckett and Eliza *not* concerned with?" This essentially means that, from the answer choices listed, they *would be* concerned with three of the four options (answer choices B–D). Because Sophia and Christian are focused on communications management, they, not Beckett and Eliza, would be responsible for developing the communications management plan and ensuring that all stakeholders receive timely and relevant information about the project. **Answer choices B–D:** These all contain facets of project quality management. These quality processes are central to ensuring that the delivered solution meets the requirements of the stakeholders.

A Guide to the Project Management Body of Knowledge (PMBOK®), Figure 8-1, "Project Quality Overview."

PMI Professional in Business Analysis (PMI-PBA)® Examination Content Outline, 2013, "Evaluation," Task 1.

150. Your organization is making a significant investment in a new enterprise resource planning (ERP) system. This system is purpose-built for your industry and is highly configurable. Your sponsor, Seraphina, has requested that you use the requirements modeling language to assist with the refinement of requirements. To what benefit?
    a. These models will focus on the system design to maximize value.
    b. These models will visually model the requirements for easy consumption by stakeholders.
    c. These models will provide the basis for business transformation.
    d. These models will focus on the project's goals and objectives for the purpose of transformation.

The correct answer is: **B**

This is a difficult question and tests your understanding of the various modeling languages and their intended purposes. The Practice Guide describes the four modeling languages: (a) Business Process Modeling Notation (BPMN), (b) Requirements Modeling Language (RML), (c) System Modeling Language (SysML), and (d) Unified Modeling Language (UML). RML is used to visually model requirements and approaches requirements from the perspective of the end user.

Answer Choices A, C, & D: These are all made-up answer choices and do not pertain to any of the modeling languages.

Business Analysis for Practitioners: A Practice Guide, Section 4.10.6, "Modeling Languages."

PMI Professional in Business Analysis (PMI-PBA)® Examination Content Outline, 2013, "Analysis," Task 2.

151. Henley is the business analyst for construction of a high-rise tower in lower Manhattan; she is accountable for providing the estimates for installing carpet and painting the walls and trim. Based on a recent project, she knows the carpet team can install 5,000 square feet per day. The current project calls for 35,000 square feet of carpet, so Henley estimates the work effort at seven days. What technique did she use to arrive at this estimate?
    a. Analogous
    b. Parametric
    c. Bottom-up
    d. Three-point

The correct answer is: **B**

Estimating is a vital competency for business analysts, and this question tests your knowledge of the four techniques that can be used to estimate activity durations. Parametric estimating uses algorithms based on historical data to forecast duration or cost for a current initiative. Because estimating is an iterative process, overtime, these estimates are further refined.

Answer Choice A: The analogous estimation technique uses data from prior projects of similar size to estimate work effort and cost when there is minimal information available

Answer Choice C: The bottom-up estimating technique totals the estimates of all lower-level components within the work breakdown structure to approximate project duration.

Answer Choice D: Three-point estimating relies on three factors to establish an approximation for the activities duration: most likely (M), optimistic (O), and pessimistic (P).

A Guide to the Project Management Body of Knowledge (PMBOK®), Section 6.5.2.3, "Parametric Estimating."

Business Analysis for Practitioners: A Practice Guide, Section 3.5.2.6, "Estimate the Work."

PMI Professional in Business Analysis (PMI-PBA)® Examination Content Outline, 2013, "Planning," Task 4.

152. Isla, a skilled business analyst, would like to use a tool that can objectively compare solution options for her subject matter experts. What can you recommend?
    a. Multi-criteria weighted ranking
    b. Nominal group technique
    c. Business Process Model and Notation (BPMN)
    d. Requirements modeling technique

The correct answer is: **A**

Business analysts can further assist stakeholders by incorporating the use of multi-criteria weighted ranking models to objectively compare solution options. When using this tool, proposed solutions are ranked based on pre-established measures. Selection preference order is then calculated as an average of the ranking.

Answer Choice B: Originally developed by Andre Delbecq and Andrew H. Van de Ven, the nominal group technique (also referred to as *NGT* or *MultiVoting*) is used to quantitatively produce consensus for a particular discussion topic, essentially refining results generated during brainstorming sessions

Answer Choice C: Business process modeling notation is used to model business processes to evaluate potential changes and optimizations.

Answer Choice D: This is a made-up answer choice, although it's worded similarly to requirements modeling language.

Business Analysis for Practitioners: A Practice Guide, Section 4.15.3, "Weighted Ranking."

PMI Professional in Business Analysis (PMI-PBA)® Examination Content Outline, 2013, "Analysis," Task 3.

153. The Executive Steering Committee for your project has just approved your business plan. Unfortunately, they've only provided you with four months to complete the requirements evaluation process and solution development. As the business analyst, how do you approach this constraint?
    a. Perform a MoSCoW analysis to determine the features for the release window.
    b. Work with the time window provided to complete the requirements and development efforts; as needed, request more time in the form of a change request.

    c. Analyze the team's capability and capacity within the defined time period, conduct a MoSCoW analysis, and prioritize the work accordingly.

    d. Conduct a feasibility assessment based on technology, system, and cost effectiveness.

The correct answer is: **C**

This question describes a very common scenario, and the answer choices require careful analysis to select the best approach to the constraint. Regardless of the lifecycle approach, most initiatives do not have the luxury of infinite time. The sponsor has provided two timeframes; it's the business analyst's responsibility to work within the fixed timelines (known as *time-boxing*), and deliver a solution aligned to the stakeholder requirements. Answer choice C offers the most complete response. The first step in the assessment would be to assess the team's capability to determine if all work could be handled internally or if external resources would be required for augmentation. Second, in terms of capacity, are there any resource, financial, or technology constraints the team would need to address. Lastly, the MoSCoW method could be used to help stakeholders prioritize the established requirements. This method will help to objectively guide the evaluation process, focusing on those items that have priority.

Answer Choice A: This answer choice is partially correct, but it doesn't address the team's capability and capacity within the time-boxed period.

Answer Choice B: Selecting this approach would lead the Executive Steering Committee to believe all the work could be completed with the defined time period. By completing the analysis, you'll have a better idea if this will be a concern.

Answer Choice D: Technology, system, and cost effectiveness feasibility assessments would not be helpful in this situation.

Business Analysis for Practitioners: A Practice Guide, Section 4.11.6.1, "Prioritization Schemes."

PMI Professional in Business Analysis (PMI-PBA)® Examination Content Outline, 2013, "Analysis," Task 2.

154. You are establishing the process for securing approval of solution requirements for a project using extreme programming (XP). Collaborating with your project manager, Sasha, what format would your offshore developers prefer?

    a. Business requirements document

    b. Functional specification

    c. User story

    d. Use case

The correct answer is: **C**

This question tests your understanding of agile and techniques used with adaptive lifecycle projects. Requirements for adaptive lifecycle projects are typically stated in the form of user stories. When properly written, they follow the INVEST acronym: **i**ndependent, **n**egotiable, **v**aluable, **e**stimable, **s**mall, and **t**estable in the format: As a <role or type of user>, I want <goal/desire>, so that <benefit>. Large user stories, known as *epics*, are decomposed into manageable user stories and added to the backlog.

Answer Choice A: Business requirements documents describe the purpose/intent of the component and metrics to evaluate its impact on the organization.

Answer Choice B: Functional specifications are decomposed from functional requirements and describe the behaviors or functions of the component; they are often categorized as *solution utility*.
Answer Choice D: Use cases are used to describe scenarios and explain complex interactions between users and systems in plain text, articulating how the system should operate, the associated flows, and the expected benefit, all from the vantage point of a lead actor.

Business Analysis for Practitioners: A Practice Guide, Section 5.4, "Approving Requirements," and Section 4.11.5, "Guidelines for Writing Requirements."

PMI Professional in Business Analysis (PMI-PBA)® Examination Content Outline, 2013, "Planning," Task 3.

155. Lauren, an experienced project manager and business analyst, is building her work plan for the construction of the new kindergarten wing at the local elementary school. She has identified a few instances in which successor activities cannot begin immediately following the completion of predecessor activities. For example, the desks cannot be moved into the classrooms until all the tile floors have cured. How can Lauren best represent a two-day delay on her work plan for this activity?
    a. FF +2
    b. FS –2
    c. FS +2
    d. SF –2

The correct answer is: **C**

This is an advanced topic regarding scheduling and the sequencing of activities, which may appear on the exam. The precedence diagramming method (PDM) includes four dependencies or logical relationships: finish-start (FS), finish-finish (FF), start-start (SS), and start-finish (SF). Leads are time increments in which successor activities can be advanced ahead of a predecessor (indicated as a "–"), and lags note delays (indicted by a "+"). In this scenario, the predecessor finish task is the curing of the tile floors and the moving of the desks into the classroom would be a start task. This would be noted as processor lag for moving the desks, represented as FS +2 in the work plan.
Answer Choice A: FF indicates a finish–finish relationship, which would not be appropriate in this scenario, regardless of lead or lag.
Answer Choice B: FS –2 is a finish–start lead suggesting that the chairs could be moved two days *before* the floors cured.
Answer Choice D: SF indicates a start–finish relationship, which would not be appropriate in this scenario, regardless of lead or lag.

A Guide to the Project Management Body of Knowledge (PMBOK®), Section 6.3.2.3, "Leads and Lags."

PMI Professional in Business Analysis (PMI-PBA)® Examination Content Outline, 2013, "Planning," Task 4.

156. While building his work plan, Jonathan noted several dependencies in the real estate module of the ERP. Based on information learned at a recent seminar by experts in the field, how can these best be catalogued?
    a. Discretionary
    b. External preferred
    c. Internal preferred
    d. Mandatory

The correct answer is: **A**

This question challenges your knowledge and experience with regard to dependencies. There are four possible combinations of dependencies for each requirement that may appear on the exam: (a) mandatory internal, (b) mandatory external, (c) discretionary internal, and (d) discretionary external. Establishing dependencies based on industry leading practice is an example of discretionary dependencies, also known as soft-logic or preferred logic.

Answer Choices B & C: This scenario does not ask if they are internal or external to the project. In addition, the qualifier of *preferred* is invalid and not associated with dependencies.

Answer Choice D: Mandatory dependencies are those determined to be bound by contractual, legal, or statutory obligations, or essential to deliver the solution.

A Guide to the Project Management Body of Knowledge (PMBOK®), 6.3.2.2, "Dependency Determination."

PMI Professional in Business Analysis (PMI-PBA)® Examination Content Outline, 2013, "Planning," Task 4.

157. Frederick, a professor at the local university, is explaining the principle of requirements containment. For a requirement document to be "self-contained," what must be true?
    a. It must contain all the requirement information for the development team to design, build, and test the development object.
    b. It must not be dependent on any other requirement document.
    c. It must be measurable, feasible, complete, consistent, and correct.
    d. It must contain all the information based on the stakeholders' requirements.

The correct answer is: **A**

Self-containment is a key principle when creating requirements documents. Although it's generally accepted practice that they stand alone, the accurate test is that they contain all the necessary information for the development team to design, build, and test the object.

Answer Choice B: Although it's true that requirement documents must not be dependent on any other requirement document, this answer choice does not correctly address the principle of requirements containment.

Answer Choice C: Although requirement documents must be measurable, feasible, complete, consistent, and correct, this answer choice does not correctly address the principle of requirements containment.

Answer Choice D: This is a distractor and is an incorrect answer choice.

Business Analysis for Practitioners: A Practice Guide, 4.11.5.1, "Functional Requirements."

PMI Professional in Business Analysis (PMI-PBA)® Examination Content Outline, 2013, "Analysis," Task 7.

158. Jinx, the VP of supply chain for a regional hospital, has established the need for several highly customized workflows. The project team has established the process for these requirements to be documented; however, before approving the onshore specialty development, what can the Project Steering Committee suggest?
    a. Request a detailed review of the solution requirements with the supply chain team and developer.
    b. Require that the supply chain lead present a cost–benefit analysis to the Change Control Board.
    c. Review the customized workflows with the VP of supply chain from a competitor, conduct a detailed review with the supply chain team, and require a cost–benefit analysis be presented to the Change Control Board.
    d. Ask the VP of supply chain from a competitor to review the business requirements documents and provide their opinion.

The correct answer is: **C**

This is a difficult question, and each answer choice requires careful consideration. The first step is to understand where you are in the process and make note of the desired outcome. In this case, because of the high degree of customization, the Project Steering Committee is looking for guidance before approving the expenditure of resources. Requirements for predictive and iterative lifecycle projects are presented in the form of business requirements documents or solution requirements documents. Once created, these artifacts may be reviewed by stakeholders or industry experts to provide their opinions. In this scenario, the most complete approach is to (a) review the customized work-flows with the VP of supply chain from a competitor, (b) conduct a detailed review with the supply chain team; concluding with (c) present a required cost–benefit analysis to the Change Control Board for approval.

Answer Choice A: This answer choice does not take into account valuable insights from outside the organization, and the options would be very one-sided.

Answer Choice B: This is a partially correct answer choice; a cost–benefit analysis should be presented to the Change Control Board for approval.

Answer Choice D: This is also a partially correct answer choice; seeking the options from industry experts can help avoid costly mistakes. Unfortunately, this is not a complete answer choice.

Business Analysis for Practitioners: A Practice Guide, Section 5.4.2, "Approval Levels," and Section 4.11.5, "Guidelines for Writing Requirements."

PMI Professional in Business Analysis (PMI-PBA)® Examination Content Outline, 2013, "Planning," Task 3.

159. You are collaborating with Colleen, your project manager, on a project artifact that outlines how requirements will be analyzed, documented, and managed. What are the required inputs to aid in creation of this document?
    a. Scope management plan, requirements management plan, stakeholder register
    b. Project charter, project management plan, scope management plan
    c. Enterprise environmental factors, organizational process assets, charter, project management plan
    d. Enterprise environmental factors, organizational process assets, scope management plan, requirements management plan, stakeholder register

The correct answer is: **C**

This is a very difficult question, and is representative of a question that is likely to appear on the exam. To answer correctly requires in-depth knowledge of the requirements management plan—a component of the overall project management plan—which outlines how requirements will be analyzed, documented, and managed over the project lifecycle. The same inputs are also required to create the scope management plan. Answer choice C establishes the required inputs to create this document: enterprise environmental factors, organizational process assets, charter, and project management plan.

Answer Choices A, B, & D: These answer choices are all incorrect and don't outline the correct inputs to create the requirements management plan.

A Guide to the Project Management Body of Knowledge (PMBOK®), Section 5.1.

PMI Professional in Business Analysis (PMI-PBA)® Examination Content Outline, 2013, "Planning," Task 3.

160. Your sponsor, Tabitha, and project manager, Braxton, are quite impressed with the significant number of ideas that have been generated during the facilitated elicitation session. To organize these ideas, Tabitha suggests:

a. You organize them using an affinity diagram.
b. You organize them using an interrelationship diagram.
c. They all be included in the work breakdown schedule (WBS), which is part of the project management plan.
d. They be outlined in the solution statement.

The correct answer is: **A**

This question tests your knowledge of affinity diagrams, which is a very useful technique to cluster and connect related themes. They are used to organize and group large amounts of data into natural relationships and are typically used following brainstorming sessions.

Answer Choice B: Interrelationship diagrams, one of the tools used in scope modeling, aid in visualizing complex problems and relationships. When using the tool, is quite common to find factors that influence each other. In cases in which there is more than one influencing factor, the team needs to determine which factor is stronger, and note only one. The tool would not be used for organizing ideas.

Answer Choices C & D: These are both distractors and incorrect answer choices.

Business Analysis for Practitioners: A Practice Guide, Section 2.4.5.2, "Affinity Diagram."

PMI Professional in Business Analysis (PMI-PBA)® Examination Content Outline, 2013, "Needs Assessment," Task 2.

---

161. Your project manager, Raelyn, is establishing a platform that will serve as the central repository for project plans, requirements documentation, and stakeholder lists that will be used over the life of the project. What is this platform considered?
    a. A project management information repository
    b. A project management collaboration site
    c. SharePoint or team wiki
    d. A project management information system

The correct answer is: **D**

On the exam, you can expect at least one question pertaining to information repositories, team collaboration tools, and knowledge management. PMI is a strong proponent of lessons learned and knowledge transfer, and this question is representative of one that may appear on the exam. Project management information systems (PMIS) do not have to be sophisticated tools; in fact, MS Excel and Shared Network Folders are the most common. Keep in mind that the key principle is that they need to serve as the central repository for all team documentation. The PMIS is also considered an enterprise environmental factor.

Answer Choices A, B, and C: These are all incorrect descriptions of the central repository for project plans, requirements documentation, and stakeholder lists that will be used over the life of the project.

A Guide to the Project Management Body of Knowledge (PMBOK®), Section 4.4.2.3, "Project Management Information System."

PMI Professional in Business Analysis (PMI-PBA)® Examination Content Outline, 2013, "Planning," Task 5.

---

162. You work for an advertising agency in a highly competitive market. In an effort to increase market share across several product lines, you've been asked to support Emily, a strategic marketing expert. During your initial meeting with Emily, you provided her with statistics from the web

development team that included number of visitors, time on each page, transaction volume, and source of traffic. Although impressed with the data, Emily also wants to understand market share, competition, industry cycle times, and an overall competitive analysis. What can you suggest?

a. Advise Emily that the Sherman Act of 1890 has very strict antitrust guidelines, and further that what she is asking for is banned by the Fair Trade Commission Act.

b. Recommend hiring a third party to conduct the competitive analysis, looking for both public and nonpublic data.

c. Advise Emily that the data cannot be reasonably gathered, much less analyzed.

d. Suggest to Emily that you conduct the analysis, looking only for data that has been made publicly available.

The correct answer is: **D**

This is a very wordy question, which during the exam can be frustrating and confusing. The first step in approaching this type of question is to isolate precisely what you're being asked and test for relevance with each of the answer choices. In this scenario, your subject matter expert, Emily, ". . . wanted to understand market share, competition, industry cycle times, and an overall competitive analysis." To further research these areas and understand the competitive landscape, the organization could conduct either a benchmarking study or a competitive analysis—in both cases, relying only on data that is publicly available.

Answer Choice A: This is a distractor; *avoid answer choices on the exam that may have technical merit but are beyond the knowledge of the business analyst.*

Answer Choice B: A competitive analysis that uses nonpublic data would be an ethical concern and could result in legal issues.

Answer Choice C: This is an incorrect answer, because the data can be gathered from a number of sources.

Business Analysis for Practitioners: A Practice Guide, Section 2.3.3, "Gather Relevant Data to Evaluate the Situation."
PMI Professional in Business Analysis (PMI-PBA)® Examination Content Outline, 2013, "Needs Assessment," Task 1.

163. Jacob is working on the development of the business case for your project. You just completed the initial needs assessment, when you were asked to provide the situation statement. To clearly communicate the nature of the problem, you construct the statement in the following format:

a. Problem–cost–benefit
b. Opportunity–effect–cost
c. Opportunity–outcome–impact
d. Problem–effect–opportunity

The correct answer is: **C**

Well-written situation statements present a complete understanding of the problem to be addressed or the opportunity to be pursued, along with the contributing effects and the overall impacts. They provide enough detail to establish the problem (or opportunity) of "x," the effect (outcome) of "y," with the impact of "z." Alternatively, the situation statement can be presented as problem–effect–impact. Should the project team be addressing a problem, remember to replace *opportunity* with *problem*. **On the exam, remember, the statement needs to be clear, concise, and unambiguous.**

Answer Choices A, B, & D: These are all incorrect formats of the situation statement.

Business Analysis for Practitioners: A Practice Guide, 2.3.4, "Draft the Situation Statement."

PMI Professional in Business Analysis (PMI-PBA)® Examination Content Outline, 2013, "Needs Assessment," Task 1.

164. Who has accountability for determining the extent to which the project will trace and track requirements?
    a. It is the project manager's responsibility to assess and determine the extent to which the project will trace and track requirements.
    b. It is the business analyst's responsibility to assess and determine the extent to which the project will trace and track requirements.
    c. The business analyst should discuss this topic with the project manager over lunch.
    d. It is the stakeholders' responsibility to assess and determine the extent to which the project will trace and track requirements.

The correct answer is: **C**

In organizations in which the project management and business analyst role are split, strong collaboration is vital to ensuring the success of the initiative. Throughout the Practice Guide are key points of collaboration; building a familiarity with them will help you approach these types of questions with relative ease. To ensure that the long-term viability of both creating and maintaining the requirements traceability matrix, this requires a collaborative conversation between the project manager and the business analyst.

Answer Choices A & B: Over the duration of the project, there are several opportunities for collaboration. These answer choices are singularly focused and incorrect.

Answer Choice D: This is a distractor; the use of stakeholders in this scenario is too generic.

Business Analysis for Practitioners: A Practice Guide, Section 5.2.1, "What is Traceability?"

PMI Professional in Business Analysis (PMI-PBA)® Examination Content Outline, 2013, "Traceability and Monitoring," Task 1.

165. Rayden is new to business analysis, and while helping you draft several key documents, she asks you to provide examples of organizational process assets. You suggest:
    a. Work authorization systems, project management information system
    b. Guidance from consultants, industry professionals, and technical associations who are available as resources to the performing organization
    c. Communication channels influenced by organizational culture and structure
    d. Artifacts from previous projects, organizational polices, procedural documents, checklists, or templates

The correct answer is: **D**

This can be very difficult question, and one that is often confused with enterprise environmental factors. Spanning project scope and time management, organizational process assets (OPAs) are referenced 12 times in the *PMBOK® Guide.* Organizational process assets are artifacts specific to the performing organization that can encompass lessons learned, templates, plans, polices, and/ or procedures. Answer choice D lists some of the most common OPAs—artifacts from previous projects, organizational polices, procedural documents, checklists, or templates.

Answer Choices A & C: Work authorization systems, project management information systems, organizational culture, and established communication channels are examples of enterprise environmental factors, which are conditions outside the control of the project team but which could impact, constrain, or influence the project.

Answer Choice B: This is a distractor answer choice and not an example of an organizational process asset.

A Guide to the Project Management Body of Knowledge (PMBOK®), Sections 2.1.4, "Organizational Process Assets," Chapter 5, "Project Scope Management," Chapter 6, "Project Time Management."

PMI Professional in Business Analysis (PMI-PBA)® Examination Content Outline, 2013, "Planning Tasks" 1–6.

166. Your organization is considering moving from an on-premise solution to one that is cloud based with Icarus Software Ltd. As the senior business analyst, you've been asked to assist with documenting the requirements for the transition. Your team is accustomed to 10-minute response time from an on-call resource, 24/365. In working with your counterpart from Icarus, how would you classify performance, continuity, and availability requirements?
    a. Nonfunctional requirements
    b. Service-level agreement
    c. Business requirements
    d. Stakeholder requirements

The correct answer is: **A**

This question tests your understanding of the requirement types. ***On the exam, you can expect several scenarios-based questions on each type and their practical application.*** Nonfunctional requirements are subset of solution requirements, which describe the environmental conditions of the component. They are typically characterized as *quality* or *additive* requirements and are described as "must haves"; they are often categorized as solution warrantee requirements. Examples of quality and behavior attributes include availability, capacity, continuity, performance, security and compliance, and service-level management.

Answer Choice B: Service-level agreements (SLAs) are contracts between service providers (either internal or external to an organization) and a customer. They quantifiably define the level of service that can be expected for a given facet. They can be defined at the customer level, a service level, or a combination (MultiLevel).

Answer Choice C: Business requirements describe the higher-level needs of the organization.

Answer Choice D: Stakeholder requirements describe the needs of a stakeholder or group of stakeholders and how they will benefit from and interact with the component.

Requirements Management: A Practice Guide, Section 5.2.2."Define Types of Requirements."

PMI Professional in Business Analysis (PMI-PBA)® Examination Content Outline, 2013, "Analysis," Task 6.

167. Tamsin, a business analyst for an international cargo carrier, has been analyzing data in an attempt to determine the leading causes of variation at the shipping port. Her hypothesis is that there are most likely only a few factors driving the majority of the defects. What tool can she use to validate her assumption?
    a. Control chart
    b. Pareto chart
    c. Process model
    d. Trend analysis

The correct answer is: **B**

On the exam, you can expect definition, situational, and interpretation style questions related to Pareto analysis and charts. A Pareto analysis, named after Italian economist Vilfredo Pareto (1895), is a technique that uses the Pareto principle, which is commonly referred to as the *80/20 rule.* The principle suggests that approximately 20% of the causes account for 80% of the work. Pareto charts are a used for visualizing the data.

Answer Choice A: Control charts are often used in Six Sigma to visually show how a process changes overtime. There are three limits that the business analyst will study: the average, the upper control limit, and the lower control limit.

Answer Choice C: Process models describe business processes; they can take the form of process flows, use cases, or user stories.

Answer Choice D: Like Pareto, trend analysis is also a data-analysis technique. However, in this scenario, it would not be used to determine the leading causes of variation.

Business Analysis for Practitioners: A Practice Guide, Section 2.3.3, "Gather Relevant Data to Evaluate the Situation."

PMI Professional in Business Analysis (PMI-PBA)® Examination Content Outline, 2013, "Needs Assessment," Task 1.

168. You are working with your IT security manager, Sienna, on a project to implement a new firewall for a pharmaceutical company whose two primary customers are the Centers for Disease Control and Prevention (CDC) and the Food and Drug Administration (FDA). While planning the kick-off meeting, you suggest collaborating on a responsibility assignment matrix. Why?
    a. A responsibility assignment matrix is necessary because the company operates in a highly regulated industry.
    b. Because of the sensitive nature of the data and the risk associated with the project, the CDC and the FDA require a responsibility assignment matrix as part of the statement of work.
    c. The RACI will identify those who are either affected or perceived to be affected by the project.
    d. The completion of a reliability and maintainability matrix is part of the Statement on Standards for Attestation Engagements (SSAE16) for auditing purposes.

The correct answer is: **C**

The culture of the organization and industry will influence a number of items over the duration of the project. During project planning, the responsibility assignment matrix (RACI) was created, which established accountability for requirements signoff. This matrix was included as part of the requirements management plan and business analysis plan. They help to clearly identify: who's responsible for completing the work; the one and only person who is accountable; experts that can be consulted; and others who need to be informed. In this scenario, in which they are operating in a highly regulated industry, the RACI will identify those who are either affected or perceived to be affected by the project.

Answer Choice A: Although this is a true statement, it doesn't properly answer the question

Answer Choices B & D: These are distractor answer choices and not valid answer selections.

Business Analysis for Practitioners: A Practice Guide, Section 2.3.1, "Identify Stakeholders."

PMI Professional in Business Analysis (PMI-PBA)® Examination Content Outline, 2013, "Needs Assessment," Task 4.

169. What tool can Amias use to ensure that his subject matter experts are creating requirements documents of high quality?

a. A checklist
b. The requirements traceability matrix
c. Planguage
d. Requirements acceptance criteria

The correct answer is: **A**

For the exam, and in practice, please remember the nine characteristics of high quality require-ments documentation: unambiguous, precise, consistent, correct, compete, measurable, feasible, traceable, and testable. The business analyst can use checklists to ensure that requirements doc-umentation satisfies these characteristics. These checklists become organizational process assets, which are archived for future reference.

Answer Choice B: The requirements traceability matrix is the essential link between the baselined scope and organizations pillars (mission, vision, goals, and objectives). There is nothing within this document that would ensure that subject matter experts are creating requirements documents of high quality.

Answer Choice C: Although Tom Gilb created the concept of a planning language (*Planguage*) to address ambiguous and incomplete nonfunctional requirements, it does not ensure that users are creating documents of high quality.

Answer Choice D: This is a made-up answer choice.

Business Analysis for Practitioners: A Practice Guide, Section 4.11.5.1, "Functional Requirements."

PMI Professional in Business Analysis (PMI-PBA)® Examination Content Outline, 2013, "Analysis," Task 7.

170. Elyse is designing an interactive website for customers at a dine-in restaurant to place food and beverage orders. Her primary stakeholder, Leonidas, requested that she create a model that plots how customers will navigate the screens and the system's responses based on their selections. What model should she use?
    a. Process-flow diagram
    b. System interface flow
    c. Interoperability diagram
    d. User-interface flow

The correct answer is: **D**

This question tests your working understanding of interface models, which can consist of (a) report tables, (b) system interface tables, (c) user-interface flows, (d) wireframes, and (e) display-action-response models. User-interface flows are models that display specific pages or screens and plot out how users navigate them based on screen prompts or selections.

Answer Choice A: Process-flow diagrams are used to describe elements of a solution, process, or project. They are powerful tools for visualizing both current (as-is) and future (to-be) states, and are often referred to as *swim-lane diagrams*.

Answer Choice B: System interface flow is a made-up answer choice.

Answer Choice C: Interoperability diagrams are used for documenting relationships and boundar-ies of interfacing systems.

Business Analysis for Practitioners: A Practice Guide, Section 4.10.11.3, "User Interface Flow."

PMI Professional in Business Analysis (PMI-PBA)® Examination Content Outline, 2013, "Analysis," Task 2.

171. It's been nearly ten years since your organization optimized its processes. Gabriel, the VP of strategic initiatives, has requested that you model the existing processes to help prepare the organization for a transformative initiative. To accomplish this task, what tool would you use?
    a. Unified Modeling Language (UML)
    b. System Modeling Language (SysML)
    c. Business Process Modeling Notation (BPMN)
    d. Process-flow diagram

The correct answer is: **C**

This is a difficult question that tests your understanding of the various modeling languages and their intended purposes. The Practice Guide describes the four modeling languages: (a) Business Process Modeling Notation (BPMN), (b) Requirements Modeling Language (RML), (c) System Modeling Language (SysML), and (d) Unified Modeling Language (UML). BPMN is similar to flow charts and process flow diagrams; the difference is that it accommodates the interest of both business and technical users by creating a unified framework for the distinct purpose of modifying/optimizing processes.

Answer Choice A: The Unified Modeling Language can be used either when specifying design models or during requirements specification.

Answer Choice B: The System Modeling Language is a subset of the Unified Modeling Language and is used when business analysts need to analyze complex systems.

Answer Choice D: Process-flow diagrams are used to describe elements of a solution, process, or project. They are powerful tools for visualizing both current (as-is) and future (to-be) states; they often referred to as *swim-lane diagrams*.

Business Analysis for Practitioners: A Practice Guide, Section 4.10.6, "Modeling Languages."

PMI Professional in Business Analysis (PMI-PBA)® Examination Content Outline, 2013, "Analysis," Task 2.

172. As an experienced business analyst, Thea suspects that there may be gaps in the business objective model that she developed for a local pet store. She would like to create a model detailing the lifecycle of an object through various conditions. What tool can Thea use to uncover any potential gaps?
    a. An entity relationship diagram further detailing the workflows from the business objective model
    b. A pair-wise table, which evaluates the initial and target states for object
    c. A process-flow diagram aligned to industry leading practice
    d. A system diagram aligned to industry leading practice

The correct answer is: **B**

This is a difficult question, because the scenario is introducing two unrelated concepts. The first is a distractor, ". . . How would you uncover any potential gaps in the business objective model?" Whereas the actual question is, "What *tool can be used* to uncover any potential gaps in a model detailing the lifecycle of an object through various conditions?" To both establish and validate object transitions, state tables and diagrams (aka *pair-wise tables*) can be used to track objects over their lifecycles. These models evaluate the initial and target states for objects. Furthermore, the models can also be used to ensure that nothing has been overlooked in the requirements definition process.

Answer Choice A: This is a made-up answer choice.

Answer Choices C & D: These are both incorrect answer choices.

Business Analysis for Practitioners: A Practice Guide, Section 4.10.10.4, "State Table and State Diagrams."

PMI Professional in Business Analysis (PMI-PBA)® Examination Content Outline, 2013, "Analysis," Task 6.

173. Stefan is having difficulty explaining the motivations, behaviors, and ultimate goals for the targeted audience of the newly designed website. To aid in understanding their characteristics, as the business analyst, you:
    a. Recommend creating a short narrative about each user group
    b. Suggest creating a user analysis based on the targeted user groups
    c. Determine models that can be used to aid in the analysis of user groups
    d. State each of the elements in the format of: As a <role or type of user>, I want <goal/desire>, so that <benefit>, to help articulate the needs of the targeted audience

The correct answer is: **A**

Personas define an archetype—a type of person that would use a system. They represent fictitious people, who are based on the team's knowledge of real users. A persona analysis is a short narrative story that describes the goals, behaviors, motivations, environment, demographics, and skills of each user group. Based on fictional characters, the tool is used to understand stakeholder needs and characteristics. (Remember, in practice, you should never use someone's actual name in a meeting or when designing a system.)

Answer Choices B & C: These are made-up answer choices.

Answer Choice D: This is an example of a user story, which would not be used to aid in understanding the targeted audience's characteristics or to explain their motivations, behaviors, or ultimate goals.

Business Analysis for Practitioners: A Practice Guide, Section 3.3.3.2, "Persona Analysis."

PMI Professional in Business Analysis (PMI-PBA)® Examination Content Outline, 2013, "Planning," Task 3.

174. You are serving as both project manager and business analyst on a small project in a heavily regulated industry. Your sponsor, Thaïs, would like to implement a formal approach to tracking changes to project artifacts. Where should this requirement be identified?
    a. In the configuration management system and version control system
    b. In the business analysis plan
    c. In the requirements management plan
    d. In the scope management plan

The correct answer is: **B**

This is a question of moderate difficulty, because the business analysis plan and requirements management plan contain similar elements. In organizations of varying size and complexity, it's common that the role of business analyst and project manager will be filled by the same person. **For the exam, you'll need to become very familiar with the contents of both artifacts.** The requirements for managing the versioning of documents (aka *version control*) is identified only within the business analysis plan. This continuum can be as simple as a table at the beginning of a document, to a formal system in which documents are checked in/out and locked while being edited.

Answer Choice A: Configuration management systems will often provide relevant and useful information to better understand the current state and can serve as repositories for historical information. The formal approach for tracking changes would not be identified within this tool.

Answer Choice C: Version control is not a section within the requirements management.

Answer Choice D: The scope management plan is used for monitoring and controlling project scope. It would not contain the requirements for managing the versioning of documents.

Business Analysis for Practitioners: A Practice Guide, Section 5.8.2.2, "Version Control Systems (VCS)."

PMI Professional in Business Analysis (PMI-PBA)® Examination Content Outline, 2013, "Planning," Task 5.

175. Your project manager, Mila, is concerned about scope creep at both the project and the product level. To address this concern, you both agreed to institute a requirements traceable matrix. After the matrix is created and approved, what can you do to set boundaries?

    a. Present the matrix to your stakeholders for approval on the approach.

    b. Perform a dependency analysis.

    c. Establish a process for monitoring and controlling scope.

    d. Establish a baseline.

The correct answer is: **D**

Although this question is a little wordy, it's simply asking, "What can be done to establish boundaries once the requirements traceable matrix is created?" To address this concern, leading practice is to establish a requirements baseline. This baseline contains all the work that was approved to be included as part of the project. It's used as a means of comparison over time as various requirements are added and others are deferred. The baseline is also referred to as a *requirements snapshot. Remember, for the exam, baseline is a measurement for future comparison.*

Answer Choices A & C: These are both distractors and incorrect answer choices.

Answer Choice B: This is an incorrect answer choice; the dependency analysis technique is used to determine whether requirements can exist without the presence of another.

Business Analysis for Practitioners: A Practice Guide, Section 5.5.1, "What is a Requirements Baseline?"

PMI Professional in Business Analysis (PMI-PBA)® Examination Content Outline, 2013, "Traceability and Monitoring," Task 2.

176. Your subject matter expert, Jennifer from the radio repair shop, recently submitted her requirements documents for the new wire reel. Although they are all very well written, they seem to be missing a key component:

    a. There were no preconditions to enable testing.

    b. They are missing the actors and personas.

    c. Some requirements were labeled TBD because the team was still addressing some elements.

    d. They should have been in the format of a user story or epic.

The correct answer is: **C**

While drafting requirements, it's acceptable to have some elements marked as "TBD," considering the dynamic nature of projects and the evolution of requirements. The key aspects of requirement documents are that they are: complete, consistent, correct, feasible, measurable, precise, testable, traceable, and unambiguous.

Answer Choice A: Preconditions are not associated with requirement documents.
Answer Choice B: Personas define an archetype—a type of person that would use a system. They represent fictitious people based on the team's knowledge of real users, whereas actors represent roles and are associated with use cases. Neither would be related to the requirement documents.
Answer Choice D: This is an incorrect answer choice. User stories and epics are associated with the agile delivery methodology and would not be applicable in this scenario.
Business Analysis for Practitioners: A Practice Guide, Section 4.11.5, "Guidelines for Writing Requirements."
PMI Professional in Business Analysis (PMI-PBA)® Examination Content Outline, 2013, "Analysis," Task 6.

177. You are an experienced project manager, and you have been asked to fill the key business analyst role on a high-profile project. Shortly after starting the requirements sessions, you've noticed considerable changes to the requirements documents. To assist with your efforts, you:
    a. Ask the subject matter experts to add versioning control to each document.
    b. Add the requirements documents to a control system.
    c. Conduct an analysis to determine the material nature of the differences.
    d. Provide your technical resources with only the most current version, based on the control date.

The correct answer is: **B**

Documentation is an essential facet of every project, and the proper management is critical for auditing, compliance, communication, and sharing lessons learned. This question tests your understanding of version control and document management. *For the exam, remember that the version control system is a subset of the configuration management system.* Aside from the more common uses, such as storing code, the version control system can track the revisions for requirements documentation.
Answer Choice A: Although adding version control to each document is a leading practice, the scenario is focused on the broader topic. Adding requirements documents to a control system addresses the broader group.
Answer Choice C: Although this may be a downstream task, the predecessor task would be to add the requirements documents to a control system
Answer Choice D: This is an incorrect answer choice.
Requirements Management: A Practice Guide, Section 5.8.2.2, "Version Control Systems."
PMI Professional in Business Analysis (PMI-PBA)® Examination Content Outline, 2013, "Traceability and Monitoring," Task 5.

178. You work for Porcine Lifesciences LLC, the leading producer of aortic pig valves and instruments for use in human heart valve replacement surgeries. To gain a better understanding of the operation and the related steps involved with surgery, a local hospital has invited you to view a surgery from the observation suite. What type of observational activity is this considered?
    a. Passive
    b. Participatory
    c. Active
    d. Simulation

The correct answer is: **A**

This is a very wordy scenario, which can lead to confusion at this point in the exam. When approaching this type of scenario, remember to isolate precisely what the question is asking—in this case, "What is the operational activity associated with viewing?" Observation allows individuals to gain a greater understanding of the situation under real-world conditions. There are four categories of observation (passive, active, participatory, and simulation). With passive, a person quietly observes others while carefully documenting the situation. Because there is zero interruption, at a later date the observer may follow up to seek clarification.

Answer Choice B: When conducting a participatory observation, the observer would jointly perform the activities with the individual performing the process.

Answer Choice C: When using active observation, the observer would ask questions of the individual performing the process.

Answer Choice D: When using the simulation technique, the observer mimics or recreates the activities from the perspective of the end user.

Business Analysis for Practitioners: A Practice Guide, Section 4.5.5.6, "Observation."

PMI Professional in Business Analysis (PMI-PBA)® Examination Content Outline, 2013, "Analysis," Task 2.

179. What is the benefit of using the iterative planning technique of rolling wave planning?
    a. Team members skilled at planning can provide their input to the plan creation.
    b. The work plan can be decomposed to varying levels of detail.
    c. Rolling wave planning is based on the WBS, which contains a hierarchical decomposition of all the work to be completed as part of the project.
    d. Rolling wave planning highlights all the significant mandatory events in the project.

The correct answer is: **B**

This is a difficult question, which tests your knowledge and understanding of the tools and techniques associated with scheduling. Rolling wave planning is a form of progressive elaboration, in which activities to be performed in the near term are planned at a much greater level of detail than activities planned for later date in the project.

Answer Choice A: This answer choice focuses on the plan creation, which doesn't properly address the question.

Answer Choice C: Although the WBS contains a hierarchical decomposition of all the work to be completed as part of the project, it's not solely based on the technique of rolling wave planning.

Answer Choice D: This is a made-up answer choice.

Business Analysis for Practitioners: A Practice Guide, Section 3.4.2.1, "Determining the Proper Level of Detail."

A Guide to the Project Management Body of Knowledge, Section 6.2.2.2, "Rolling Wave Planning."

PMI Professional in Business Analysis (PMI-PBA)® Examination Content Outline, 2013, "Planning," Task 3.

180. Skyler holds certifications from the Project Management Institute in both project management and business analysis. He is presently working to estimate construction times for a multi-seasonal resort in New Hampshire. Due to the weather uncertainty, he is using a technique that takes into account pessimistic, most likely, and optimistic estimates from his subject matter experts. What estimating technique is Skyler using?
    a. Analogous
    b. Three-point
    c. Parametric
    d. Bottom-up

The correct answer is: **B**

Estimating is a vital competency for business analysts, and this question tests your knowledge of the four techniques that can be used to estimate activity durations. To consider uncertainties that may influence a project, three-point estimates based on pessimistic, most likely, and optimistic durations provide an approximate range for the activities duration. Because estimating is an iterative process, over time, these estimates are further refined.

Answer Choice A: The analogous estimation technique uses data from prior projects of similar size to estimate work effort and cost when there is minimal information available. Because estimating is an iterative process, over time, these estimates are further refined.

Answer Choice C: Parametric estimating considers the relationship between variables for estimating cost and duration.

Answer Choice D: Bottom-up estimating totals the estimates of all lower-level components within the work breakdown structure to approximate project duration.

A Guide to the Project Management Body of Knowledge (PMBOK®), Section 6.5.2.4, "Three Point Estimating."
Business Analysis for Practitioners: A Practice Guide, Section 3.5.2.6, "Estimate the Work."
PMI Professional in Business Analysis (PMI-PBA)® Examination Content Outline, 2013, "Planning," Task 4.

181. Cassian is working for a digital advertising agency and is attempting to articulate how customers would interact with a timed advertisement that would load prior to free content on a client's website. How could Cassian best explain the concepts and interactions in an attempt to solicit feedback?
    a. Cassian could conduct a series of elicitation sessions focused on soliciting feedback.
    b. Cassian's team could create a high-fidelity prototype of the finished product.
    c. Cassian's team could create an evolutionary prototype of the finished product.
    d. Cassian could create graphical illustrations, which would eventually be discarded.

The correct answer is: **D**

This is a tricky question, providing a detailed back story with similar answer choices. After isolating the question, the next step to answering confidently is to find a correlation to a possible answer choice. Storyboards (aka *graphical illustrations*) are a type of low-fidelity prototype. Although most commonly associated with movie and television production, they are very versatile tools that focus on the user experience via a series of cartoons or drawings.

Answer Choice A: This is an incorrect answer choice.

Answer Choices B & C: Although both answer choices B and C appear to be reasonable options, both are incorrect.

Requirements Management: A Practice Guide, Section 4.5.5.7, "Prototyping."
PMI Professional in Business Analysis (PMI-PBA)® Examination Content Outline, 2013, "Analysis," Task 1.

182. To ensure project consistency and completeness, your project manager, Tom, suggests that he will manage the requirements traceability matrix, because he has greater insight into the tasks and deliverables associated with each project phase. Is this approach acceptable?
    a. Yes. For fundamental accuracy, Tom as the PM should maintain the matrix.
    b. No. Muriel, as the organizational change management lead, should maintain the matrix to ensure consistency and completeness. Her utilization is at 50%.
    c. No. The business analyst, Woodrow, who is a part-time resource and new to the organization, should manage the matrix to ensure consistency and completeness.
    d. Yes, the approach is acceptable.

The correct answer is: **B**

This is a difficult scenario with several plausible answer choices. When available to a project, it is the business analyst's responsibly to maintain the requirements traceability matrix. However, please keep in mind, the overall effort may not require a full-time resource. Because the requirements traceability matrix must be maintained for the entire project lifecycle, it may be appropriate at some point to transition the maintenance to another team member. To realize the benefits, there must be an organizational commitment to maintain the matrix. In this scenario, by the organizational change management lead maintaining the matrix, they will have a better insight into future state and planning/communication pertaining to the changes.

Answer Choices A & D: These are incorrect answer choices; whenever available to a project, someone other than the project manager should maintain the requirements traceability matrix.

Answer Choice C: This is a viable answer choice; however, because Woodrow is new to the organization, he would not be the most obvious choice to maintain the requirements traceability matrix. Answer choice B is the better option.

Requirements Management: A Practice Guide, Section 5.1, "Overview."

PMI Professional in Business Analysis (PMI-PBA)® Examination Content Outline, "Traceability and Monitoring," Task 1.

183. Brock is a well-respected business analyst for a Fortune 200 chemical company. He receives praise from senior leadership for providing support, and he's very successful in terms of mediation and conflict resolution. While conducting his 360-degree peer review, you were mostly in agreement with leadership, just adding the following comments:

a. It is a pleasure to work with Brock; he is both a friend and mentor.

b. Although it's a pleasure to work with Brock, if he were to focus more on understanding the motivations, goals, and objectives of the team, he would be more successful in several aspects of business analysis.

c. Although Brock is skilled in many areas, he needs to exhibit less authority and more influence; in addition, focusing on the business processes could help identify opportunities for further improvement in the organization.

d. It's a pleasure to work with Brock. In terms of improvement, he would be more valuable to the organization if he were to focus more on the business processes, which in turn could help identify opportunities for further improvement. As a leader, he needs to demonstrate slightly more consigned authority.

The correct answer is: **C**

Elicitation throughout the lifecycle, from conception to evaluation, is a critical activity. Business analysts who are most successful support leadership with information that enables effective decision making; apply influence, without authority; are successful in negotiating or mediating when resolving conflict; and define problems or opportunities, which enable the organization to grow and thrive. In this scenario, the individual should exhibit less authority and more influence; in addition, focusing on the business processes could help identify opportunities for further improvement in the organization.

Answer Choices A, B, & D: These all made up and incorrect answer choices.

Requirements Management: A Practice Guide, Section 4.2.2, "Importance of Eliciting Information."

PMI Professional in Business Analysis (PMI-PBA)® Examination Content Outline, 2013, "Analysis," Task 1.

184. Your project manager, Prescott, has recently successfully completed a $150 million USD software project for a major healthcare customer, for which it won the PMI Project of the Year Award. When talking about your new $50,000 USD project, he offers to share his requirements traceability matrix, stressing that you must use this tool on your project. You decide:
    a. To use his requirements traceability matrix—if it helped to win Project of the Year, it's a tool and technique to use.
    b. To use his requirements traceability matrix—it's a PMO process asset.
    c. To review his matrix, but adapt it for your needs to trace requirements to the business mandate.
    d. To review his matrix, but adapt it for your needs to trace a list of actions.

> **The correct answer is: D**
>
> PMI is a big proponent of sharing lessons learned, advancing knowledge, and collaboration. The tools and techniques available to business analysts are robust and highly flexible. In this scenario, the template can be easily adapted to suit the specific needs of the project. *Remember, for the exam, the requirements traceability matrix can be used to trace and track use cases to features, and test cases to deliverables.* Answer choice D is an example of actors performing steps or interacting with a system, as outlined in a use case.
>
> **Answer Choices A, B, & C:** These are all incorrect answer choices, as they don't provide a valid rationale.
>
> *Requirements Management: A Practice Guide*, Section 5.2.1, "What is Traceability?"
>
> *PMI Professional in Business Analysis (PMI-PBA)® Examination Content Outline*, 2013, "Traceability and Monitoring," Task 1.

185. Koda, your project manager, is working to decompose the elements of the project into logical and manageable portions so that you can begin creating the requirements traceability matrix. Once complete, what will she have created?
    a. The project management plan
    b. The details of all project and product work
    c. An outline of the requirements traceability matrix
    d. The scope baseline

> **The correct answer is: B**
>
> This is both an artifact- and sequence-related question. The first observation is that the artifact will be an input to creating the requirements traceability matrix; this helps us to understand sequence. Second, the work effort will result in a logical decomposition of project work into manageable portions. The work breakdown structure (WBS) is a fundamental project management deliverable; it is a logical decomposition, represented in the form of a hierarchy, representing the total scope of the project. It is often accompanied by a WBS dictionary, which is a detailed description of each WBS package. *For the exam, remember, the work breakdown structure represents all the effort for the initiative, including project and product work.*
>
> **Answer Choice A:** The project management plan is the aggregation of all the subsidiary project plans.
>
> **Answer Choice C:** The requirements traceability matrix can be used to track product scope and, more specifically, work breakdown structure deliverables.

Answer Choice D: The scope baseline is a significant project deliverable and key artifact. It's not one single document; rather, it includes the scope statement, the work breakdown structure, and the WBS dictionary. *On the exam, remember, the baseline is a measurement for future comparison.*

A Guide to the Project Management Body of Knowledge (PMBOK®), Section 5.4.2, "Create WBS: Tools and Techniques."

PMI Professional in Business Analysis (PMI-PBA)® Examination Content Outline, "Traceability and Monitoring," Task 1.

186. Achilles is an experienced business analyst and an expert facilitator. While working with Hayley, a subject matter expert for the seed division of the Ministry of Agriculture & Farmers, to uncover objects that are related to or may have dependencies with other objects, he inquires as to the cardinality relationships. To what is Achilles referring?
    a. The minimum number of times an instance in one entity can be associated with instances in a related entity
    b. A line notation
    c. The maximum number of times an object in one entity can be associated with objects in a related entity
    d. Business data objects

The correct answer is: **C**

Although this a lengthy and wordy question, it's asking a simple question: "What is cardinality?" *During the exam, isolate precisely what the question is asking, and don't extrapolate.* When creating entity relationship diagrams or business data diagrams, cardinality indicates the maximum number of times an object in one entity can be associated with objects in a related entity. Objects can include concepts, individuals, or whatever else the business analyst considers important; it is often modeled using the crow's foot notation.

Answer Choice A: Ordinality is used to uncover mandatory and optional relationships. It represents the *minimum* number of times an instance in one entity can be associated with instances in a related entity.

Answer Choice B: Line notation is a made-up answer. *For the exam, remember, when using entity relationship diagrams, the crow's foot notation is the most commonly used style to indicate both cardinality and ordinality.*

Answer Choice D: Entity relationship diagrams visualize business data objects; this is not directly related to cardinality.

Requirements Management: A Practice Guide, Section 4.10.10.1, "Entity Relationship Diagram."

PMI Professional in Business Analysis (PMI-PBA)® Examination Content Outline, 2013, "Analysis," Task 2.

187. Cecelia is in the process of developing her elicitation plan for a meeting with Gitchi Goo, the owner of a local gallery and a famous artist. What are some of the elements that should be included in the plan?
    a. What does Cecelia need to know to answer the question, what methods will be used to elicit information from Gitchi, in what order should activities be conducted?
    b. How and when elicitation will be conducted with Gitchi.
    c. In what order should communication activities be conducted, and what methods will be used to elicit information from Gitchi regarding her paintings?
    d. The source of information, activities to be conducted, and methods of communication.

The correct answer is: **A**

This question challenges your experience with elicitation techniques and the guiding plan for elicitation activities. As the business analyst develops the elicitation plan, there are several elements to consider: (a) What is the information they are seeking? (b) Where can the information be found? (c) How can the information be sourced? and (d) What is the appropriate sequence of elicitation activities? Answer choice A offers the most complete response.

Answer Choice B: The "how and when" refers to the plan itself, not the elements.

Answer Choice C: This is a partially correct answer choice; it is missing, "What information are they seeking?" Or, in the context of the question, "What does Cecelia need to know to answer the question?"

Answer Choice D: The source of information, activities be conducted, and methods of communication are all elements of the communication management plan.

Business Analysis for Practitioners: A Practice Guide, Section 4.3.1, "Develop the Elicitation Plan."

PMI Professional in Business Analysis (PMI-PBA)® Examination Content Outline, 2013, "Analysis," Task 1.

188. Aries was contracted to develop several queries and reports for the legal department of a utility company. To ensure that requirements were appropriately addressed, your subject matter expert, Kiera, created a prototype along with a model that establishes the details for each of the reports. What type of model did Kiera provide Aries?

   a. Interface model
   b. Business intelligence model
   c. Laszlo report table
   d. Report table

The correct answer is: **D**

This question tests your working understanding of interface models, which can consist of (a) report tables, (b) system interface tables, (c) user-interface flows, (d) wireframes, and (e) display-action-response models. Report tables are models containing the attributes and the details for each report at the object/field level. They specify requirements to ensure that requirements are appropriately addressed. When creating a prototype, attributes should be noted, because they add context for the textual information within the table.

Answer Choice A: Although interface models consist of report tables, the answer is too generic; they also include system interface tables, user-interface flows, wireframes, and display-action-response models.

Answer Choice B: Business intelligence model is a made-up answer choice.

Answer Choice C: Laszlo report table is a made-up answer choice.

Business Analysis for Practitioners: A Practice Guide, Section 4.10.11.1, "Report Table."

PMI Professional in Business Analysis (PMI-PBA)® Examination Content Outline, 2013, "Analysis," Task 2.

189. Although Tristan received approval from his sponsor and primary stakeholders for his business analysis plan, when the time arrives to conduct a test cycle, the lead resources were unable to perform their assigned tasks. What could be the root cause?

   a. The system wasn't sufficiently complete to conduct testing.
   b. Those performing the testing didn't participate in the review of the plan.
   c. The plan didn't outline the testing components.
   d. Test scripts were not prepared for the cycle.

The correct answer is: **B**

If team members and essential resources aren't trained on the delivery methodology, the approach to business analysis, and the overall plans, it's very hard to hold them accountable and expect that they'll be successful. Aside from the project sponsor and key stakeholders, team members who have lead roles should be included in the plan review. Individuals who are actively involved with planning are more inclined to support the activities and deliverables.

Answer Choices A, C, & D: These are all distractor answer choices.

Requirements Management: A Practice Guide, Section 3.5.5, "Review the Business Analysis Plan with Stakeholders."

PMI Professional in Business Analysis (PMI-PBA)® Examination Content Outline, 2013, "Planning," Task 4.

190. Upon returning from a PMI-PBA® Boot Camp, your sponsor, Alina, has requested that the team begin using the Planguage tool. What is the benefit of using this tool, which was developed by Tom Gilb?
    a. The tool is used to address concerns with ambiguous and incomplete functional requirements.
    b. The tool is used to address concerns with ambiguous and incomplete nonfunctional requirements.
    c. The tool provides a standard set of templates for solution requirements.
    d. The tool ensures consistency across all requirement types.

The correct answer is: **B**

This is an example of an advanced question that is likely to appear on the exam. The topic is covered in the *Examination Content Outline* but not the Practice Guides. In Tom Gilb's book *Competitive Engineering: A Handbook for Systems Engineering, Requirements Engineering, and Software Engineering Using Planguage* (Gilb, 2005), he introduces the concept of a planning language, what he calls "Planguage," to address ambiguous and incomplete nonfunctional requirements. The tool uses defined identifiers (e.g., tags) to qualify and quantify the quality elements of requirements. Examples include: *tag,* a unique identifier for each statement; *gist,* a summary of details of the nonfunctional requirements; *requirement,* a condition or capability described by a subject matter expert to address a concern or achieve an objective; *rationale,* justification for the requirement; *priority level,* the priority in relation to the other identified requirements (wish, stretch, fail, goal, survival); *stakeholders,* individuals or groups of individuals who are either interested in or impacted by the requirement; *status,* which can be undetermined, under revision, exited, approved, validated, or verified; and *owner,* the individual or department responsible for the requirement.

Answer Choices A, C, & D: These are all incorrect associations of the tool.

PMI Professional in Business Analysis (PMI-PBA)® Examination Content Outline, Knowledge and Skills, #20, "Measurement Skills and Techniques."

PMI Professional in Business Analysis (PMI-PBA)® Examination Content Outline, 2013, "Analysis," Task 6.

191. Aurelia is in the process of auditing the project's quality requirements to ensure that the output meets the stakeholder requirements and expectations. What process is she performing?
    a. Perform quality assurance
    b. Monitor and control quality
    c. Update quality checklists
    d. Conduct quality control review

The correct answer is: **A**

On the exam, you can expect a number of questions pertaining to both quality assurance (QA) and quality control (QC). Quality assurance is focused on prevention; with quality control, the emphasis is on testing and inspection. Quality assurance tasks are outlined in the project's quality management plan; the focus is on preventing defects through planning and by inspecting work in progress (WIP). *On the exam, remember QA is proactive, focusing on (a) defect prevention, (b) processes, (c) 7-step continual improvement.*

Answer Choices B: Monitor and control quality is not a process. From the *PMBOK® Guide*, you'll recall the process is called *control quality*, covered in the Project Quality Management Knowledge Area.

Answer Choices C & D: These are all made-up answer choices.

A Guide to the Project Management Body of Knowledge (PMBOK®), Section 8.2, "Quality Assurance."

PMI Professional in Business Analysis (PMI-PBA)® Examination Content Outline, 2013, "Evaluation," Task 2.

192. You are the program manager for a new line of nanorobotic vehicles. You ask Lou, a senior business analyst and board-certified occupational therapist, to conduct a series of sessions with potential customers, suppliers, and strategic business partners to better understand their needs. Lou is going to conduct a:
    a. Needs assessment workshop
    b. Stakeholder requirements session
    c. Business requirements session
    d. Nominal group technique workshop

The correct answer is: **B**

This scenario tests your understanding of elicitation sessions, which are vital to ensure that business analysts properly understand stakeholder needs. There are many methods that can be used to elicit information from stakeholders: interviews, facilitated sessions, focus groups, observations, questionnaires, surveys, and brainstorming, to name a few; the goal early on with these sessions is to thoroughly understand stakeholder needs so they can be traced and tracked over the life of solution development.

Answer Choices A, C, & D: These are all made-up answer choices.

Business Analysis for Practitioners: A Practice Guide, Section 1.7.2, "Requirement Types."

PMI Professional in Business Analysis (PMI-PBA)® Examination Content Outline, 2013, "Needs Assessment, Task 4."

193. Your sponsor has just recommended a functional requirement that was not previously identified during the needs assessment or included in the approved requirements snapshot. Your project manager suggests you consult:
    a. The scope management system (SMS)
    b. The configuration management system (CMS)
    c. The requirements management system (RMS)
    d. The service knowledge management system (SKMS)

The correct answer is: **B**

This question can be challenging if you're not familiar with the tools supporting business analysis. The configuration management system (CMS) is used to ensure conformity. It stores all project-related documents and changes, including those related to requirements. Furthermore, the CMS contains the approval levels for authorizing changes.

Answer Choices A & C: The scope management system and requirements management system are made up tools. Some organizations may use project management information systems to manage scope and requirements.

Answer Choice D: In some organizations, the configuration management system is a subset of the service knowledge management system, which represents the total body of knowledge for an IT organization.

Business Analysis for Practitioners: A Practice Guide, Section 5.8.2.1, "Configuration Management System (CMS)."

PMI Professional in Business Analysis (PMI-PBA)® Examination Content Outline, 2013, "Traceability and Monitoring," Task 5.

194. You're meeting with Jackie, the project manager for XYZ Soap Company. In your role as business analyst, she's asking that you document the product quality conditions required for the new line of fragrance-free soap to be effective. You agree, stating that these qualities will be included as part of the:
    a. Business quality requirements
    b. Stakeholder requirements
    c. Solution and nonfunctional requirements
    d. Functional and solution requirements

The correct answer is: **C**

Nonfunctional requirements are subset of solution requirements that describe the environmental conditions of the component. They are typically characterized as quality or additive requirements and are described as "must haves" and are often categorized as *solution warrantee requirements.* Common IT examples include: availability, capacity, continuity, and performance. ***On the exam, remember, "nonfunctional = quality, additive, or warrantee."***

Answer Choice A: Business quality requirements is a made-up requirement type.

Answer Choice B: Stakeholder requirements describe the needs of a stakeholder or group of stakeholders and how they will benefit from and interact with the component.

Answer Choice D: Functional requirements are a subset of solution requirements that describe the behaviors or functions of the component—a statement of conformity. Described in Yes/No terms, they are often categorized as *solution utility.*

Business Analysis for Practitioners: A Practice Guide, Section 1.7.2, "Requirement Types."

PMI Professional in Business Analysis (PMI-PBA)® Examination Content Outline, 2013, "Needs Assessment," Task 1.

195. You work for the Blue Ball Point Pen Company. Grayson is in the process of prioritizing the features for a new line of translucent pens, based on user stories submitted by the team members. What role does Grayson have?
    a. The business analyst
    b. Your product owner
    c. The scrum master
    d. The project manager

The correct answer is: **B**

This question challenges your understanding of roles associated the agile delivery methodology, of which there are only four: team member, product owner, scrum master, and stakeholder. When using agile, the product owner has the overall accountability for the product requirements. Based on business need, the established priority of a requirement can change over the life of a project. This would be reflected in the backlog.

Answer Choice A: When using agile, there is no role *business analyst.*

Answer Choice C: The scrum master is the facilitator for the agile development team.

Answer Choice D: When using agile, there is no role *project manager*, because the team is self-directed, guided by the scrum master.

Business Analysis for Practitioners: A Practice Guide, Section 5.5.3, "Maintaining the Product Backlog."

PMI Professional in Business Analysis (PMI-PBA)® Examination Content Outline, 2013, "Traceability and Monitoring," Task 1

196. Shortly after go-live, your internal auditor, Maddox, identified a list of concerns and observations. Management agreed with all the audit findings, which have been classified as high-priority defects. Who has accountability to see these concerns to resolution?

   a. The project team, because they will be correcting the defective code
   b. The business analyst, despite rolling off to another project
   c. The sponsor, who has yet to sign off on the project and release final payment
   d. The stakeholders, as they are the ones most impacted by the defective code

The correct answer is: **B**

The business analyst, barring extenuating circumstances, would monitor the defect remediation efforts and report status to all interested stakeholders. The business analyst would remain engaged through the entire evaluation process: (a) measuring solution performance, (b) analyzing results, (c) determining gaps, (d) improving the solution, (e) concluding with sunsetting or decommissioning of the solution.

Answer Choices A, C, & D: These are all distractor answer choices. Unless otherwise noted in the RACI, although the others may be involved to some extent, it's the business analyst who is "accountable" to see these concerns to resolution.

Business Analysis for Practitioners: A Practice Guide, Section 5.8.4, "Controlling Changes Related to Defects."

PMI Professional in Business Analysis (PMI-PBA)® Examination Content Outline, 2013, "Traceability and Monitoring," Task 5.

197. Malakai, an experienced application developer, is working on a complex interface between two electronic health record (EHR) systems. Because he's a contractor working for your system integrator and will roll off once you're live, you'd like him to create an artifact that notes all the attributes and subsequent details of each interface. What template can you suggest?

   a. Interface flow table
   b. System interface table
   c. Interface schematic
   d. Interface element table

The correct answer is: **B**

This question tests your working understanding of interface models, which can consist of (a) report tables, (b) system interface tables, (c) user interface flows, (d) wireframes, and (e) display-action-response models. System interface tables are models containing the attributes and the details for each interface at the object/field level. They specify requirements for both the target and source systems, to ensure that requirements are appropriately addressed.

Answer Choices A, C, & D: These are all made-up answer choices, worded to closely resemble the names of interface models.

Business Analysis for Practitioners: A Practice Guide, Section 4.10.11.2, "System Interface Table."

PMI Professional in Business Analysis (PMI-PBA)® Examination Content Outline, 2013, "Analysis," Task 2.

198. Your project intern, Miles, hears you discussing the traceability matrix with Karen, the lead assigned to manage requirements for the ERP initiative. After the discussion, he inquires as to the benefit of the tool. How would you respond?

 a. The requirements traceability matrix helps to control and monitor project scope.
 b. The requirements traceability matrix helps to manage sponsor expectations.
 c. The requirements traceability matrix helps to control and monitor product scope.
 d. The requirements traceability matrix helps to manage customer expectations.

The correct answer is: **C**

As requirements are identified, they are added to the requirements traceability matrix, and attributes are assigned to help monitor and control product scope. Once requirements have been approved, a baseline can be established and status reported to stakeholders, as defined in the business analysis plan. The baseline is a measurement for future comparison.

Answer Choice A: The scope management plan is used for monitoring and controlling project scope.

Answer Choices B & D: The stakeholder engagement plan would provide guidance on how to best manage sponsor and customer expectations.

Requirements Management: A Practice Guide, Section 5.1, "Overview."

PMI Professional in Business Analysis (PMI-PBA)® Examination Content Outline, 2013, "Traceability and Monitoring," Task 1

199. As a business analyst, what project artifact can Kaison refer to in order to plan and assess project performance?

 a. The requirements management plan
 b. The work breakdown structure
 c. The requirements traceability matrix
 d. The time management plan

The correct answer is: **B**

The work breakdown structure (WBS) is a fundamental project management deliverable; it is a logical decomposition, represented in the form of a hierarchy, representing the total scope of the project. It is often accompanied by a WBS dictionary, which is a detailed description of each WBS package. The status of the WBS deliverables can gauge project performance.

Answer Choice A: The requirements management plan provides the overarching direction through-out the project, formally establishing the *how* and *when* for all solution development activities.

Answer Choice C: The requirements traceability matrix is not the best answer choice, although it is used for tracking product requirements from their origin through Evaluation. *On the exam, be careful of the slight nuances in words.*

Answer Choice D: The requirements management plan will cover elements of both the project and product, identifying stakeholders and their roles; establishing the framework for communications; and articulating the guidelines for managing requirements. It cannot be used for the planning and assessment of project performance.

Business Analysis for Practitioners: A Practice Guide, Section 5.2, "Traceability."

PMI Professional in Business Analysis (PMI-PBA)® Examination Content Outline, 2013, "Planning," Task 2.

200. Chanel is working with stakeholders to determine the metrics that will be used during solution evaluation, and also to establish the definition of success. Which of the elements below can influence project costs?
    a. Service-level agreements
    b. Acceptance criteria
    c. Scope baseline
    d. Operational-level agreements

The correct answer is: B

Project costs are in large part influenced by acceptance criteria. During the Planning activities, the business analyst will work closely with the stakeholders and the project manager to ensure that all quality requirements are thoroughly documented, agreed to, and communicated to all interested parties.

Answer Choice A: Service level agreements (SLAs) are contracts between service organizations and customers that establish performance metrics and response times. Once established, these metrics can continually be validated during Evaluation.

Answer Choice C: The scope baseline is a significant project deliverable and key artifact, it's not one single document; rather, it includes the scope statement, the work breakdown structure, and the WBS dictionary.

Answer Choice D: Operational-level agreements are contracts within the service organization.

A Guide to the Project Management Body of Knowledge (PMBOK®), Section 8.1.1, "Plan Quality Management: Inputs."

PMI Professional in Business Analysis (PMI-PBA)® Examination Content Outline, 2013, "Planning," Task 6.

Practice Test: Study Matrix

Question	Domain	Study Notes
1	Planning	
2	Needs Assessment	
3	Needs Assessment	
4	Needs Assessment	
5	Analysis	
6	Needs Assessment	
7	Analysis	
8	Analysis	
9	Planning	
10	Analysis	
11	Needs Assessment	
12	Traceability and Monitoring	
13	Evaluation	
14	Evaluation	
15	Traceability and Monitoring	
16	Analysis	
17	Analysis	
18	Analysis	
19	Planning	
20	Analysis	

Practice Test: Study Matrix

Question	Domain	Study Notes
21	Evaluation	
22	Evaluation	
23	Analysis	
24	Analysis	
25	Analysis	
26	Needs Assessment	
27	Analysis	
28	Analysis	
29	Evaluation	
30	Analysis	
31	Planning	
32	Evaluation	
33	Planning	
34	Planning	
35	Traceability and Monitoring	
36	Analysis	
37	Needs Assessment	
38	Traceability and Monitoring	
39	Planning	
40	Analysis	

Practice Test: Study Matrix

Question	Domain	Study Notes
41	Traceability and Monitoring	
42	Analysis	
43	Needs Assessment	
44	Traceability and Monitoring	
45	Needs Assessment	
46	Planning	
47	Evaluation	
48	Evaluation	
49	Analysis	
50	Analysis	
51	Traceability and Monitoring	
52	Planning	
53	Planning	
54	Needs Assessment	
55	Analysis	
56	Planning	
57	Analysis	
58	Analysis	
59	Analysis	
60	Needs Assessment	

Practice Test: Study Matrix

Question	Domain	Study Notes
61	Analysis	
62	Planning	
63	Analysis	
64	Needs Assessment	
65	Traceability and Monitoring	
66	Traceability and Monitoring	
67	Planning	
68	Analysis	
69	Needs Assessment	
70	Needs Assessment	
71	Planning	
72	Analysis	
73	Analysis	
74	Analysis	
75	Analysis	
76	Analysis	
77	Needs Assessment	
78	Analysis	
79	Traceability and Monitoring	
80	Traceability and Monitoring	

Practice Test: Study Matrix

Question	Domain	Study Notes
81	Analysis	
82	Planning	
83	Planning	
84	Analysis	
85	Needs Assessment	
86	Analysis	
87	Analysis	
88	Needs Assessment	
89	Analysis	
90	Analysis	
91	Needs Assessment	
92	Traceability and Monitoring	
93	Planning	
94	Analysis	
95	Planning	
96	Evaluation	
97	Planning	
98	Analysis	
99	Needs Assessment	
100	Traceability and Monitoring	

Practice Test: Study Matrix

Question	Domain	Study Notes
101	Evaluation	
102	Planning	
103	Analysis	
104	Evaluation	
105	Analysis	
106	Analysis	
107	Analysis	
108	Traceability and Monitoring	
109	Analysis	
110	Traceability and Monitoring	
111	Analysis	
112	Analysis	
113	Analysis	
114	Needs Assessment	
115	Analysis	
116	Evaluation	
117	Traceability and Monitoring	
118	Needs Assessment	
119	Evaluation	
120	Analysis	

Practice Test: Study Matrix

Question	Domain	Study Notes
121	Analysis	
122	Needs Assessment	
123	Planning	
124	Traceability and Monitoring	
125	Planning	
126	Traceability and Monitoring	
127	Needs Assessment	
128	Evaluation	
129	Needs Assessment	
130	Planning	
131	Evaluation	
132	Planning	
133	Planning	
134	Planning	
135	Evaluation	
136	Needs Assessment	
137	Planning	
138	Needs Assessment	
139	Traceability and Monitoring	
140	Evaluation	

Practice Test: Study Matrix

Question	Domain	Study Notes
141	Needs Assessment	
142	Planning	
143	Traceability and Monitoring	
144	Needs Assessment	
145	Analysis	
146	Evaluation	
147	Planning	
148	Needs Assessment	
149	Evaluation	
150	Analysis	
151	Planning	
152	Analysis	
153	Analysis	
154	Planning	
155	Planning	
156	Planning	
157	Analysis	
158	Planning	
159	Planning	
160	Needs Assessment	

Practice Test: Study Matrix

Question	Domain	Study Notes
161	Planning	
162	Needs Assessment	
163	Needs Assessment	
164	Traceability and Monitoring	
165	Planning	
166	Analysis	
167	Needs Assessment	
168	Needs Assessment	
169	Analysis	
170	Analysis	
171	Analysis	
172	Analysis	
173	Planning	
174	Planning	
175	Traceability and Monitoring	
176	Analysis	
177	Traceability and Monitoring	
178	Analysis	
179	Planning	
180	Planning	

Practice Test: Study Matrix

Question	Domain	Study Notes
181	Analysis	
182	Traceability and Monitoring	
183	Analysis	
184	Traceability and Monitoring	
185	Traceability and Monitoring	
186	Analysis	
187	Analysis	
188	Analysis	
189	Planning	
190	Analysis	
191	Evaluation	
192	Needs Assessment	
193	Traceability and Monitoring	
194	Needs Assessment	
195	Traceability and Monitoring	
196	Traceability and Monitoring	
197	Analysis	
198	Traceability and Monitoring	
199	Planning	
200	Planning	

Practice Test: Study Matrix by Domain

Question	Domain	Study Notes
2	Needs Assessment	
3	Needs Assessment	
4	Needs Assessment	
6	Needs Assessment	
11	Needs Assessment	
26	Needs Assessment	
37	Needs Assessment	
43	Needs Assessment	
45	Needs Assessment	
54	Needs Assessment	
60	Needs Assessment	
64	Needs Assessment	
69	Needs Assessment	
70	Needs Assessment	
77	Needs Assessment	
85	Needs Assessment	
88	Needs Assessment	
91	Needs Assessment	
99	Needs Assessment	
114	Needs Assessment	

Practice Test: Study Matrix by Domain

Question	Domain	Study Notes
118	Needs Assessment	
122	Needs Assessment	
127	Needs Assessment	
129	Needs Assessment	
136	Needs Assessment	
138	Needs Assessment	
141	Needs Assessment	
144	Needs Assessment	
148	Needs Assessment	
160	Needs Assessment	
162	Needs Assessment	
163	Needs Assessment	
167	Needs Assessment	
168	Needs Assessment	
192	Needs Assessment	
194	Needs Assessment	
1	Planning	
9	Planning	
19	Planning	
31	Planning	

Practice Test: Study Matrix by Domain

Question	Domain	Study Notes
33	Planning	
34	Planning	
39	Planning	
46	Planning	
52	Planning	
53	Planning	
56	Planning	
62	Planning	
67	Planning	
71	Planning	
82	Planning	
83	Planning	
93	Planning	
95	Planning	
97	Planning	
102	Planning	
123	Planning	
125	Planning	
130	Planning	
132	Planning	

Practice Test: Study Matrix by Domain

Question	Domain	Study Notes
133	Planning	
134	Planning	
137	Planning	
142	Planning	
147	Planning	
151	Planning	
154	Planning	
155	Planning	
156	Planning	
158	Planning	
159	Planning	
161	Planning	
165	Planning	
173	Planning	
174	Planning	
179	Planning	
180	Planning	
189	Planning	
199	Planning	
200	Planning	

Practice Test: Study Matrix by Domain

Question	Domain	Study Notes
5	Analysis	
7	Analysis	
8	Analysis	
10	Analysis	
16	Analysis	
17	Analysis	
18	Analysis	
20	Analysis	
23	Analysis	
24	Analysis	
25	Analysis	
27	Analysis	
28	Analysis	
30	Analysis	
36	Analysis	
40	Analysis	
42	Analysis	
49	Analysis	
50	Analysis	
55	Analysis	

Practice Test: Study Matrix by Domain

Question	Domain	Study Notes
57	Analysis	
58	Analysis	
59	Analysis	
61	Analysis	
63	Analysis	
68	Analysis	
72	Analysis	
73	Analysis	
74	Analysis	
75	Analysis	
76	Analysis	
78	Analysis	
81	Analysis	
84	Analysis	
86	Analysis	
87	Analysis	
89	Analysis	
90	Analysis	
94	Analysis	
98	Analysis	

Practice Test: Study Matrix by Domain

Question	Domain	Study Notes
103	Analysis	
105	Analysis	
106	Analysis	
107	Analysis	
109	Analysis	
111	Analysis	
112	Analysis	
113	Analysis	
115	Analysis	
120	Analysis	
121	Analysis	
145	Analysis	
150	Analysis	
152	Analysis	
153	Analysis	
157	Analysis	
166	Analysis	
169	Analysis	
170	Analysis	
171	Analysis	

Practice Test: Study Matrix by Domain

Question	Domain	Study Notes
172	Analysis	
176	Analysis	
178	Analysis	
181	Analysis	
183	Analysis	
186	Analysis	
187	Analysis	
188	Analysis	
190	Analysis	
197	Analysis	
12	Traceability and Monitoring	
15	Traceability and Monitoring	
35	Traceability and Monitoring	
38	Traceability and Monitoring	
41	Traceability and Monitoring	
44	Traceability and Monitoring	
51	Traceability and Monitoring	
65	Traceability and Monitoring	
66	Traceability and Monitoring	
79	Traceability and Monitoring	

Practice Test: Study Matrix by Domain

Question	Domain	Study Notes
80	Traceability and Monitoring	
92	Traceability and Monitoring	
100	Traceability and Monitoring	
110	Traceability and Monitoring	
108	Traceability and Monitoring	
117	Traceability and Monitoring	
124	Traceability and Monitoring	
126	Traceability and Monitoring	
139	Traceability and Monitoring	
143	Traceability and Monitoring	
164	Traceability and Monitoring	
175	Traceability and Monitoring	
177	Traceability and Monitoring	
182	Traceability and Monitoring	
184	Traceability and Monitoring	
185	Traceability and Monitoring	
193	Traceability and Monitoring	
195	Traceability and Monitoring	
196	Traceability and Monitoring	
198	Traceability and Monitoring	

Practice Test: Study Matrix by Domain

Question	Domain	Study Notes
13	Evaluation	
14	Evaluation	
21	Evaluation	
22	Evaluation	
29	Evaluation	
32	Evaluation	
47	Evaluation	
48	Evaluation	
96	Evaluation	
101	Evaluation	
104	Evaluation	
116	Evaluation	
119	Evaluation	
128	Evaluation	
131	Evaluation	
135	Evaluation	
140	Evaluation	
146	Evaluation	
149	Evaluation	
191	Evaluation	

Final Thoughts

Business analysts serve a vital role in organizations, often functioning as the intermediary between the business areas that they support and operations or project teams supporting the product or solution. To be effective in their role, business analysts should have a fundamental understanding of both the industry and the business for which they are aligned, along with the vernacular, nomenclature, and industry leading practices.

Business analysts directly support most of the activities related to the creation of product scope. In cases of IT software development or the implementation of enterprise software applications, they are the bridge between assessing the need of the organization and delivering a viable solution. Business analysts help to guide the conversation as to functions that the system *can* and ***should*** perform to deliver value.

If the delivered or supported solution doesn't meet the business requirements, as business analysts, we've failed, regardless of how well the project was managed or the total cost of ownership. This Exam Practice Test and Study Guide covers the most important topics in each of the domains required for business analysts to be successful and deliver value to organizations. In addition, it will help you pass the very challenging exam with relative ease.

By putting this knowledge into action, practitioners of business analysis will be better equipped to:

- ✓ Successfully align product requirements with the organizations pillars
- ✓ Effectively engage with stakeholders
- ✓ Navigate situations of ambiguity with greater ease
- ✓ Improve communication with stakeholders
- ✓ Manage change in an integrated manner
- ✓ Understand the rationale for decisions
- ✓ Evaluate whether the delivered solution fulfills the value proposition

*"**Knowing What To Do** must be translated into **Doing What You Know**"*

— Brian Williamson

Acronyms

#

7QC	7 Basic Quality Tools

A

AC	Actual Cost
ACWP	Actual Cost of Work Performed
AD	Activity Description
ADM	Arrow Diagramming Method
AE	Apportioned Effort
AF	Actual Finish Date
AOA	Activity – On – Arrow
AON	Activity – On – Node
API	Application Programming Interface
AS	Actual Start Date
ASP	Application Service Provider
ATTD	Acceptance-Test-Driven Development

B

B2B	Business-to-Business
B2C	Business-to-Consumer
B2G	Business-to-Government
BA	Business Analyst
BAC	Budget at Completion
BAP	Business Analysis Plan

B (cont.)

BCWP	Budgeted Cost of Work Performed
BCWS	Budgeted Cost of Work Scheduled
BOM	Bill of Materials
BPMN	Business Process Modeling Notation
BRD	Business Requirements Document
BU	Business Unit

C

CA	Control Account
CAB	Change Advisory Board
CAP	Control Account Plan
CAPEX	Capital Expenditure
CBA	Cost–Benefit Analysis
CBT	Computer-Based Test
CCB	Change Control Board
CCM	Critical Chain Method
CCR	Continuing Certification Requirements
CEO	Chief Executive Officer
CFO	Chief Financial Officer
CI	Configuration Item
CIO	Chief Information Officer
CM	Configuration Management
CMDB	Configuration Management Database

CMS	Configuration Management System
COA	Chart of Accounts
COB	Close of Business
COE	Center of Excellence
COO	Chief Operating Officer
COQ	Cost of Quality
COTS	Commercial-Off-the-Shelf
CPAF	Cost-Plus Award Fee
CPF	Cost-Plus Fee
CPFF	Cost-Plus Fixed Fee
CPI	Cost Performance Index
CPIF	Cost-Plus Incentive Fee
CPM	Critical Path Method
CPPC	Cost-Plus-Percentage of Cost
CRM	Customer Relationship Management
CRP	Conference Room Pilot
CRUD	Create, Read, Update, and Delete
CSF	Critical Success Factor
CT	Continuity Testing
CTO	Chief Technology Officer
CV	Cost Variance
CWBS	Contract Work Breakdown Structure

D

DD	Data Date
DFD	Data Flow Diagram
DI	Didactic Interaction
DITL	Day-in-the-Life Test
DMAIC	Define, Measure, Analyze, Improve, and Control
DR	Disaster Recovery
DU	Duration
DUR	Duration

E

EAC	Estimate at Completion
eCAB	Emergency Change Advisory Board
ECO	Examination Content Outline

EF	Early Finish Date
EMV	Expected Monetary Value
EOD	End of Day
EOW	End of Week
ERD	Entity Relationship Diagram
ERP	Enterprise Resource Planning
ES	Early Start Date
ESA	Enterprise Software Application
ESC	Executive Steering Committee
ET	Exploratory Testing
ETC	Estimate to Complete
EUT	End User Testing / Training
EV	Earned Value
EVM	Earned Valued Management
EVT	Earned Valued Technique

F

FF	Finish-to-Finish or Free Float
FFP	Firm Fixed Price
FIFO	First In–First Out
FM	Financial Management
FMEA	Failure Mode and Effect Analysis
FPIF	Fixed-Price Incentive Fee
FS	Finish-to-Start
FTE	Full-Time Equivalent Employee
FV	Future Value

G

GERT	Graphical Evaluation and Review Technique
GPoG	Great Pyramid of Giza

H

HCM	Human Capital Management

I

IaaS	Infrastructure-as-a-Service
IFB	Invitation for Bid

INVEST	Independent, Negotiable, Valuable, Estimable, Small, Testable	**NPS**	Net Promoter Score
IRR	Internal Rate of Return	**NPV**	Net Present Value
ITIL	Information Technology Infrastructure Library	**NVT**	Non-Value Added Time

O

OSA	Operational Support and Analysis
RCV	Release Control and Validation
SO	Service Operation
SOA	Service Offerings and Agreements
MALC	Managing Across the Lifecycle
IV&V	Independent Verification and Validation

OBS	Organizational Breakdown Structure
OCR	Optical Character Recognition
OD	Original Duration
OLA	Operational-Level Agreement
OPA	Organizational Process Asset
OPEX	Operating Expenditure

J

JAD	Joint Application Development

P

PaaS	Platform-as-a-Service
PBP	Payback Period
PBT	Paper-Based Test
PC, PCT	Percent Complete
PDCA	Plan, Do, Check, Act
PDM	Precedence Diagramming Method
PDU	Professional Development Unit
PERT	Program Evaluation Review Technique
PEST	Political, Economic, Social, Technological
PF	Planned Finish Date
PfMP®	Portfolio Management Professional
PgMP®	Program Management Professional
PM	Project Management or Project Manager
PMBOK®	*A Guide to the Project Management Body of Knowledge*
PMI	Project Management Institute
PMI-ACP®	PMI Agile Certified Professional
PMI-PBA®	PMI Professional in Business Analysis
PMIS	Project Management Information System
PMI-RMP®	PMI Risk Management Professional

K

KA	Knowledge Area
KPI	Key Performance Indicators
KT	Knowledge Transfer

L

LF	Late Finish date
LIFO	Last In, First Out
LLC	Limited Liability Company
LOE	Level of Effort
LRC	Linear Responsibility Charts
LS	Late Start date
LTV	Life Time Value

M

MMF	Minimum Marketable Features
MoSCoW	Must Have, Should Have, Could Have, Won't Have

N

NAV	Net Asset Value
NDA	Non-Disclosure Agreement
NGT	Nominal Group Technique

PMI-SP®	PMI Schedule Management Professional
PMO	Portfolio, Program, or Project Management Office
PMP®	Project Management Professional
PPA	Project or Program Process Assets
PS	Planned Start Date
PSC	Project Steering Committee
PSWBS	Project Summary Work Breakdown Structure
PT	Performance Test
PV	Planned Value or Present Value

Q

QA	Quality Assurance or Quality Analysis
QC	Quality Control
QFD	Quality Function Deployment

R

RACI	Responsible, Accountable, Consulted, Informed
RAM	Responsibility Assignment Matrix (aka RACI)
RBS	Resource Breakdown Structure or Risk Breakdown Structure
RCA	Root Cause Analysis
RD	Remaining Duration
RDL	Role Delineation Study
RFI	Request for Information
RFP	Request for Proposal
RFQ	Request for Quote
RICE-BW	Reports, Interfaces, Conversions, Extensions or Enhancements, Build, Workflow (Product Elements to Trace)
RM	Requirements Management
RMP	Requirements Management Plan
ROE	Net Present Value
ROI	Return on Investment

S

SaaS	Software-as-a-Service
SCM	Supply Chain Management
SDLC	Software Development Life Cycle
SF	Scheduled Finish Date or Start-to-Finish
SIPOC	Suppliers, Inputs, Process, Outputs, Customers
SIT	System Integrated Test
SKMS	Service Knowledge Management System
SLA	Service-Level Agreement
SMART	Specific, Measurable, Agreed To, Realistic, Time-bound
SME	Subject Matter Expert
SMS	Scope Management System
SKMS	Service Knowledge Management System
SOP	Standard Operating Procedures
SOW	Statement of Work
SPI	Scheduled Performance Index
SS	Scheduled Start Date or Start-to-Start
ST	System Test
STEEP	Social, Technological, Economic, Environmental, Political
STEEPLE	Social, Technological, Economic, Environmental, Political, Legal, Ethics
SV	Schedule Variance
SWOT	Strengths, Weakness, Opportunities, and Threads
SysML	System Modeling Language

T

T&M	Time and Material
TC	Target Completion date
TCO	Total Cost of Ownership
TDD	Test-Driven Development
TF	Target Finish Date or Total Float

TQM	Total Quality Management		**VOC**	Voice of the Customer
TS	Target Start Date			

U

			WIIFM	"What's in it for me?"
UAT	User Acceptance Test		**WBS**	Work Breakdown Structure
UI	User Interface		**WIP**	Work in Process
UML	Unified Modeling Language			
UT	Unit Testing			
UX	User Experience			

W

X

XP	Extreme Programming

V

Y

VAT	Value Added Time		**YoY**	Year-over-Year
VCS	Version Control Systems		**YTD**	Year-to-Date
VE	Value Engineering			

Bibliography

Blais, S. P. (2012). *Business Analysis: Best Practices for Success*. Hoboken, NJ: Wiley.

Cadle, J., D. Paul, and P. Turner (2011). *Business Analysis Techniques: 72 Essential Tools For Success*. Swindon, UK: British Informatics Society Limited.

Carkenord, B. A. (2009). *Seven Steps to Mastering Business Analysis*. Fort Lauderdale, FL: Ross.

Gilb, T. (n.d.). "Quantifying the Qualitative." Available at http://www.resultplanning.com

Gilb, T. (1997). "Requirements-Driven Management: A Planning Language," Crosstalk, June 1997, pp. 18–42. Unavailable online, contact Crosstalk at http://www.crosstalkonline.org/contact-us/ to arrange for reprints.

Gilb, T. (2001). *A Handbook for Systems & Software Engineering Management Using Planguage*. Boston, MA: Addison-Wesley.

Hossenlopp, R., and K. B. Hass (2008). *Unearthing Business Requirements: Elicitation Tools and Techniques*. Vienna, VA: Management Concepts.

Hooks, I. F., and K. A. Farry (2001). *Customer-Centered Products: Creating Successful Products Through Smart Requirements Management*. New York, NY: AMACOM.

Ivan (n.d.). "25 Facts About the Great Pyramid of Giza." Retrieved November 20, 2016, from http://www.ancient-code.com/25-facts-about-the-great-pyramid-of-giza/

PMI (n.d.). "Ethics in Project Management." Newtown Square, PA: Project Management Institute, Inc. Retrieved October 28, 2016, from https://www.pmi.org/about/ethics/code

PMI (n.d.). "Learn About PMI." Newtown Square, PA: Project Management Institute, Inc. Retrieved October 16, 2016, from http://www.pmi.org/about/learn-about-pmi/founders

PMI (2013). *A Guide to the Project Management Body of Knowledge (PMBOK®)*. Newtown Square, PA: Project Management Institute, Inc. Information (and free download for PMI members) available at https://www.pmi.org/pmbok-guide-standards/foundational/pmbok

PMI (2014). "Requirements Management: A Core Competency For Project and Program Success." *Pulse of the Profession®*. Newtown Square, PA: Project Management Institute, Inc. Available at http://www.pmi.org/-/media/pmi/documents/public/pdf/learning/thought-leadership/pulse/requirements-management.pdf

PMI (2015). *Business Analysis for Practitioners: A Practice Guide*. Newtown Square, PA: Project Management Institute, Inc. Information available at https://www.pmi.org/pmbok-guide-standards/practice-guides/business-analysis

PMI (2016). "About PMI's Certification Program." *PMI-PBA Handbook*, p. 2. Newtown Square, PA: Project Management Institute, Inc. Available in PDF format at http://www.pmi.org/-/media/pmi/documents/public/pdf/certifications/professional-business-analysis-handbook.pdf

PMI (2016). *Requirements Management: A Practice Guide*. Newtown Square, PA: Project Management Institute, Inc. Information (and free download for PMI members) available at https://www.pmi.org/pmbok-guide-standards/practice-guides/requirements-management

PMI (2017). Online Credential Registry. Newtown Square, PA: Project Management Institute, Inc. Retrieved January, 2017, from https://www.pmi.org/certifications/registry

PMI (2017). "PMI Fact File." *PMI Today*®. Newtown Square, PA: Project Management Institute, Inc. Information available at https://www.pmi.org/learning/publications/pmi-today

PMI (2017). *PMI Professional in Business Analysis (PMI-PBA)*® *Handbook*. Newtown Square, PA: Project Management Institute, Inc. Available at http://www.pmi.org/-/media/pmi/documents/public/pdf/certifications/professional-business-analysis-handbook.pdf

Podeswa, H. (2007). *The Business Analyst's Handbook*. Independence, KY: Cengage Learning PTR. Information available at http://www.cengage.com/c/the-business-analyst-s-handbook-1e-podeswa

Reichheld, F. F. (2003). "The One Number You Need to Grow." *Harvard Business Review*, December 2003.

Robertson, S., and J. Robertson (2013). *Mastering the Requirements Process: Getting Requirements Right*. Upper Saddle River, NJ: Addison-Wesley.

Wood, M. A. (1997). "Toward a Theory of Stakeholder Identification and Salience: Defining the Principle of Who." *Academy of Management Review*, 22(4), 853–886.

Resources

Great Pyramid of Giza

    https://discoveringegypt.com/pyramids-temples-of-egypt/pyramids-of-giza/

    http://www.nationalgeographic.com/pyramids/khufu.html

    http://www.ancient-code.com/25-facts-about-the-great-pyramid-of-giza/

    https://www.ancient.eu/Great_Pyramid_of_Giza/

Project Management Collaboration Tools

    https://www.pmi.org/learning/library/project-management-collaboration-tools-virtual-7040

The United States Patent and Trademark Office (USPTO) Facilitator's Tool Kit

    https://www.uspto.gov/web/offices/com/oqm-old/Facilitation.pdf

Index

S